VASCULAR PLANTS OF OHIO

A Manual for Use in Field and Laboratory

•

CLARA G. WEISHAUPT

Department of Botany
The Ohio State University

THIRD EDITION

KENDALL/HUNT PUBLISHING COMPANY
DUBUQUE, IOWA

Copyright © 1960, 1968, 1971 by
Clara Gertrude Weishaupt

Library of Congress Catalog Card Number: 75-96816

ISBN 0—8403—0084—0

Printed in the United States of America

Foreword

In revising this manual, I have made changes in keys only when it seemed possible for me to make them more accurate or easier for the student to use. I have learned of a few species, not included in the first edition, that are part of the Ohio flora; these species have been added. A few names have been changed; in each instance, the name from the first edition has been added as a synonym. No other synonyms have been included.

An asterisk before the name of a plant indicates that the species is not native in Ohio. Doubtfully native species are not marked. Most of the plants marked with an asterisk grow wild in Ohio; others do not, but are included because it has been learned from experience that they are often available for study when wild plants of the family or genus are not.

In the keys, degree of union of carpels is, at least in most instances, stated rather exactly. In descriptions, however, the unqualified statement that carpels are united means that they are united approximately to summit of ovularies. Additional statements about styles and stigmas indicate whether the upper parts of the carpels (styles and stigmas) are united or not.

I am grateful to my colleagues at the Department of Botany, The Ohio State University, for their help and criticism in the preparation of this manual, as well as to many others who have made suggestions. I am grateful, also, to my former students whose ease or difficulty in using the first edition has helped me to make what I hope are easier and better keys.

Columbus, Ohio   Clara G. Weishaupt
October, 1967

Since 1967 I have learned that plants of several species not included in the previous edition have been found in Ohio; these species have been added. I have made a number of revisions, including changes in several keys in the hope of making them clearer and easier to use. The manual has been enlarged by the addition of diagrams, explanations supplementary to the diagrams, and winter-keys to some genera of deciduous woody plants.

I express my thanks to those who have called my attention to omissions and errors, and to my colleagues in Botany at The Ohio State University whose criticisms of the revisions and additions have been very helpful.

Columbus, Ohio   Clara G. Weishaupt
March, 1971

Key to Families of Vascular Plants of Ohio and Vicinity

I. KEY TO FAMILIES OF PTERIDOPHYTES (SEEDLESS VASCULAR PLANTS)

a. Plants aquatic, floating, small; leaves at most 1.5 cm long, papillose or hairy, actually or appearing 2-ranked. Salviniaceae, p. 21

a. Plants aquatic or terrestrial, not floating.

    b. Aerial stems ridged and grooved, the main one with central cavity, leafless except for toothed sheaths at nodes; sporophylls in terminal cones. Equisetaceae, p. 15

    b. Aerial stems absent or, if present, leafy, not ridged, not grooved, sometimes horizontal.

        c. Leaves quill-like, in a tuft from a very short thick stem, expanded bases containing sporangia; growing in mud. Isoetaceae, p. 16

        c. Leaves 4-foliolate, petioles long, erect; stems horizontal; sporangia in sporocarps; growing in mud. Marsileaceae, p. 21 (See third c.)

        c. Leaves scalelike or small and linear, at most 1.5 cm long, in spirals or 4-ranked. (See fourth c.)

            d. Sporophylls in terminal 4-angled cones; leaf with a ligule. Selaginellaceae, p. 16

            d. Sporophylls in terminal terete cones or in zones alternating with zones of foliage leaves; leaf without a ligule. Lycopodiaceae, p. 15

        c. Leaves not quill-like, not 4-foliolate, not scalelike, not small and linear; usually large, usually compound, from rhizomes.

            d. Leaves twining, climbing; leaflets 2-forked, each fork with a palmately lobed blade. Schizaeaceae, p. 17

            d. Leaves neither twining nor climbing.

                e. Sporangia borne on leaves or portions of leaves that are without flat green blades.

                    f. Sporangia coherent in 2 rows in a stalked unbranched spikelike cluster; leaf consisting of 2 segments on a common petiole, one an entire green blade, the other the cluster of sporangia. Ophioglossaceae, p. 16

                    f. Sporangia not in an unbranched spikelike cluster.

                        g. Sporangia in beadlike globular clusters, the clusters enfolded by modified pinnules, or in chainlike rows. Polypodiaceae, p. 17

                        g. Sporangia separate, opening with a terminal longitudinal cleft into 2 valves.

                            h. Either leaves or leaflets dimorphic, some green and foliaceous, others bearing sporangia and not foliaceous; leaves large, lanceolate to oblong. Osmundaceae, p. 17

                            h. Leaf consisting of 2 segments on a common petiole, one green and foliaceous, compound or divided, the other bearing a cluster of sporangia. Ophioglossaceae, p. 16

                e. Sporangia in sori at margin of or distributed on under surface of flat green leaf-blades.

                    f. Leaves delicate, 1 cell thick between veins, pinnate-pinnatifid; sporangia at base of an elongate axis in a cuplike indusium; rare in Ohio. Hymenophyllaceae, p. 17

                    f. Leaves thicker; sori with or without indusia, marginal or variously distributed over the surface; indusia various; sporangia opening transversely. Polypodiaceae, p. 17

II. KEY TO FAMILIES OF SPERMATOPHYTES (SEED PLANTS)

GENERAL SECTION

a. Ovules and seeds naked, attached to open carpels, without stigmas, styles, and ovularies; plants woody. GYMNOSPERMS, SECTION A, p. 2

a. Ovules and seeds enclosed in an ovulary formed from a closed carpel or group of united carpels; stigma or stigmas present, and often style or styles; plants herbaceous or woody. ANGIOSPERMS, SECTION B, p. 2

SECTION A. GYMNOSPERMS

a. Sporophylls on dwarf branches, not in cones; leaves with expanded blades, dichotomously veined. Ginkgoaceae, p. 21
a. Sporophylls in cones (rarely solitary); leaves needlelike, flat and linear, or scalelike.
  b. Leaves flat and linear, yellow-green beneath; staminate cone of 5-8 peltate stamens; ovule solitary, surrounded by fleshy red aril. Taxaceae, p. 21
  b. Both kinds of sporophylls in cones; ovules more than 1.
    c. Leaves needlelike or flat and linear; leaves and sporophylls in spirals, foliage leaves sometimes in fascicles or clusters on dwarf branches. Pinaceae, p. 21
    c. Leaves scalelike or awl-shaped; leaves and sporophylls opposite or whorled. Cupressaceae, p. 22

SECTION B. ANGIOSPERMS

a. Usually flower parts 3 or a multiple of 3 and leaves parallel-veined; vascular bundles closed, usually not in a circle but scattered through the pith; annual rings of growth not present in stems; cotyledon 1. (Included are small floating aquatics with thalluslike body, about 1 cm long or less, seldom observed in flower.) MONOCOTYLEDONS, SECTION I-B, p. 2
a. Usually flower parts 4 or 5 or a multiple of 4 or 5 and leaves not parallel-veined; vascular bundles open, usually in a circle around central pith; annual growth-rings forming from cambium when stems are perennial; cotyledons 2. DICOTYLEDONS, SECTION II-B, p. 3

SECTION I-B. MONOCOTYLEDONS

a. Flowers without typical perianth (sometimes with perianth of scales or bristles); if with minute perianth, then on a spadix, with or without a spathe.
  b. Plant body thalluslike, about 1 cm long and wide or less, with or without roots; thalli solitary or in colonies; flowers seldom observed; floating aquatics. Lemnaceae, p. 63
  b. Plants with stems and leaves; aquatic or terrestrial.
    c. Flowers on a spadix subtended by a spathe. Araceae, p. 62
    c. Flowers not on a spadix subtended by a spathe.
      d. Flowers subtended by or enclosed by scales or glumes; leaves sheathing.
        e. Flowers enclosed by 2 glumes (lemma and palea); perianth represented by 2 or 3 scales (lodicules); leaf-sheaths usually open. Gramineae, p. 27
        e. Flowers in axil of a single glume, sometimes a second glume forming an enclosing sac around the ovulary; leaf-sheaths usually closed. Cyperaceae, p. 50
      d. Flowers not subtended by or enclosed by glumes.
        e. Erect herbs of aquatic or mud habitats; if aquatic, stems and leaves emersed; flowers in dense spikes or heads; leaves linear.
          f. Flowers in a solitary dense terminal spike, staminate flowers above, carpellate flowers below. Typhaceae, p. 23
          f. Flowers in heads scattered along upper part of stem or its branches, staminate heads above, carpellate below. Sparganiaceae, p. 23 (See third f.)
          f. Flowers in a dense spike (spadix) borne laterally on edge of flattened scape; rhizome aromatic. Acorus in Araceae, p. 62
        e. Submersed aquatics; floating leaves sometimes present also.
          f. Leaves alternate or, if opposite, then entire; flowers in elongate or capitate spikes or in axillary clusters. Zosteraceae, p. 23
          f. Leaves opposite, usually toothed, linear, widened at base; flowers solitary, axillary, sessile. Najadaceae, p. 25
a. Flowers with typical perianth.
  b. Flowers hypogynous.
    c. Carpels 2 or more, separate or slightly united at base.
      d. Carpels 3 or 6, becoming follicles.
        e. Flowers in umbels; carpels 6; stamens 9. Butomaceae, p. 27
        e. Flowers in racemes; carpels 3 or 6; stamens 3-6. Juncaginaceae, p. 25

d. Carpels more than 6, becoming achenes; flowers whorled on axis of raceme or panicle. <u>Alismaceae</u>, p. 26

c. Carpels 2 or more, ovularies united; styles and/or stigmas may be separate.

  d. Flowers small, in small solitary heads on leafless scapes, each flower in axil of a bract; leaves all basal.

    e. Flowers bisporangiate; petals and anther-bearing stamens 3; calyx of 3 sepals, irregular; petals yellow. <u>Xyridaceae</u>, p. 64

    e. Flowers monosporangiate; petals 3 or 2; stamens 4 or 6; perianth white. <u>Eriocaulaceae</u>, p. 64

  d. Flowers not in solitary heads at end of leafless scapes; if flowers in heads, then heads not solitary, or flowers not as above.

    e. Flowers with a glumaceous perianth of 6 similar divisions; stamens 3 or 6; leaves linear, flat or terete, sometimes septate. <u>Juncaceae</u>, p. 65

    e. Perianth petaloid or of 3 green sepals and 3 colored or white petals.

      f. Terrestrial plants.

        g. Perianth of 6 (rarely 4) similar parts; stamens 6 (rarely 4). <u>Liliaceae</u>, p. 67

        g. Perianth of 3 green sepals and 3 white or colored petals.

          h. Leaves alternate, parallel-veined; corolla regular or zygomorphic, ephemeral. <u>Commelinaceae</u>, p. 64

          h. Leaves in whorls of 3, usually not parallel-veined; corolla regular, not ephemeral. <u>Trillium</u> in <u>Liliaceae</u>, p. 67

      f. Aquatic or marsh plants.

        g. Perianth tubular or funnelform, limb 2-lipped or equally 6-lobed. <u>Pontederiaceae</u>, p. 65

        g. Perianth divided to base or nearly so.

          h. Flowers in a raceme; stamens 3 or 6. <u>Juncaginaceae</u>, p. 25

          h. Flowers in an umbel; stamens 9. <u>Butomaceae</u>, p. 27

b. Flowers epigynous.

  c. Aquatic, submersed; leaves short, sessile, in crowded whorls, or long and ribbonlike; carpellate flowers eventually floating as a result of elongation of stipe or of hypanthium base. <u>Hydrocharitaceae</u>, p. 27

  c. Not aquatic or, if rarely aquatic, then stem and leaves emersed.

    d. Diecious; vines with longitudinally-ribbed whorled or alternate leaves and small greenish-white flowers in axillary panicles or spikes. <u>Dioscoreaceae</u>, p. 73

    d. Not diecious; not vines.

      e. Stamens 6.

        f. Ovulary wholly inferior. <u>Amaryllidaceae</u>, p. 73

        f. Ovulary partly inferior. A few genera of <u>Liliaceae</u>, p. 67

      e. Stamens 3. <u>Iridaceae</u>, p. 74

      e. Stamens 1 or 2, united with upper portion of gynecium. <u>Orchidaceae</u>, p. 75

SECTION II-B. DICOTYLEDONS

A FEW FAMILIES IN WHICH FLOWER PARTS ARE DIFFICULT TO INTERPRET

a. Perianth parts many, intergrading, sometimes intergrading with stamens.

  b. Shrubs; flowers perigynous; perianth parts similar, maroon, lanceolate, on concave hypanthium; stamens several to many; carpels several to many, separate. <u>Calycanthaceae</u>, p. 109

  b. Aquatic herbs; blades 1-7 dm wide, floating or emersed, ovate to circular, peltate or with a sinus; perianth parts, stamens, and carpels many; flowers hypogynous or epigynous. <u>Nymphaeaceae</u>, p. 102

a. Perianth parts about 15 or fewer; flowers hypogynous.

  b. Trees and shrubs; perianth parts similar, 9-15; stamens many; carpels many, separate or somewhat united; receptacle elongate. <u>Magnoliaceae</u>, p. 108

  b. Herbs; perianth zygomorphic, of 6 or 8 parts, largest and most conspicuous part cornucopia-shaped or saccate, spurred at base; stamens 5; ovulary 5-loculed. <u>Balsaminaceae</u>, p. 151

KEY TO SECTIONS

a. Plants woody; flowers monosporangiate, the two kinds in separate clusters, one or both kinds in aments or heads; perianth absent or calyx present. SECTION II-B-1, p. 4

a. Plants herbaceous or woody; if plants woody, then flowers bisporangiate, or clusters not as above, or both calyx and corolla present.
  b. Perianth none. SECTION II-B-2, p. 4
  b. Perianth present (rarely absent in carpellate flowers but present in staminate flowers).
    c. Flowers hypogynous or perigynous.
      d. Either calyx or corolla present (usually calyx) but not both. SECTION II-B-3, p. 5
      d. Both calyx and corolla present.
        e. Petals separate, at least below. SECTION II-B-4, p. 6
        e. Petals more or less united. SECTION II-B-5, p. 10
    c. Flowers epigynous (ovulary wholly or partly inferior). SECTION II-B-6, p. 12

SECTION II-B-1. PLANTS WOODY; FLOWERS MONOSPORANGIATE, ONE OR BOTH KINDS IN AMENTS OR HEADS; COROLLA ABSENT

a. Shrubs; foliage and twigs resin-dotted, fragrant; calyx absent; aments short-cylindric or globose; carpellate flowers single under each ament-bract. Myricaceae, p. 83
a. Trees or, if shrubs, not as above.
  b. Both carpellate and staminate flowers in cylindric aments.
    c. Lowest scale of winter bud directly above leaf-scar, or scale 1; fruit a capsule, seeds many, comate; perianth a small cup, 1-2 glands, or none; diecious. Salicaceae, p. 79
    c. Scales of winter bud not as above; fruit multiple or 1-seeded.
      d. Carpellate flowers with 4-parted perianth which enlarges in fruit; bracts small or absent; whole carpellate ament becoming a multiple fruit; diecious. Moraceae, p. 89
      d. Carpellate flowers 2-3 under each ament-bract, perianth absent or minute; bracts obvious, sometimes closed sacs; fruit a nutlet or a samara; monecious. Corylaceae, p. 84
  b. Carpellate flowers not in cylindric aments.
    c. Carpellate flowers in dense heads.
      d. Staminate flowers in cylindric aments and stem thornless, or in loose globular aments and stem usually thorny; style and stigma 1; fruit multiple. Moraceae, p. 89
      d. Staminate flowers in heads; carpellate heads peduncled and drooping; stem thornless.
        e. Staminate heads in erect racemes; stipule-rings absent. Hamamelidaceae, p. 124
        e. Staminate heads solitary, drooping; stipule-rings present. Platanaceae, p. 125
    c. Carpellate flowers not in dense heads, but solitary or few together or in budlike clusters or in few-flowered heads.
      d. Style and stigma 1; carpellate flowers solitary or in few-flowered heads, the staminate in umbels or short racemes; fruit a drupe. Nyssaceae, p. 171
      d. Styles or stigmas or both 2 or more; fruit a nut; monecious.
        e. Flowers appearing before leaves, the carpellate in small budlike clusters, 2 flowers under each bract, the staminate in cylindric aments. Corylaceae, p. 84
        e. Flowers appearing with or after leaves; carpellate flowers in an involucre or bur.
          f. Leaves pinnately compound; carpellate flowers solitary or in short spikes terminal on branches of current season; staminate aments elongate. Juglandaceae, p. 83
          f. Leaves simple; carpellate flowers axillary, solitary or few together, each in a cuplike involucre or 2-3 together in an involucre or bur; staminate flowers in erect spikes or in drooping slender or globular aments. Fagaceae, p. 86

SECTION II-B-2. PERIANTH NONE

a. Trees; leaves opposite; flowers monosporangiate or bisporangiate, in panicles. Oleaceae, p. 176
a. Herbs.
  b. Sap milky; flowers monosporangiate, several staminate flowers and 1 carpellate flower within a calyxlike structure (cyathium); staminate flower of 1 stamen, carpellate of 3 united carpels. Euphorbiaceae, p. 145
  b. Sap not milky; flowers not in a cyathium.
    c. Erect marsh herbs with ovate alternate leaves; flowers small, perfect, in nodding spikes. Saururaceae, p. 79
    c. Aquatic herbs with tiny axillary solitary or clustered flowers.
      d. Leaves opposite, entire; flowers monosporangiate, 1-3 in axils; carpels 2, ovulary 4-loculed. Callitrichaceae, p. 147
      d. Leaves alternate, mostly forked into narrow lobes; flowers perfect, carpels and locules 2. Podostemaceae, p. 121 (See third d.)

4

      d. Leaves whorled, dichotomously dissected; flowers monosporangiate, solitary; carpel and locule 1. Ceratophyllaceae, p. 101

SECTION II-B-3. PERIANTH OF EITHER CALYX OR COROLLA BUT NOT BOTH

a. Carpels more than 1, separate.
    b. Vines; leaves opposite, mostly compound; usually diecious. Ranunculaceae, p. 103
    b. Prickly shrubs or small trees; flowers in axillary clusters appearing before or with pinnately compound leaves; corolla present, calyx absent. Rutaceae, p. 143 (See third b)
    b. Herbs or rarely non-prickly shrubs, not vines; carpels becoming achenes, follicles, or berries.
      c. Flowers hypogynous. Ranunculaceae, p. 103
      c. Flowers perigynous. Rosaceae, p. 125
a. Carpels 2 or more, at least the ovularies united or partly united, or carpel 1.
    b. Herbs, sometimes herbaceous vines.
      c. Plants non-green, without chlorophyll; leaves small and scalelike. Pyrolaceae, p. 171
      c. Plants green, with chlorophyll.
        d. Small decumbent plants of wet places; blades round-ovate; sepals 4, stamens 8, carpels 2; flower tinged with red or yellow. Saxifragaceae, p. 122
        d. Without the above set of characters.
          e. Ovulary 1-loculed.
            f. Leaves 3-foliolate, pinnately compound, decompound, or whorled and dichotomously dissected.
              g. Submersed aquatics; leaves whorled, dissected. Ceratophyllaceae, p. 101
              g. Not submersed aquatics or, if so, then leaves not as above.
                h. Leaves 3-foliolate; stamens diadelphous. Leguminosae, p. 133
                h. Leaves not 3-foliolate; stamens separate.
                  i. Flower hypogynous; leaves decompound. Ranunculaceae, p. 103
                  i. Flower perigynous; leaves pinnately compound. Rosaceae, p. 125
            f. Leaves simple or compound but not as above.
              g. Some or all leaves opposite, palmately lobed or palmately compound; stigmas 2; locule 1; diecious herbs; erect, or vines. Cannabinaceae, p. 89
              g. Without the above set of characters.
                h. Leaves opposite.
                  i. Calyx corollalike, investing ovulary in such way that flower appears epigynous; bracts below flower or flower-cluster. Nyctaginaceae, p. 96
                  i. Calyx not as above.
                    j. Filaments united in a tube; flowers in spikes or heads, often woolly; fruit a utricle. Amaranthaceae, p. 96
                    j. Filaments not united in a tube.
                      k. Leaves minute and scalelike. Chenopodiaceae, p. 94
                      k. Leaves not as above.
                        l. Style 1, elongate, or stigma a brushlike tuft. Urticaceae, p. 90
                        l. Styles or stigmas or both 2 or more; stigma not brushlike.
                          m. Flowers monosporangiate, the carpellate without perianth, enclosed by 2 bracts; fruit 1-seeded. Chenopodiaceae, p. 94
                          m. Flowers perfect or monosporangiate, not enclosed in bracts; calyx present; fruit 1-seeded or placenta free central. Caryophyllaceae, p. 97
                h. Leaves alternate or all basal.
                  i. Fruit a capsule, seeds more than 1; placentae 3, parietal.
                    j. Plant stellate-pubescent; without stipules. Cistaceae, p. 156
                    j. Plant glabrous or hairs simple; with stipules. Violaceae, p. 157
                  i. Fruit 1-seeded, usually an achene or a utricle.
                    j. Stipules sheathing; stems erect or trailing or twining, rarely prickly; fruit an achene. Polygonaceae, p. 91
                    j. Stipules absent or not sheathing; flowers small, often greenish.
                      k. Style 1, sometimes very short, or absent; stigma 1 or a brushlike tuft. Urticaceae, p. 90
                      k. Styles or stigmas or both 2-3; stigma not brushlike.

l. Leaves spine-tipped, narrow. Chenopodiaceae, p. 94
l. Leaves not spine-tipped.
    m. Plants often mealy (with whitish scales); calyx herbaceous or fleshy; flowers bractless or the carpellate enclosed by 2 non-scarious bracts. Chenopodiaceae, p. 94
    m. Plants not mealy; calyx usually scarious; each flower subtended, but not enclosed, by 1 or more dry or scarious, sometimes spiny, bracts. Amaranthaceae, p. 96
  e. Ovulary with 2 or more locules.
    f. Flowers monosporangiate (if rarely some perfect ones present, then blades large, peltate, lobed); styles 2-3, each often lobed or dissected; capsule 2-3-loculed, 1-2 seeds in each locule. Euphorbiaceae, p. 145
    f. Flowers bisporangiate or other characters not as above.
      g. Leaves opposite or whorled.
        h. Leaves whorled; prostrate small herbs; flowers in small clusters; sepals 5; stamens 3-4; stigmas and locules 3. Aizoaceae, p. 97
        h. Leaves opposite; septa of ovulary sometimes incomplete.
          i. Flower perigynous; style 1; sepals 4. Lythraceae, p. 160
          i. Flower hypogynous; styles 2-5; sepals 5. Caryophyllaceae, p. 97
      g. Leaves alternate or all basal.
        h. Locules of ovulary 2.
          i. Sepals 4; stamens 6, 4, or 2. Cruciferae, p. 112
          i. Sepals 2; stamens many; blades palmately lobed. Papaveraceae, p. 110
        h. Locules of ovulary more than 2; sepals usually 5; stamens usually 10.
          i. Locules usually 5; carpels united below (about halfway), divergent above, the beaks dehiscing crosswise. Crassulaceae, p. 121
          i. Locules about 10; berry dark purple. Phytolaccaceae, p. 97
b. Woody plants.
  c. Leaves and young stems covered with brownish or silvery peltate scales; carpel 1. Elaeagnaceae, p. 160
  c. Leaves and stems without such scales.
    d. Shrubby green plants partially parasitic on trees; leaf-blades entire, thick; flowers in axillary spikes; fruit a berry. Loranthaceae, p. 91
    d. Plants not growing on other plants, not parasitic.
      e. Anthers opening by uplifted lids; plants aromatic. Lauraceae, p. 109
      e. Anthers not opening by uplifted lids.
        f. Leaves or leaf-scars opposite; flowers sometimes monosporangiate.
          g. Style 1; stamens usually 2; fruit a samara or drupe; calyx minute, 4-cleft or -toothed. Oleaceae, p. 176
          g. Styles 2 (rarely more); stamens 4-10, often 8; fruit of 2 (rarely 3 or more) samaras; calyx evident or minute. Aceraceae, p. 149
        f. Leaves or leaf-scars alternate.
          g. Vines; sepals minute; petals sometimes coming off as a cap; stamens opposite petals; leaves palmately veined or compound. Vitaceae, p. 152
          g. Not vines.
            h. Trees; leaves 2-ranked; stamens and sepals 5 or stamens and calyx-lobes 3-9; stigmas 2; fruit a samara or a drupe. Ulmaceae, p. 88
            h. Shrubs; leaves not 2-ranked or, if so, then stigma 1.
              i. Leaves 2-ranked, stamens 8, sepals 4, flowers yellow, or leaves not 2-ranked, flowers pink; fruit a drupe. Thymelaeaceae, p. 160
              i. Leaves not 2-ranked; flowers not pink.
                j. Flowers greenish-white; stamens and sepals 5. Rhamnaceae, p. 151
                j. Flowers yellowish; stamens and petals 4-5; sepals minute and deciduous. Aquifoliaceae, p. 148

SECTION II-B-4. FLOWERS HYPOGYNOUS OR PERIGYNOUS; PERIANTH OF BOTH CALYX AND COROLLA; PETALS SEPARATE

a. Stamens more than twice as many as petals.
  b. Carpels 2 or more, separate or slightly united at base of ovularies.

c. Carpels 5 (6), united about halfway, upper portions divergent as horns that dehisce crosswise in fruit; herbs; blades simple, serrate, elliptic. Crassulaceae, p. 121
c. Plants not as above.
    d. Sepals, petals, and stamens attached to hypanthium (flower perigynous); sepals and petals usually 5; stipules usually present; herbaceous and woody. Rosaceae, p. 125
    d. Sepals, petals, and stamens attached to receptacle (flower hypogynous).
        e. Vines.
            f. Leaves simple, alternate, palmately veined; diecious. Menispermaceae, p. 108
            f. Leaves compound, opposite; styles long, plumose in fruit. Ranunculaceae, p. 103
        e. Not vines.
            f. Herbs.
                g. Aquatics with floating entire peltate blades; petals and sepals usually 3, purple. Nymphaeaceae, p. 102
                g. Not aquatics or, if rarely so, blades not as above. Ranunculaceae, p. 103
            f. Trees and shrubs.
                g. Twigs with complete stipule-rings; perianth-segments 9-18, similar, or sepals and petals somewhat differentiated. Magnoliaceae, p. 108
                g. Twigs without stipule-rings; sepals 3, petals 6, maroon. Annonaceae, p. 109
b. Carpels 2 or more, at least the ovularies united, or carpel 1.
    c. Herbs and shrubs; leaves opposite, blades entire, gland-dotted; petals yellow, 4-5; styles and carpels 2-5; ovulary 1-loculed, placentae parietal, or locules as many as carpels and placentae axile. Hypericaceae, p. 155
    c. Without the above set of characters.
        d. Stamens many, filaments united in a sheath around styles; locules and styles or style-branches 5-many; corolla regular, petals 5; shrubs and herbs. Malvaceae, p. 153
        d. Without the above set of characters.
            e. Woody plants.
                f. Hoary-tomentose evergreen shrubs with scalelike leaves. Cistaceae, p. 156
                f. Not evergreen; leaves not scalelike.
                    g. Leaves pinnately compound or decompound, present at anthesis; petals 1-5; ovulary 1-loculed, parietal placenta 1. Leguminosae, p. 133
                    g. Leaves simple; flowers sometimes appearing before leaves.
                        h. Anthers opening by uplifted lids; plants aromatic. Lauraceae, p. 109
                        h. Anthers opening otherwise.
                          i. Flower perigynous; ovulary 1-loculed; leaves sometimes appearing after flowers; peduncle not adnate to bract. Rosaceae, p. 125
                          i. Flower hypogynous; ovulary 5-loculed; leaves present before flowers; peduncle adnate to large bract. Tiliaceae, p. 153
        e. Herbs.
             f. Ovulary 1-loculed.
                g. Placenta 1, parietal, or seed or seeds basally attached.
                    h. Stipules sheathing; fruit an achene. Polygonaceae, p. 91
                    h. Stipules absent or not sheathing.
                        i. Flowers to 5 cm wide, usually solitary; corolla regular; flowering plant with 2 half-circular lobed leaves. Berberidaceae, p. 107
                        i. Flowers much smaller than above or spurred, irregular, in racemes or panicles; leaves not as above. Ranunculaceae, p. 103
                g. Placentae 2 or more, parietal.
                    h. Leaves palmately compound, leaflets 3-7; petals and sepals 4; placentae 2. Capparidaceae, p. 111
                    h. Leaves not palmately compound.
                      i. Corolla irregular; placentae 3 or more. Resedaceae, p. 121
                      i. Corolla regular, or each whorl of petals regular.
                          j. Blades entire, lanceolate or very small; sepals 5, 2 smaller, or sepals 3, persistent; petals mostly 3-5. Cistaceae, p. 156
                          j. Blades not as above; sepals usually 2, early deciduous; petals 4 to 8 to many; juice colored or milky. Papaveraceae, p. 110
            f. Ovulary with 2 or more locules.
                g. Leaves pitcher-shaped; flowers large; bog herbs. Sarraceniaceae, p. 121
                g. Leaves not pitcher-shaped.
                    h. Flowers monosporangiate; plants stellate-hairy. Euphorbiaceae, p. 145

        h. Flowers perfect, zygomorphic; filaments united in a split sheath; locules 2; sepals 5, 2 larger; petals 3. <u>Polygalaceae</u>, p. 144 (See third h)

        h. Flowers perfect; filaments separate; locules 5. <u>Crassulaceae</u>, p. 121

a. Stamens not more than twice the petals.

  b. Flower with a fringed corona; vines with tendrils; blades palmately lobed. <u>Passifloraceae</u>, p. 159

  b. Flower without a fringed corona.

    c. Stamens the same number as the petals and opposite them.

      d. Woody vines, usually with tendrils; calyx minute; ovulary 2-loculed; blades simple and broadly ovate, or compound, usually palmately veined. <u>Vitaceae</u>, p. 152

      d. Shrubs and small trees; calyx evident; ovulary 2-5-loculed; blades not as above. <u>Rhamnaceae</u>, p. 151 (See third d)

      d. Shrubs and herbs; ovulary 1-loculed.

        e. Anthers opening by uplifted lids; ovules basal or placenta 1, parietal. <u>Berberidaceae</u>, p. 107

        e. Anthers opening otherwise.

          f. Calyx usually 5-parted (4-7); placenta free central. <u>Primulaceae</u>, p. 174

          f. Sepals 2; placenta free central or basal. <u>Portulacaceae</u>, p. 97

    c. Stamens the same number as petals and alternate with them, or more or fewer than petals.

      d. Carpels 2 or more, separate or only slightly united.

        e. Styles, or stigmas, or both, united; ovularies separate.

          f. Trees; leaves pinnately compound; flowers in large panicles. <u>Simaroubaceae</u>, p. 144

          f. Herbs.

            g. Carpels 2, stigmas united, styles separate; stamens usually united with stigma; pollen in pollinia. <u>Asclepiadaceae</u>, p. 179

            g. Carpels 2-3, bases of styles united; blades pinnately divided; flowers axillary; diffuse small herbs. <u>Limnanthaceae</u>, p. 147

        e. Styles and stigmas separate; ovularies separate or slightly united at base.

          f. Aquatics with submersed opposite palmately dissected leaves and few floating entire peltate leaves; sepals and petals 3; stamens 6. <u>Nymphaeaceae</u>, p. 102

          f. Plants not as above.

            g. Trees or vines; flowers monosporangiate; leaves simple, palmately veined.

              h. Vines; diecious; flowers in racemes or panicles. <u>Menispermaceae</u>, p. 108

              h. Trees; monecious; flowers in spherical heads. <u>Platanaceae</u>, p. 125

            g. Plants herbaceous or flowers perfect.

              h. Leaves thick and fleshy and petals 4-5, or carpels dehiscing crosswise; sepals and carpels 4-5; stamens 8-10; herbs; blades simple. <u>Crassulaceae</u>, p. 121

              h. Plants not as above.

                i. Carpels 2, united below, summits separate and often divergent; herbs; leaves simple; fruit a capsule or 2 follicles. <u>Saxifragaceae</u>, p. 122

                i. Carpels more than 2, or leaves compound, or both; herbs and shrubs; fruit an aggregate of achenes, berries, drupes, or follicles.

                  j. Flowers hypogynous. <u>Ranunculaceae</u>, p. 103

                  j. Flowers perigynous. <u>Rosaceae</u>, p. 125

      d. Carpels 2 or more, at least the ovularies united, or carpel 1.

        e. Leaves opposite, simple, entire, gland-dotted; petals yellow or pink; rarely blades not dotted, then petals pink and plants marsh herbs; locules 3-5 or placentae parietal and locule 1; herbs and shrubs. <u>Hypericaceae</u>, p. 155

        e. Leaves not opposite, simple, entire, and gland-dotted, or one or more other characters not as above.

          f. Trees, shrubs, and woody vines.

            g. Leaves opposite.

              h. Corolla not regular; leaves palmately compound. <u>Hippocastanaceae</u>, p. 150

              h. Corolla regular.

                i. Flower with prominent disk which appears to cover ovulary; locules 2-5; small trees and prostrate or erect shrubs, sometimes evergreen; leaves simple, pinnately veined; seeds with arils. <u>Celastraceae</u>, p. 148

                i. Plants not as above.

                  j. Petals 4, linear; sepals 4, minute; stamens 2, rarely more; locules 2; fruit a drupe; blades entire. <u>Oleaceae</u>, p. 176

8

j. Petals, sepals, and stamens 5; locules 3; fruit a bladdery capsule; blades 3-foliolate. Staphyleaceae, p. 149 (See third j)

j. Petals and sepals usually about 5; stamens 4-10, often 8; locules 2, rarely more; fruit of 2 (rarely more) united samaras; blades not entire, usually palmately lobed. Aceraceae, p. 149

g. Leaves alternate.

   h. Evergreen shrubs; blades scalelike, or entire and revolute-margined.

      i. Leaves small, scalelike; hoary-tomentose. Cistaceae, p. 156

      i. Leaves not scalelike; blades entire, revolute-margined; anthers opening by terminal pores. Ericaceae, p. 171

   h. Leaves deciduous, or blades not as above.

      i. Vines; leaves simple; flowers in a terminal panicle; fruit an orange capsule, locules 3, seeds with arils. Celastraceae, p. 148

      i. Not vines or, if so, then leaves compound.

         j. Ovulary 1-loculed; trees, shrubs, and vines.

            k. Stigmas 3, styles 3 or partly united or short; petals 5, small, corolla regular; fruit a drupe. Anacardiaceae, p. 147

            k. Stigma and style 1; petals usually 5 (3-5); corolla regular or zygomorphic; fruit a legume. Leguminosae, p. 133

         j. Ovulary with 2 or more locules.

            k. Leaves simple; locules 4-8; flowers axillary. Aquifoliaceae, p. 148

            k. Leaves 3-foliolate, bladelets dotted; locules 2; flowers in terminal clusters. Rutaceae, p. 143

f. Herbs, including herbaceous vines and small plants slightly woody at base.

  g. Petals 5, corolla usually zygomorphic; stamens 5 or 10 (9); stigma, style, and locule 1; placenta 1, parietal; fruit 1-many-seeded, a legume or indehiscent; leaves usually compound, with pulvini. Leguminosae, p. 133

  g. Without the above set of characters.

   h. Fleshy non-green plants or small evergreens; stamens 8-10, twice the petals; locules 4-5; style 1; stigma large. Pyrolaceae, p. 171

   h. Plants green, not evergreen or, if so, not as above.

      i. Hypanthium globose, tubular, or urn-shaped.

         j. Anthers opening by terminal pores, each with a spur; flowers 4-merous. Melastomaceae, p. 161

         j. Anthers not as above; flowers 4-7-merous. Lythraceae, p. 160

      i. Hypanthium absent or not as above.

         j. Blades circular or spatulate, covered with reddish glandular hairs; small bog herbs. Droseraceae, p. 121

         j. Plants not as above.

            k. Ovulary 1-loculed.

               l. Placenta free central or basal; leaves opposite or whorled, entire. Caryophyllaceae, p. 97

               l. Placenta or placentae parietal, or leaves not as above.

                  m. Placenta 1, or indehiscent fruit 1-seeded, or both.

                     n. Placenta 1, parietal; seeds more than 1.

                        o. Sepals, petals, and stamens 5. Leguminosae, p. 133

                        o. Sepals 6, falling early; petals 6-9; flowering plant with 2 half-circular leaves. Berberidaceae, p. 107

                     n. Fruit 1-seeded.

                        o. With sheathing stipules. Polygonaceae, p. 91

                        o. Stipules absent or not sheathing; petals 4.

                           p. Sepals 2; corolla irregular. Fumariaceae, p. 111

                           p. Sepals 4; corolla regular. Cruciferae, p. 112

                  m. Placentae 2; fruit with 2 or more seeds. (See third m)

                     n. Petals 5 or more.

                        o. Petals 8 or more; leaf 1, basal, circular; juice red; style 1. Papaveraceae, p. 110

                        o. Petals 5; leaves more than 1, mostly basal; styles 2. Saxifragaceae, p. 122

                     n. Petals 4.

9

o. Leaves palmately compound; petals clawed. <u>Capparida-ceae</u>, p. 111

o. Leaves decompound or dissected; outer petals unlike inner, 1 or 2 spurred or saccate. <u>Fumariaceae</u>, p. 111

m. Placentae 3-6; fruit a capsule.

n. Corolla regular; blades simple, entire.

o. Placentae 4; petals white. <u>Saxifragaceae</u>, p. 122

o. Placentae 3; petals yellow or red. <u>Cistaceae</u>, p. 156

n. Corolla zygomorphic; placentae 3-6.

o. Flower-stalks axillary or basal, 1-few-flowered; petals 5; placentae 3. <u>Violaceae</u>, p. 157

o. Flowers in dense racemes; petals cleft. <u>Resedaceae</u>, p. 121

k. Ovulary 2-4-loculed. (See third k)

l. Styles 2-3, each forked; stellate-hairy. <u>Euphorbiaceae</u>, p. 145

l. Plants not as above.

m. Corolla zygomorphic; filaments united.

n. Leaves compound; petals 5. <u>Leguminosae</u>, p. 133

n. Leaves simple; petals 3. <u>Polygalaceae</u>, p. 144

m. Corolla regular, or filaments separate, or both.

n. Tiny herbs of water and mud; petals, stamens, and locules 2-4; leaves opposite. <u>Elatinaceae</u>, p. 156

n. Plants of other habitats, usually larger.

o. Leaves opposite or whorled, entire; locules 2-4, septa often incomplete. <u>Caryophyllaceae</u>, p. 97

o. Leaves alternate, or all basal, or not entire.

p. Sepals and petals 3; ovularies 2-3, almost separate; leaves pinnately divided. <u>Limnanthaceae</u>, p. 147

p. Sepals and petals 5; styles 2; stamens 5-10; leaves mostly basal. <u>Saxifragaceae</u>, p. 122 (See third p)

p. Sepals and petals 4; style 1; stamens 6, two shorter, or stamens 2 or 4. <u>Cruciferae</u>, p. 112

k. Ovulary 5-10-loculed.

l. Blades simple, not lobed.

m. Blades serrate. <u>Crassulaceae</u>, p. 121

m. Blades entire; stamens 5; petals 5; styles 5 or united at base. <u>Linaceae</u>, p. 142

l. Blades lobed or compound; petals 5; ovulary 5-loculed.

m. Plants prostrate; blades pinnately compound; flowers axillary, solitary; fruit spiny. <u>Zygophyllaceae</u>, p. 143

m. Plants not as above.

n. Blades 3-foliolate; stamens 10. <u>Oxalidaceae</u>, p. 142

n. Blades lobed to compound but not 3-foliolate; stamens 5 or 10; fruit with long beak. <u>Geraniaceae</u>, p. 143

SECTION II-B-5. FLOWERS HYPOGYNOUS OR PERIGYNOUS; PERIANTH OF BOTH CALYX AND COROLLA; PETALS UNITED

a. Stamens more numerous than corolla-lobes.

b. Ovulary 1-loculed; style 1; placentation parietal; leaves alternate or basal.

c. Placenta and stigma 1; petals 5 (3-5); corolla zygomorphic or regular; stamens 10 (3-10); herbaceous or woody; leaves mostly 1-2 times compound. <u>Leguminosae</u>, p. 133

c. Placentae 2, stigma 2-lobed; petals 4; corolla zygomorphic or isobilateral; stamens 6; herbs; leaves decompound or dissected. <u>Fumariaceae</u>, p. 111

b. Ovulary with 2 or more locules.

c. Styles or style-branches 5 to many; stamens many, filaments united in a column around style or styles; locules 5 to many; herbs or shrubs. <u>Malvaceae</u>, p. 153

c. Plants not as above.

d. Trees and shrubs, including some small evergreen scarcely woody plants.

e. Styles 4 or united below; deciduous trees; flowers monosporangiate or some perfect; mostly diecious; petals 4, stamens 16 in staminate flowers. <u>Ebenaceae</u>, p. 175

10

e. Style 1 or absent; often evergreen, sometimes small and scarcely woody; flowers usually perfect; petals 5 (4); stamens to twice as many. Ericaceae, p. 171
  d. Herbs.
    e. Corolla regular, petals 5; locules 5; blades 3-foliolate. Oxalidaceae, p. 142
    e. Corolla zygomorphic, petals 3; locules 2; blades simple. Polygalaceae, p. 144
a. Stamens as many as or fewer than corolla-lobes.
  b. Stamens opposite corolla-lobes and as many as corolla-lobes; corolla regular, the lobes obvious; style 1; placenta free central; fruit a capsule. Primulaceae, p. 174
  b. Stamens alternate with corolla-lobes or fewer than corolla-lobes; corolla regular or not.
    c. Corolla scarious, veinless, usually 4-lobed, persistent on fruit; flowers small, in spikes or heads; herbs; usually, leaves basal and ribbed lengthwise. Plantaginaceae, p. 206
    c. Plants not as above.
      d. Plants without chlorophyll; leaves small or bractlike.
        e. White or yellow twining stem-parasites; styles and locules 2. Convolvulaceae, p. 180
        e. Not twining; root-parasites; style and locule 1. Orobanchaceae, p. 204
      d. Plants green, with ordinary leaves.
        e. Carpels 2, ovularies separate; stigmas, or styles, or both, united; blades entire.
          f. Styles and stigmas united; stamens separate; pollen of simple grains. Apocynaceae, p. 178
          f. Styles separate, stigmas united; stamens usually united and adnate to stigma; pollen in pollinia. Asclepiadaceae, p. 179
        e. Carpels 2 or more, at least the ovularies united, or carpel 1; ovulary sometimes 4-lobed and having the appearance of 4 separate ovularies.
          f. Ovulary actually or appearing 4-loculed, separating at maturity into four 1-seeded (or rarely two 2-seeded) nutlets; corolla 4-5-toothed or -lobed.
            g. Leaves alternate; stamens 5. Boraginaceae, p. 183
            g. Leaves opposite; stamens 4 or 2; style-tip 2-lobed; stem often square.
              h. Ovulary usually not deeply 4-lobed; style apical, only 1 lobe stigmatic; plants often not aromatic; stamens 4. Verbenaceae, p. 186
              h. Ovulary usually deeply 4-lobed; style usually basal; foliage nearly always gland-dotted and aromatic; stamens 4 or 2. Labiatae, p. 186
          f. Ovulary not as above.
            g. Style, stigma, locule, and parietal placenta 1; herbs and woody plants; leaves compound or decompound, alternate, with pulvini. Leguminosae, p. 133
            g. Without the above set of characters.
              h. Trees, shrubs, and woody vines.
                i. Anthers opening by terminal pores; not vines; deciduous or evergreen. Ericaceae, p. 171
                i. Anthers not opening by terminal pores or, if so, then plants vines.
                  j. Leaves opposite or whorled.
                    k. Corolla regular; petals 4; stamens usually 2. Oleaceae, p. 176
                    k. Corolla somewhat zygomorphic; petals 5; fertile stamens 2-4.
                      l. Trees, shrubs, and vines; capsule linear or lance-linear. Bignoniaceae, p. 203
                      l. Trees; capsule ovoid. Scrophulariaceae, p. 197
                  j. Leaves alternate.
                    k. Flowers about 1 cm wide; berry red; ovules many; locules 2; shrubs and somewhat woody vines, deciduous. Solanaceae, p. 195
                    k. Flowers smaller than above; fruit a drupe.
                      l. Shrubs and vines; deciduous; ovulary 1-loculed; stigmas 3. Anacardiaceae, p. 147
                      l. Shrubs and small trees; sometimes evergreen; ovulary with 4-8 locules. Aquifoliaceae, p. 148
              h. Herbaceous plants, sometimes herbaceous vines.
                i. Aquatic or mud plants; leaves linear or dissected, bearing tiny bladders; corolla zygomorphic stamens 2. Lentibulariaceae, p. 205
                i. Plants not as above.
                  j. Ovulary 1-loculed.
                    k. Corolla and calyx 2-lipped; flowers 6-8 mm long, in spikes; fruit reflexed, 1-seeded; leaves opposite. Phyrmaceae, p. 206

11

        k. Plants not as above; placentae usually 2, parietal.
           l. Fertile stamens 4; corolla 5-lobed, 2–3 cm long; placentae 2,
               T-shaped; capsule 2-horned. Martyniaceae, p. 204
           l. Fertile stamens as many as corolla-lobes.
              m. Leaves alternate, opposite, or whorled; blades scalelike, or
                 floating, or entire, or 3-foliolate; placentae 2 or rarely
                 ovules on most of inner surface of ovulary; mostly glabrous.
                 Gentianaceae, p. 177
              m. Leaves alternate; blades lobed to compound, not scalelike, not
                 floating, not entire, not 3-foliolate; placentae 2; usually
                 hairy. Hydrophyllaceae, p. 182
     j. Ovulary with 2 or more locules.
        k. Flowers very small, in small axillary peduncled heads; mature fruit
           separating into two 1-seeded nutlets. Verbenaceae, p. 186
        k. Plants not as above.
           l. Fertile stamens 5 (or nearly always 5), as many as petals; corolla
              lobed or entire, regular or nearly so, not 2-lipped.
              m. Vines or, if not, then 2 large bracts beneath the calyx, more
                 or less covering it; corolla usually large and trumpet-shaped,
                 often entire. Convolvulaceae, p. 180
              m. Not vines; without such bracts beneath the calyx.
                n. Anthers opening by terminal slits or pores, or connate or
                   connivent in a cone, or both. Solanaceae, p. 195
                n. Anthers not as above.
                  o. Stigmas 3; ovulary 3-loculed; leaves simple and entire
                   or compound with entire leaflets or pinnately parted into
                   linear segments. Polemoniaceae, p. 181
                  o. Stigmas and locules not 3, or leaves not as above.
                    p. Blades compound or pinnatifid or palmately lobed;
                      style cleft at apex or styles 2; locules 2.
                      Hydrophyllaceae, p. 182
                  p. Blades entire, toothed, or shallowly pinnately lobed;
                      style 1; stigma 1 or somewhat lobed.
                    q. Flowers in elongate single or panicled racemes or
                      spikes; locules 2; corolla rotate or saucer-
                      shaped, not plaited. Scrophulariaceae, p. 197
                    q. Flower-clusters not as above; locules 2–4, if 2,
                      then corolla tubular, funnelform, or salverform,
                      plaited and often twisted in bud; calyx often en-
                      larged in fruit. Solanaceae, p. 195
           l. Fertile stamens 4 or 2, fewer than corolla-lobes or, if the same
            number as corolla-lobes, then corolla irregular; locules 2.
              m. Seeds few (about 2–8) in each locule; leaves simple, opposite,
                blade-margins usually entire or nearly so; seeds on curved
                projections; capsule splitting to base elastically; if stamens
                are 4, then corolla is not 2-lipped. Acanthaceae, p. 205
              m. Seeds many in each locule; leaves alternate or opposite, blade-
                margins various; seed-stalks and capsule not as above; if,
                rarely, locules are few-seeded, then usually stamens are 4
                and corolla is 2-lipped. Scrophulariaceae, p. 197

SECTION II-B-6. FLOWERS WHOLLY OR PARTIALLY EPIGYNOUS

a. Stems of fleshy spiny segments jointed together; leaves small or absent; perianth parts and stamens
   many, intergrading, on epigynous hypanthium. Cactaceae, p. 159
a. Stems not as above; leaves present, evident.
   b. Herbs.
      c. Flowers in involucrate heads; ovulary 1-loculed; fruit an achene.
        d. Stamens 2–4, separate; flowers perfect; corollas tubular; calyx small, cup-shaped; leaves
           opposite; awns of bracts of receptacle exceeding flowers. Dipsacaceae, p. 212

d. Stamens 5, filaments separate, anthers usually united in ring around style; flowers bisporangiate, monosporangiate, or neutral; corollas all tubular, all ligulate, or both kinds in same head; leaves opposite, alternate, or whorled, sometimes all basal. Compositae, p. 214

c. Flowers not in involucrate heads or, if so, then ovulary not 1-loculed.
  d. Aquatic, some or all leaves submersed, dissected; flowers emersed.
    e. Flowers axillary or in spikes; sepals 3-4; petals 4 or none. Haloragaceae, p. 163
    e. Flowers in umbels; perianth parts in 5's. Sium in Umbelliferae, p. 165
  d. Not aquatics as described above.
    e. Either calyx or corolla present, but not both.
      f. Small plant with decumbent stem, blades round-ovate; carpels 2, united below, ovulary 1-loculed below; sepals usually 4; stamens at edge of disk, anthers yellow to red. Saxifragaceae, p. 122
      f. Plants not as above.
        g. Ovulary 1-loculed; sepals 5; stamens 5. Santalaceae, p. 91
        g. Ovulary with more than 1 locule, or calyx minute or pappuslike.
          h. Perianth parts (petals) 5.
            i. Petals united; stamens usually 1-3 (-5); locules 1, or locules 3, 2 empty; calyx minute or expanding late and appearing pappuslike. Valerianaceae, p. 211
            i. Petals separate; flowers in umbels or rarely in heads.
              j. Styles 2, with stylopodium; petiole-bases sheathing; fruit dry, the two 1-seeded carpels separating at maturity. Umbelliferae, p. 165
              j. Styles 2-5, without stylopodium; if only 2, then plant with a single whorl of palmately compound leaves; fruit a berry or drupe. Araliaceae, p. 164
          h. Perianth parts (petals or sepals) 4 or 3.
            i. Leaves alternate, sometimes basal; ovulary 6-loculed; sepals 3. Aristolochiaceae, p. 91
            i. Leaves opposite, whorled, or alternate; locules fewer.
              j. Locules of ovulary 4; sepals and stamens 4; leaves not whorled. Onagraceae, p. 161
              j. Locules of ovulary 2; petals and stamens 3 or 4; leaves whorled. Rubiaceae, p. 206
    e. Both calyx and corolla present.
      f. Petals separate.
        g. Sepals 2; ovulary 1-loculed.
          h. Leaves entire, fleshy; capsule circumscissile. Portulacaceae, p. 97
          h. Leaves toothed, not fleshy; fruit indehiscent. Onagraceae, p. 161
        g. Sepals more than 2 or ovulary more than 1-loculed or both.
          h. Ovulary only partly inferior; carpels 2, more or less united below; ovulary 2-loculed or 1-loculed with 2 parietal placentae; petals 5, stamens 5-10. Saxifragaceae, p. 122
          h. Ovulary wholly inferior or nearly so.
            i. Flowers in dense terminal cluster subtended by 4 large petaloid bracts; petals, sepals, and stamens 4. Cornaceae, p. 170
            i. Flower clusters not subtended by large petaloid bracts.
              j. Flowers not in umbels or heads; petals usually 4 (2-6), stamens as many or twice as many. Onagraceae, p. 161
              j. Flowers in umbels or rarely in heads; petals and stamens usually 5.
                k. Styles 2, with stylopodia; petiole bases sheathing; fruit dry, the two 1-seeded carpels separating at maturity. Umbelliferae, p. 165
                k. Styles 2-5, without stylopodia; if only 2, then plant with a single whorl of palmately compound leaves; fruit a berry or drupe. Araliaceae, p. 164
      f. Petals united.
        g. Leaves alternate on the stem or basal.
          h. Stamens usually 3 in ours; flowers monosporangiate; vines with tendrils. Cucurbitaceae, p. 212
          h. Stamens 5; flowers bisporangiate; not vines.

13

          i. Stamens opposite the petals; placenta free, central; ovulary only partly inferior. <u>Primulaceae</u>, p. 174
          i. Stamens alternate with the petals; ovulary 2-several-loculed, wholly inferior. <u>Campanulaceae</u>, p. 213
      g. Leaves opposite or whorled.
        h. Corolla-lobes 4, rarely 3; stamens 4, rarely 3; ovulary 2-loculed <u>or</u>, if 4-loculed, then plant a small prostrate evergreen; leaves opposite with stipules or whorled without apparent stipules. <u>Rubiaceae</u>, p. 206
        h. Corolla-lobes 5; leaves opposite, stipules present or absent.
          i. Ovulary 3-5-loculed; stamens 5; calyx deeply parted. <u>Caprifoliaceae</u>, p. 209
          i. Ovulary 3-loculed, 2 locules empty, or 1-loculed; ovule 1; stamens usually 1-3 (5); calyx minute or expanding late and becoming pappuslike. <u>Valerianaceae</u>, p. 211
b. Woody plants, including a few small scarcely woody evergreens.
  c. Leaves opposite or whorled.
    d. Shrubby evergreen plants parasitic on trees; leaf-blades entire, coriaceous; flowers in short axillary spikes. <u>Loranthaceae</u>, p. 91
    d. Plants not growing on other plants, not parasitic; sepals sometimes minute.
      e. Petals separate.
        f. Stamens 4; ovulary 2-loculed, 1 ovule in each locule; fruit a drupe. <u>Cornaceae</u>, p. 170
        f. Stamens 8 to many; ovulary 2-4-loculed, 2 or more ovules in each locule; fruit a capsule. <u>Saxifragaceae</u>, p. 122
      e. Petals united.
        f. Corolla 4-lobed; erect shrubs or small trees with flowers in heads, or small evergreens with prostrate stems. <u>Rubiaceae</u>, 206
        f. Corolla 5-lobed or, if rarely 4-lobed, then plants erect shrubs with flowers not in heads. <u>Caprifoliaceae</u>, p. 209
  c. Leaves alternate.
    d. Flowers monosporangiate, the staminate in loose globular clusters or in erect spikes, carpellate flowers and fruits within a bur. <u>Fagaceae</u>, p. 86
    d. Flowers bisporangiate, or clusters not as above.
      e. Ovulary 1-loculed.
        f. Shrubs, sometimes prickly; petals, sepals, and stamens 5; epigynous hypanthium present; blades palmately lobed. <u>Saxifragaceae</u>, p. 122
        f. Trees, without prickles; hypanthium absent; sepals and petals about 5, small or petals none; stamens about 10; leaves pinnately veined, entire or nearly so. <u>Nyssaceae</u>, p. 171
      e. Ovulary 2-5 loculed; petals present; sepals present or absent, sometimes minute or represented by only a rim.
        f. Petals separate.
          g. Petals, sepals, and stamens 4; ovulary 2-loculed.
            h. Petals white; flowers in terminal clusters; style 1. <u>Cornaceae</u>, p. 170
            h. Petals yellow; flowers in axillary clusters; styles 2. <u>Hamamelidaceae</u>, p. 124
          g. Petals and sepals 5; stamens 5-many.
            h. Stamens as many as petals and opposite them; only base of ovulary inferior; leaves simple. <u>Rhamnaceae</u>, p. 151
            h. Stamens alternate with petals or more than petals; ovulary wholly inferior or nearly so.
              i. Leaves decompound; stems and leaves with prickles; flowers in umbels; calyx small. <u>Araliaceae</u>, p. 164
              i. Leaves simple or once compound; stamens 5 to many; flower clusters various; calyx evident. <u>Rosaceae</u>, p. 125
        f. Petals united; stamens as many as petals to twice as many or more.
          g. Pollen sacs opening by terminal pores, or plants small trailing leafy evergreens; stamens not united in a ring at base. <u>Ericaceae</u>, p. 171
          g. Pollen sacs not opening by pores; deciduous trees and shrubs; stamens united in a ring; ovulary sometimes 1-loculed above. <u>Styracaceae</u>, p. 175

EQUISETACEAE, Horsetail Family

Equisetum L., Horsetail, Scouring-rush

Stems erect or decumbent, from rhizomes, ridged and grooved, deposits of silica on the ridges; stomates in the grooves in definite rows, in bands, or scattered; leaves small, united except at their tips, forming sheaths at the nodes, the free tips persistent or deciduous; sporophylls in cones, each sporophyll with a stalk and a peltate top beneath which several sporangia are borne.

a. Stomates not in 2 lines in the grooves; stems containing chlorophyll and branched or stems with little or no chlorophyll, fertile, usually unbranched.
  b. Branches sometimes few, not again branched; stems green, not dimorphic, central cavity 4/5 the diameter; teeth of sheaths not deciduous. E. fluviatile L., Water H.
  b. Sterile stems with many branches, green; fertile stems often unbranched, brown or pink, with little or no chlorophyll.
    c. Branches of sterile stem again branched, slender, spreading; teeth of sheaths coherent as 3-4 lobes, translucent; green branches developing on fertile stems; ridges flat-topped or concave, with short hairs or spinules on edges. E. silvaticum L.
    c. Branches of sterile stem usually not again branched; teeth usually separate; fertile stems soon withering; ridges not as above. E. arvense L., Field H.
a. Stomates in 2 lines in the grooves (except in E. fluviatile); stems not dimorphic, unbranched or with few branches, containing chlorophyll.
  b. Teeth of sheaths persistent; stem 5-12-ridged, ridges usually with 2 rows of tubercles. E. variegatum Schleich.
  b. Teeth persistent or deciduous; stem 10-many-ridged, ridges with 1 row of tubercles.
    c. Ridges of sheath not wider than grooves; stem smooth; teeth persistent, separate, lanceolate, not hyaline-margined. E. fluviatile L., Water H.
    c. Ridges of sheath more or less flattened, wider than grooves; teeth hyaline-margined, sometimes deciduous.
      d. Sheath longer than wide, flaring at summit, usually with 1 dark band at top; teeth deciduous. E. laevigatum A. Br. (E. hyemale var. intermedium A. A. Eat.; including E. kansanum Schaffn.)
      d. Sheath about as long as wide, cylindric, usually with dark band at top and at base; teeth deciduous or persistent; ridges rough. E. hyemale L.

LYCOPODIACEAE, Club-moss Family

Lycopodium L., Club-moss

Perennial, evergreen; main stem trailing on ground or subterranean, dichotomously branched, one branch usually erect; leaves small, scalelike, sometimes with decurrent base adnate to stem, or linear, 4-many-ranked; sporophylls in cones or in zones alternating with zones of foliage leaves; sporangia globose or reniform, solitary on base of sporophyll or in its axil.

a. Sporophylls in zones alternating with zones of foliage leaves.
  b. Leaves minutely toothed, widest above middle. L. lucidulum Michx., Shining C.
  b. Leaves scarcely toothed, widest at base. L. selago L. var. patens (Beauv.) Desv.
a. Sporophylls in terminal cones, sometimes differing in shape from foliage leaves.
  b. Stem flat; foliage leaves 4-ranked, scalelike; cones peduncled, often clustered; erect branches fanlike, forming funnels.
    c. Horizontal branches underground; tip of each leaf on lower surface of stem reaching to or almost to base of one next above. L. tristachyum Pursh
    c. Horizontal branches mostly on surface of ground; tip of leaf on lower surface of stem not reaching base of one next above. L. complanatum L. var. flabelliforme Fern., Trailing C.
  b. Stems not flat; foliage leaves 6-10-ranked, linear to linear-lanceolate.
    c. Erect stems unbranched; cone usually solitary, wider than stem (plus leaves) below it; sporophylls linear, base widened. L. inundatum L., Bog C.
    c. Erect stems branched; sporophylls ovate.
      d. Foliage leaves with long hairlike tips; cones 1-several, bracts of long peduncle remote, yellowish. L. clavatum L.

d. Foliage leaves with acute tips; cones solitary and sessile at ends of leafy branches; erect branches treelike. L. obscurum L., Ground-pine.

SELAGINELLACEAE, Selaginella Family

Selaginella Beauv.

Small herbs; stems erect or prostrate, branched; leaves small, each bearing on upper side near base a small membranous flap, the ligule; microsporophylls and megasporophylls 4-ranked, in same cones; cones sessile on leafy branches.

a. Sterile leaves 4-ranked, ovate, those in the lateral rows the larger; plants small, delicate, light green, of moist habitats. S. apoda (L.) Fern.
a. Sterile leaves many-ranked, uniform in size, narrow, awn-tipped; plants gray-green, of dry habitats. S. rupestris (L.) Spring

ISOETACEAE, Quillwort Family

Isoetes L., Quillwort

Growing in water or mud; leaves from flattened stems, quill-like or rushlike, with 4 lengthwise transversely-septate cavities; sporangia solitary in enlarged leaf-bases, the two kinds in different leaf-bases; ligule borne just above sporangium.

a. Surface of megaspore spinulose. I. echinospora Dur. (I. muricata Dur.)
a. Surface of megaspore reticulate. I. engelmanii A. Br.

OPHIOGLOSSACEAE, Adder's-tongue Family

With erect rhizomes and fleshy roots; leaf (portion of plant above ground) consisting of branched petiole, one branch sterile, bearing a green foliaceous structure (called blade, in the keys below), the other branch fertile, bearing sporangia.

a. Blade lobed or compound; veins ending free. Botrychium
a. Blade simple, entire; veins reticulate. Ophioglossum

1. Botrychium Sw., Grape Fern

Blade lobed or compound, sessile or stalked above common petiole; fertile branch stalked or sessile, 1-3-pinnate.

a. Blade evergreen, on stalk usually 5 cm long or more above common petiole, borne near base of plant; spores ripening in late summer or autumn.
  b. Pinnules mostly 2-4 times as long as wide.
    c. Pinnules acute at tip; blade usually bronze in autumn. B. dissectum Spreng.
      d. Pinnules cut into linear teeth or lobes. var. dissectum
      d. Pinnules lobed or unlobed, margins serrulate-dentate. var. obliquum (Muhl.) Clute
    c. Pinnules rounded to obtuse; blade green in autumn. B. oneidense (Gilbert) House
  b. Pinnules mostly less than twice as long as wide, rounded to obtuse; blade green in autumn.
    B. multifidum (Gmel.) Rupr.
a. Blades deciduous, sessile or the stalk to 3 cm long, usually borne near or above middle of plant; spores ripening in spring or summer.
  b. Blade bipinnate-pinnatifid to tripinnate, deltoid, usually more than 10 cm wide. B. virginianum (L.) Sw., Rattlesnake Fern
  b. Blade entire to pinnate-pinnatifid, less than 10 cm wide.
    c. Blade simple or with 1-4 pairs of rounded pinnae. B. simplex E. Hitchc.
    c. Blade pinnate-pinnatifid.
      d. Blade deltoid, sessile. B. lanceolatum (Gmel.) Rupr.
      d. Blade oblong or ovate, sessile or short-stalked. B. matricariaefolium A. Br.

2. Ophioglossum L., Adder's-tongue

Blade entire, elliptic, sessile on the common petiole; sporangia coherent in 2 rows in a spikelike cluster with stalk about as long as common petiole.

a. Islets formed by principal veins not enclosing smaller islets. O. vulgatum L.
a. Islets formed by principal veins enclosing smaller islets; rare. O. engelmanni Prantl

OSMUNDACEAE, Royal Fern Family

Osmunda L.

Tall; leaves pinnate to bipinnate, dimorphic, fertile leaves or fertile pinnae not foliaceous; sporangium opening by longitudinal (vertical) cleft.

a. Leaves bipinnate; lower pinnae of fertile leaves foliaceous, only upper pinnae bearing sporangia.
  O. regalis L., Royal Fern.
a. Sterile leaves pinnate-pinnatifid.
  b. All pinnae of fertile leaves bearing sporangia, none foliaceous; mature pinnae of sterile leaves brown-tomentose at base and ciliate at margin. O. cinnamomea L., Cinnamon Fern.
  b. Middle pinnae of fertile leaves bearing sporangia, upper and lower foliaceous; mature sterile pinnae not or slightly brown-tomentose at base, not ciliate at margin. O. claytoniana L., Interrupted Fern.

SCHIZAEACEAE, Curly-grass Family

Lygodium Sw., Climbing Fern

L. palmatum (Bernh.) Sw. Leaves flexible, twining, of indeterminate growth; pinnae stalked, in pairs on branches of rachis, the sterile deeply palmately lobed, the fertile apical, several times forked, smaller. Rare.

HYMENOPHYLLACEAE, Filmy Fern Family

Trichomanes L., Filmy Fern

T. boschianum Sturm Small, delicate; blades thin, pinnate-pinnatifid; sporangia on base of an elongate axis within a cup-shaped indusium, the apical portion of the axis projecting beyond the cup.

POLYPODIACEAE, Polypody Family

Leaves simple to decompound, often bearing scales, especially toward base; sporangia in sori on backs or margins of blades; indusia present or absent.

In some species, apical part of blade is less divided than the remainder; when this is true, the phrase "except at tip" is implied in statements of degree of division of blade. When position of indusium is stated, it is implied that leaf is lying with under side upward.

a. Leaf-blade simple, entire. 9. Camptosorus
a. Leaf-blades compound or pinnatifid.
  b. Fertile leaves without flat green blades, unlike the sterile.
    c. Rachis of sterile blade not winged. 3. Pteretis
    c. Rachis of sterile blade winged.
      d. Sterile blade ovate or deltoid; groups of sori enclosed in globose pinnules. 4. Onoclea
      d. Sterile blade lanceolate; sori oblong, in rows, not enclosed. 11. Woodwardia
  b. Fertile and sterile leaves (or pinnae) both with flat green blades, but not always equal in size.
    c. Sori neither with true indusia nor covered by reflexed blade-margin.
      d. Blades longer than wide, once pinnatifid. 15. Polypodium
      d. Blades about as long as wide, pinnate-pinnatifid to tripinnate-pinnatifid. 5. Dryopteris
    c. Sori marginal, covered by reflexed blade-margin. (See third c.)
      d. Sori discontinuous, covered by reflexed pinnule-teeth. 13. Adiantum
      d. Sori continuous, covered by reflexed pinna- or pinnule-margin.

  e. Blades lanceolate, pinnate or bipinnate. 12. <u>Pellaea</u>
  e. Blades deltoid, ternate, each part bipinnate. 14. <u>Pteridium</u>
 c. Sori with true indusia.
  d. Indusium wholly or partly below sporangia (between sporangia and blade).
   e. Sori marginal; indusium cuplike. 7. <u>Dennstaedtia</u>
   e. Sori not marginal.
    f. Indusium hoodlike, attached at side, partly under sporangia, arching over them at
     least when young. 2. <u>Cystopteris</u>
    f. Indusium disk- or cuplike, attached at center, edge folded over sporangia, later
     separating into lobes and spreading. 1. <u>Woodsia</u>
  d. Indusium above sporangia (sporangia between indusium and blade), covering them at least
   when young.
   e. Indusium elongate, straight or curved, attached along one side.
    f. Sori in chainlike rows parallel to midveins of pinnae or their divisions.
     11. <u>Woodwardia</u>
    f. Sori oblique to midveins of pinnae or their divisions.
     g. Small ferns; leaves mostly evergreen, mostly less than 5 cm wide at widest part.
      10. <u>Asplenium</u>
     g. Large ferns; leaves deciduous, 10-30 cm wide at widest part. 8. <u>Athyrium</u>
   e. Indusium circular or reniform, attached at center.
    f. Indusium without a sinus; sori on only smaller upper pinnae of fertile blades; blades
     pinnate, leaflets auricled at base. 6. <u>Polystichum</u>
    f. Indusium with a sinus, circular or reniform; sori not confined to upper pinnae of
     fertile leaves; blades pinnate-pinnatifid to tripinnate. 5. <u>Dryopteris</u>

1. Woodsia R. Br.

  <u>W</u>. <u>obtusa</u> (Spreng.) Torr. Low tufted ferns; blades lanceolate, pinnate-pinnatifid to bipinnate-pinnatifid; rachis and petiole glandular; sorus globose; indusium attached beneath the sporangia, the edge folded over and enclosing them when young, later splitting into several toothed segments and spreading.

2. Cystopteris Bernh., Bladder Fern

  Leaves deciduous; blades thin, lanceolate to ovate, pinnate-pinnatifid to tripinnate; indusium hoodlike, partly beneath the sporangia, covering them when young.

a. Blades widest at base, the fertile much prolonged, often bearing bulblets; veins ending mostly between teeth. <u>C</u>. <u>bulbifera</u> (L.) Bernh., Bulblet Fern
a. Blades widest just above base, not prolonged, without bulblets; veins ending mostly in teeth.
 <u>C</u>. <u>fragilis</u> (L.) Bernh., Fragile Fern

3. Pteretis Raf., Ostrich Fern

  <u>P</u>. <u>pensylvanica</u> (Willd.) Fern. Leaves dimorphic, rachis and petiole grooved; sterile blades large, oblanceolate, pinnate-pinnatifid; fertile blades shorter, pinnae smaller, ascending, pinnules beadlike, edges inrolled about groups of sori.

4. Onoclea L., Sensitive Fern

  <u>O</u>. <u>sensibilis</u> L. Leaves arising singly from spreading rhizomes, dimorphic; sterile leaves deciduous, blade deltoid, deeply pinnatifid, rachis winged, wing wider toward tip, veins anastomosing, ending free; fertile blades twice pinnate, pinnae smaller, pinnules beadlike, margins inrolled about groups of sori.

5. Dryopteris Adans., Shield Fern

  Leaves evergreen or deciduous, bipinnatifid to tripinnate; sori circular, with or without indusia.

a. Sori without indusia; blades triangular, length and width nearly equal.
 b. Lowest pair of pinnae stalked, compound at base, widest at base; rare. <u>D</u>. <u>disjuncta</u> (Ledeb.)
 C. V. Mort., Oak Fern

b. Lowest pair of pinnae sessile, narrowed at base.
   c. Lowest pair of pinnae joined by wing to next above; blades usually not longer than wide.
      D. hexagonoptera (Michx.) Christens., Broad Beech Fern
   c. Lowest pair of pinnae not thus joined; blades usually longer than wide. D. phegopteris (L.)
      Christens., Long Beech Fern
a. Sori with indusia; blades longer than wide.
  b. Blades pinnate-pinnatifid, at least the sterile ones thin; veins in base of petiole 2, free or united;
    deciduous.
   c. Lower pinnae very short; blades delicate. D. noveboracensis (L.) Gray, New York Fern
   c. Lower pinnae little shorter than middle ones; fertile blades somewhat thicker than sterile
      ones, pinnae narrower, revolute-margined. D. thelypteris (L.) Gray var. pubescens (Law-
      son) Nakai, Marsh Fern
  b. Blades pinnate-pinnatifid to tripinnate, of firmer texture than above; veins in base of petiole 3-5;
    usually evergreen.
   c. Sori near margin; blades gray- or blue-green, firm, pinnate-pinnatifid to bipinnate-
      pinnatifid. D. marginalis (L.) Gray, Marginal S.
   c. Sori not at margin; blades green.
    d. Blades bipinnate to tripinnate; lower pinnae strongly asymmetric; lobes or teeth with mu-
      cronate or spinulose tips. D. spinulosa (O. F. Muell.) Watt., Spinulose S.
     e. Basal pinnule on lower side of lowest pinna longer than one next to it; pinnae ascending;
       blade glabrous. var. spinulosa
     e. Basal pinnule, described above, not longer than one next to it; pinnae spreading at al-
       most a right angle to rachis; blade glandular. var. intermedia (Muhl.) Underw.
    d. Blades pinnate-pinnatifid to bipinnate; lower pinnae almost symmetric; lobes or teeth not
      or only slightly mucronate.
     e. Blades ovate, abruptly narrowed at tip; pinnae narrowed slightly to base. D. goldiana
       (Hook.) Gray, Goldie's Fern
     e. Blades lanceolate, gradually narrowed to tip; pinnae widest at their bases. D. cristata
       (L.) Gray
      f. Basal pinnae of fertile leaves twisted on rachis, as wide or almost as wide as long.
        var. cristata. D. bootii (Tuckerm.) Underw. is a hybrid with D. spinulosa var.
        intermedia.
      f. Basal pinnae of fertile leaves flat or nearly so, twice as long as wide or more. var.
        clintoniana (D. C. Eat.) Underw.

6. Polystichum Roth

    P. acrostichoides (Michx.) Schott, Christmas Fern. Leaves evergreen; blades pinnate, pinnae with basal auricle on upper side; fertile blades longer than sterile, pinnae on upper half smaller and bearing sori; indusia circular, peltate.

7. Dennstaedtia Bernh.

    D. punctilobula (Michx.) Moore, Hay-scented Fern. Rhizomes extensive; leaves scattered, deciduous; blades delicate, fragrant, bipinnate-pinnatifid, lanceolate; indusium cuplike, formed partly from a recurved tooth of leaf-margin.

8. Athyrium Roth

    Woodland ferns; leaves large, deciduous; blades thin, pinnate to tripinnate; sori linear or oblong, straight, curved, or U-shaped, at side of, and sometimes crossing, small veins; indusium attached along one side.

a. Blades pinnate; pinnae linear-lanceolate, entire, fertile slightly narrower than sterile. A. pycno-
   carpon (Spreng.) Tidestr., Narrow-leaved Spleenwort
a. Blades pinnate-pinnatifid to tripinnate.
  b. Blades pinnate-pinnatifid; ultimate segments blunt, entire or crenate. A. thelypteroides (Michx.)
    Desv., Silvery Spleenwort
  b. Blades mostly bipinnate, pinnules toothed to deeply pinnatifid. A. filix-femina (L.) Roth, Lady
    Fern

c. Blade widest near middle; cilia of indusia glandless. var. michauxii (Spreng.) Farw.
c. Blade widest just above base; cilia of indusia glandular. var. asplenioides (Michx.) Farw.

9. Camptosorus Link, Walking Fern

C. rhizophyllus (L.) Link   Small evergreen fern; leaves clustered, spreading; blades lanceolate, cordate at base, entire or undulate, long-attenuate to freely-rooting tips; sori elongate, irregularly scattered, sometimes paired.

10. Asplenium L. Spleenwort

Small, evergreen; blades pinnatifid to bipinnate; sori elongate, straight or slightly curved, indusium attached along one edge. Several known or reputed interspecific and intergeneric (with Camptosorus) hybrids have been collected or reported in Ohio.

a. Blades once pinnatifid or once pinnate.
  b. Blades pinnatifid, or base sometimes pinnate, long-attenuate at apex. A. pinnatifidum Nutt.
    A. trudellii Wherry is a hybrid with A. montanum.
  b. Blades once pinnate; petiole and rachis dark.
    c. Pinnae lobed or auricled at base.
      d. Pinnae usually opposite, about 1 cm long; fertile and sterile leaves similar; rare.
        A. resiliens Kunze, Small S.
      d. Pinnae usually alternate, those of the fertile blade to 3 cm long; sterile leaves spreading, shorter than the fertile erect ones. A. platyneuron (L.) Oakes, Ebony S.
    c. Pinnae not lobed or auricled at base, roundish-oblong, inequilateral. A. trichomanes L., Maidenhair S.
a. Blades pinnate-pinnatifid to bipinnate-pinnatifid.
  b. Rachis dark, at least at base; blades pinnate-pinnatifid. A. bradleyi D. C. Eat.
  b. Rachis green throughout, but petiole may be dark.
    c. Petiole dark at base; blades once pinnate, pinnae sometimes deeply cleft or pinnate at base, 6 or more on each side of rachis. A. montanum Willd., Mountain S.
    c. Petiole green to base; blades once pinnate above, twice pinnate below; pinnae fewer than 6 on each side; ultimate leaflets rhombic, cuneate at base, cleft or toothed. A. ruta-muraria L. (A. cryptolepis Fern.)

11. Woodwardia Sm., Chain Fern

Bog ferns; leaves scattered, deciduous; blades pinnatifid to pinnate-pinnatifid; venation areolate near the midveins, free near the margins; sori oblong to linear, in chainlike rows parallel to midveins of pinnae or of segments of pinnae; indusium attached by outer edge, opening at side toward midvein.

a. Fertile and sterile blades similar, pinnate-pinnatifid, rachis not winged. W. virginica (L.) Smith
a. Sterile blades pinnatifid, rachis broadly winged toward tip, segments lanceolate; fertile blades pinnatifid or almost pinnate, segments linear. W. areolata (L.) Moore

12. Pellaea Link, Cliff Brake

Small ferns growing on rocks; leaves clustered, evergreen; blades pinnate to bipinnate; fertile leaves somewhat taller than sterile, pinnae narrower; petioles dark; sori marginal, covered by reflexed margins of pinnae or pinnules.

a. Petiole and rachis pubescent; lower pinnae of fertile leaves pinnate. P. atropurpurea (L.) Link, Purple C.
a. Petiole and rachis glabrous or nearly so; lower pinnae of fertile leaves usually trifoliolate. P. glabella Mett., Smooth C.

13. Adiantum L.

Graceful, delicate; leaves decompound, pinnules lobed, dichotomously veined; rachis dark, lustrous; sori marginal, covered by reflexed pinnule-lobes.

a. Main rachis dichotomous, the 2 branches arching, the elongate compound pinnae borne along convex side of the curves. A. pedatum L., Maidenhair Fern

a. Main rachis not dichotomous, the leaf pinnately compound or decompound. *A. capillus-veneris L.,
   Venus's-hair Fern

14. Pteridium Gleditsch, Bracken

P. aquilinum (L.) Kuhn   Leaves deciduous, scattered along the extensive forking rhizomes; glands in axils of lower pinnae; blades sometimes several feet wide, firm, bipinnate-pinnatifid to tripinnate-pinnatifid; sori marginal, covered by pinnule-margins.

15. Polypodium L., Polypody

Often growing on rocks or epiphytic on trees; leaves evergreen, subcoriaceous; blades pinnatifid; sori circular, without indusia.

a. Lower surface of blades green and glabrous. P. vulgare L., Common P.
a. Lower surface of blades gray or brownish, covered with peltate scales. P. polypodioides (L.) Watt, Gray P., Resurrection Fern

MARSILEACEAE, Marsilea Family

Marsilea L.

*M. quadrifolia L. Aquatic, rooting in mud; slender petioles arising from rhizomes, blades 4-foliolate; megasporangia and microsporangia together in sporocarps borne on the petioles.

SALVINIACEAE, Salvinia Family

Small floating water ferns; microsporangia and megasporangia in separate sporocarps.

a. Leaves about 0.5 mm long. Azolla
a. Leaves 1 cm long or more. Salvinia

Azolla Lam.

A. caroliniana Willd. Minute plants 1/2 to 1 cm across; leaves alternate, papillose, 2-lobed, upper lobe emersed, lower one submersed; roots delicate.

Salvinia Adans.

*S. natans (L.) All. Leaves in whorls of 3, the 2 upper oval, spreading, floating, papillose-hairy, the lower dissected, rootlike, submersed, bearing sporocarps.

GYMNOSPERMS

GINKGOACEAE, Ginkgo Family

Ginkgo L.

*G. biloba L., Maidenhair Tree. Diecious; leaves deciduous, alternate, fan-shaped, often lobed or bifid, dichotomously-veined, crowded on dwarf branches or scattered on elongate stems; stamens and carpels on dwarf branches, appearing with the leaves.

TAXACEAE, Yew Family

Taxus L., Yew

T. canadensis Marsh., American Y. Evergreen shrub; leaves flat, linear, 1-2.5 cm long, spirally arranged, spreading and appearing 2-ranked; stamens peltate, several on a short axillary stem subtended by scales, each stamen with 4-6 microsporangia; carpellate cone a single ovule surrounded at maturity by a fleshy red aril.

PINACEAE, Pine Family

Leaves needlelike or flat and narrowly linear, alternate, sometimes crowded on dwarf branches; leaf-buds scaly; ovulate cones becoming coriaceous or woody.

a. Leaves not in fascicles or clusters on dwarf branches, those on dorsal side of stem-segment shorter than those spreading laterally.   1. <u>Tsuga</u>
a. At least some of the leaves in fascicles or clusters on dwarf branches.
  b. Deciduous; leaves in clusters of more than 5 on dwarf branches and also some leaves scattered on elongate stems.   2. <u>Larix</u>
  b. Evergreen; leaves in fascicles of 2-5 on dwarf branches.   3. <u>Pinus</u>

1. <u>Tsuga</u> (Endl.) Carr., Hemlock

<u>T</u>. <u>canadensis</u> (L.) Carr., Canadian H. Evergreen tree; leaves flat, linear, 5-15 mm long, short-petioled, with 2 white bands of stomates beneath, those lateral and ventral on the stem spreading in one plane, those dorsal on the stem shorter and appressed; monecious; cones on twigs of previous season, staminate globose, very small, carpellate ellipsoid, 1-2 cm long; seeds winged.

2. <u>Larix</u> Mill., Larch

<u>L</u>. <u>laricina</u> (DuRoi) K. Koch, Tamarack. Tall deciduous tree; some branches dwarf, tipped with leaf-clusters, some elongate with leaves scattered; leaves soft, slender, 1-2.5 cm long; cones on dwarf branches of a previous year, staminate globose, very small, carpellate 1-2 cm long; seeds winged.

3. <u>Pinus</u> L., Pine

Trees and shrubs with needle leaves in fascicles of 2-5 on dwarf branches that are borne in axils of scale-leaves; young cones on stem-segments of current year, staminate cylindric or oblong, near base of segment, carpellate above base; ovulate cones maturing at end of second or third season, becoming woody; seeds winged.

a. Leaves 5 in a fascicle; mature ovulate cones cylindric or slightly curved, drooping. <u>P</u>. <u>strobus</u> L., White P.
a. Leaves 2 or 3 in a fascicle.
  b. Leaves 3 in a fascicle, 7-14 cm long; young stems light brown or orange, not glaucous; cone-scales with sharp prickle. <u>P</u>. <u>rigida</u> Mill., Pitch P.
  b. Leaves 2 or 3 in a fascicle on same tree, 7-12 cm long; young stems purple or red-brown, glaucous. <u>P</u>. <u>echinata</u> Mill., Shortleaf or Yellow P.
  b. Leaves 2 in a fascicle.
    c. Leaves 9 cm long or more.
      d. Leaves stiff, often curved, not breaking sharply when bent; cone-scales usually with short prickle; young stems dark. *<u>P</u>. <u>nigra</u> Arnold, Austrian P.
      d. Leaves flexible, often breaking sharply when bent; cone-scales without a prickle; young stems yellow or orange; staminate cones red or purple. *<u>P</u>. <u>resinosa</u> Ait., Red P.
    c. Leaves 8 cm long or less.
      d. Young stems glaucous, purple; leaves usually twisted, 4-8 cm long; cone-scales with persistent prickle. <u>P</u>. <u>virginiana</u> Mill., Scrub P.
      d. Young stems not glaucous; prickle of cone-scales minute or none.
        e. Leaves 2-4 cm long; young stems purple or yellow-brown; cone curved, remaining closed for years. *<u>P</u>. <u>banksiana</u> Lamb., Jack P.
        e. Leaves 3-7 cm long; young stems grayish-yellow; cone reflexed; bark of upper trunk and branches yellow. *<u>P</u>. <u>sylvestris</u> L., Scots P.

CUPRESSACEAE (PINACEAE, in part), Cypress Family

Leaves opposite or whorled, scalelike or needlelike; leaf-buds not scaly; sporophylls few, cones small, the ovulate becoming coriaceous or berrylike.

a. Leaves opposite, 4-ranked, scalelike; ovulate cone becoming coriaceous. <u>Thuja</u>
a. Leaves opposite or in whorls of 3, some scalelike, some awl-shaped, or all awl-shaped; ovulate cone becoming berrylike. <u>Juniperus</u>

Thuja L., Arbor Vitae

T. occidentalis L. Evergreen small tree; monecious; twigs flat, branched, in fanlike sprays; leaves scalelike, appressed, 4-ranked; cones terminal, the 2 kinds on different branches, staminate very small, carpellate about 1 cm long, with 4-6 scales, middle scales bearing 2 ovules each; seeds winged all around.

Juniperus L., Juniper

Evergreen trees and shrubs; diecious or monecious; leaves opposite or in whorls of 3, awl-shaped, or some scalelike and some awl-shaped; both kinds of cones small; carpellate scales becoming fleshy and coalescent, mature cone dark blue, berrylike; seeds wingless.

a. Young plants with awl-shaped leaves, older plants with mostly scalelike leaves; cones terminal.
    b. Erect tree. J. virginiana L., Red Cedar.
    b. Prostrate shrub. *J. horizontalis Moench, Shrubby J.
a. Leaves all awl-shaped, jointed at base; cones axillary. J. communis L.

ANGIOSPERMS

MONOCOTYLEDONS

TYPHACEAE, Cat-tail Family

Typha L., Cat-tail

Tall marsh or aquatic herbs with rhizomes; leaves linear, 2-ranked, basal and cauline, bases sheathing; flowers monosporangiate, in cylindric spikes, upper part of spikes staminate, lower part carpellate; staminate flowers of usually 3 united stamens, bractlets hairlike; carpellate flowers of 1 carpel, style elongate, stigma one-sided, stipe of ovulary long-villous; ovule 1; sterile flowers among the fertile carpellate ones.

a. Staminate and carpellate parts of spike usually contiguous; pollen in tetrads; stigma lance-ovate, flat; tip of sterile flowers ellipsoid, exceeded by the hairs of the fertile flowers and concealed among them; mature spike greenish-brown. T. latifolia L., Broad-leaved C.
a. Staminate and carpellate parts of spike usually separated; pollen not in tetrads; stigma linear; tip of sterile flowers an inverted cone, visible at surface of spike after stigmas have fallen; mature spike cinnamon-brown. T. angustifolia L., Narrow-leaved C.

SPARGANIACEAE, Bur-reed Family

Sparganium L., Bur-reed

In mud or shallow water; stems erect; leaves alternate, 2-ranked, linear; flowers monosporangiate, in globose sessile or subsessile heads in or above axils of bracts the lower of which are usually leaflike; staminate heads above the carpellate, in spikes; staminate flowers of 3 or more stamens mixed with small scales; carpellate flowers with perianth of 3-6 scalelike parts, carpels 2, united, locules 1 or 2, style 1, fruit 1-2-seeded.

a. Stigmas 2. S. eurycarpum Engelm.
a. Stigma 1.
    b. Some carpellate heads supra-axillary; rare. S. chlorocarpum Rydb.
    b. Carpellate heads axillary.
        c. Bracts ascending; middle leaves keeled; stigma 2-4 mm long; body of fruit 5 mm long or more, lustrous. S. androcladum (Engelm.) Morong
        c. Bracts spreading; leaves not or weakly keeled; stigma 1-2 mm long; body of fruit 5 mm long or less, dull. S. americanum Nutt.

ZOSTERACEAE, Pondweed Family

Aquatic herbs with submersed stems and leaves and sometimes also floating leaves; perianth none; stamens 1-4; carpels 2-5, separate.

a. Leaves alternate or the upper opposite; flowers bisporangiate, in peduncled elongate or capitate spikes. Potamogeton

a. Leaves opposite; flowers monosporangiate, sessile or nearly so, both kinds clustered in the same axil. Zannichellia

Potamogeton L. , Pondweed

Some species with only submersed leaves, other species with both submersed and floating leaves; flowers small, perianth none; stamens 4, with a sepal-like appendage; carpels 4, separate, becoming achenes.

a. Floating leaves present.
  b. Submersed blades linear.
    c. Floating blades ovate, 2-4 cm wide, usually shorter than petiole, base often cordate, attachment to petiole curving and flexible; submersed leaves linear, not flat, blade and petiole not differentiated; stipules 4-10 cm long. P. natans L.
    c. Floating blades not as above; submersed blades flat.
      d. Floating blades tapering to tip; submersed blades very narrow or to 10 mm wide.
        e. Submersed blades 2-10 mm wide, long and ribbonlike, with obvious lengthwise median reticulate band of paler green (lacunae); stem flat. P. epihydrus Raf.
        e. Submersed blades 1-3 mm wide, lacunae from margin to midrib. P. tennesseensis Fern.
      d. Floating blades rounded at tip, 1 cm wide or less, 2-3 times as long.
        e. Stipules free from blade-base; submersed blades bristle-tipped, 0.5 mm wide or less, 1-3-nerved; spikes cylindric, peduncles 1-2 cm long. P. vaseyi Robbins
        e. Stipules adnate to blade-base; submersed blades 0.5-1.5 mm wide.
          f. Free tip of stipules twice length of adnate portion; submersed blades 1-3-nerved, acute or obtuse; fruit 3-keeled. P. diversifolius Raf.
          f. Free tip of stipules shorter than adnate portion; submersed blades 3-nerved, obtuse; fruit 1-keeled, lateral keels obscure. P. spirillus Tuckerm.
  b. Submersed blades not linear; floating blades 2-6 cm wide; peduncle often wider than stem.
    c. Petiole and stem black-spotted; submersed blades lanceolate, floating ones ovate to rounded, mostly cordate; stipules to 6 cm long, acuminate. P. pulcher Tuckerm.
    c. Submersed blades falcate-folded, lanceolate to ovate; floating ones ovate to elliptic; stipules to 12 cm long, acuminate. P. amplifolius Tuckerm. (See third c)
    c. Petiole and stem not black-spotted; submersed blades not falcate-folded.
      d. All blades long-petioled, base and apex usually acute, floating blades lance-elliptic, submersed ones lance-linear, often very long; stipules 4-10 cm long. P. nodosus Poir.
      d. Submersed blades sessile or short-petioled, lance-elliptic, margins with deciduous minute teeth; floating blades elliptic to ovate, petioled; stipules obtuse.
        e. Submersed blades 3-10 mm wide; stipules about 2 cm long. P. gramineus L.
        e. Submersed blades 1.5-4.5 cm wide, sessile or petioles to 4 cm; stipules 4-10 cm long. P. illinoensis Morong
a. Floating leaves not present.
  b. Blades serrate, often ruffled, not cartilaginous-margined, oblong to broadly linear; fruit 5-6 mm long, beak 2-3 mm long. *P. crispus L.
  b. Blades not serrate but sometimes with toothlike hairs on edge; fruit not as above.
    c. Blades auricled at base or clasping with cordate or rounded bases.
      d. Blades with rounded basal auricles, linear, often 2-ranked, margin cartilaginous and with toothlike hairs, adnate to stipule-base. P. robbinsii Oakes
      d. Blades clasping, lanceolate or wider; stipules white, with many strong nerves.
        e. Stipules conspicuous, persistent; blades lanceolate. P. praelongus Wulfen
        e. Stipules soon disintegrating; blades sometimes ovate. P. richardsonii (Benn.) Rydb.
    c. Blades not auricled, not clasping, linear, mostly less than 4 mm wide.
      d. Nerves of blade many; stem flat; stipules white, fibrous. P. zosteriformis Fern.
      d. Nerves of blade 1-7; stem terete or nearly so.
        e. Blades septate (with short cross-veins); stipules connate; flowers in 2-6 whorls.
          f. Stipule-sheath obviously wider than stem; blades 1-2 mm wide. P. vaginatus Turcz.
          f. Stipule-sheath not wider than stem; blades 1.5 mm wide or less.
            g. Stem freely branched, branches and leaves spreading; common. P. pectinatus L.
            g. Stem branched near base, branches and leaves ascending. P. filiformis Pers.
        e. Blades not septate.
          f. Stipules adnate to blade. P. spirillus and P. diversifolius. (See first pair of f's in key to distinguish these two species).
          f. Stipules free from blade but sometimes connate.

24

g. Stipules whitish, fibrous, disintegrating with age into fibers.
   h. Stipules not connate; blades 3-nerved, bristle-tipped; spikes capitate, of 1-5 flowers; peduncle to 1.5 cm long. P. hillii Morong
   h. Stipules connate; spike with 3-4 remote whorls; peduncles longer than above.
      i. Blades 5-7-nerved; peduncles flattened, to 5 cm long. P. friesii Rupr.
      i. Blades 3-nerved, attenuate at tip. P. strictifolius Benn.
g. Stipules delicate, not fibrous.
   h. Stipules not connate.
      i. Blades to 0.5 mm wide, curving outward or recurved, 1-3-nerved, without glands at base. P. vaseyi Robbins
      i. Blades to 1.5 mm wide, 3-5-nerved, not curved, glands present at base; peduncles to 4 cm long. P. berchtoldii Fieber
   h. Stipules connate; blades to 3.5 mm wide.
      i. Blades 3-5-nerved; spike capitate, peduncle clavate, 1 cm long or less; glands usually present. P. foliosus Raf.
      i. Blades 3-nerved, the laterals often obscure; spike cylindric, interrupted, peduncle usually elongate; glands usually present. P. pusillus L.

Zannichellia L., Horned Pondweed

Z. palustris L. Slender branching submersed herb; leaves linear; staminate flower of 1 stamen; carpellate flower of 2-5 separate carpels subtended by a membranous sheath; fruit an aggregate of achenes with beaklike persistent styles.

NAJADACEAE, Naiad Family

Najas L., Naiad

Submersed aquatics; monecious or diecious; stems slender, rooting from lower nodes; leaves linear, wider and sheathing at base; flowers monosporangiate, minute, axillary; staminate flower of a single stamen; carpellate flower a single carpel with usually 3 stigmas and 1 ovule; markings on seed visible through thin fruit-wall.

a. Leaves 2-4 mm wide including the coarse teeth; sheath (wide base of blade) truncate or broadly rounded at summit, entire or with 1 or 2 teeth on each side; fruit 4 mm or more long. N. marina L.
a. Leaves less than 2 mm wide; sheath (wide base of blade) toothed; fruit usually less than 4 mm long.
   b. Sheath truncate or each half rounded at top and like an erect lobe or auricle; areolae of seed rectangular.
      c. Narrow portion of blade toothed, tip often recurving; top of sheath finely toothed; areolae of seeds wider than long, in regular vertical rows. *N. minor All.
      c. Narrow portion of blade minutely toothed except near base, tip not recurving; top of sheath jagged-toothed; areolae of seeds longer than wide. N. gracillima (A. Br.) Magnus
   b. Sheath gradually rounded, longest at point where it joins narrow portion of blade; areolae of seed square or hexagonal.
      c. Seed shining, with 30 or more rows of minute hexagonal areolae; style 1 mm or more long. N. flexilis (Willd.) Rostk. & Schmidt
      c. Seed dull, with 20 or fewer rows of square areolae; style 1/2 mm long or less. N. guadalupensis (Spreng.) Magnus

JUNCAGINACEAE, Arrow-grass Family

Of marshes and bogs; flowers regular, bisporangiate in ours, small, greenish; sepals and petals 3 in ours; stamens 6, attached at base to perianth; carpels 3 or 6, separating at maturity, becoming follicles.

a. Leaves all basal; flowers in a bractless raceme; carpels 3 or 6, united until mature; ovule solitary in each carpel. Triglochin
a. Stem leafy; flowers few, in a bracted raceme; carpels 3, united only slightly at base; ovules 2 in each carpel. Scheuchzeria

Triglochin L., Arrow-grass

Leaves slender, basal, with membranous sheaths; anthers sessile.

a. Locules 6; fruit ovoid-ellipsoid, the beaks recurving at maturity. T. maritima L.
a. Locules 3; fruit linear-clavate, the carpels at maturity separating from base upward, each carpel
    pointed at base. T. palustris L.

Scheuchzeria L.

S. palustris L. Bog herbs with zigzag stems; leaves alternate, linear, tubular at apex, with
wider basal sheaths; follicles spreading at maturity.

ALISMACEAE (ALISMATACEAE), Water-plantain Family

Aquatic or marsh herbs; monecious or diecious; leaves simple, basal, long-petioled; flowers
hypogynous, monosporangiate or bisporangiate, regular, bracted, verticillate in racemes or panicles;
sepals 3, separate; petals 3, separate; stamens 6-many; carpels few to many, separate; fruit an aggregate of achenes.

a. Carpels in one circle; flowers bisporangiate. 1. Alisma
a. Carpels in spirals, in fruit becoming a globular cluster of achenes; flowers bisporangiate or mono-
    sporangiate.
    b. Flowers all bisporangiate; achenes ridged, not winged. 2. Echinodorus
    b. Upper flowers usually staminate, lower usually carpellate or bisporangiate; achenes winged.
        3. Sagittaria

1. Alisma L., Water-plantain

Leaf-blades elliptic to ovate, base cordate, rounded, or tapering; inflorescence a large panicle;
petals white or pink; stamens 6-9; carpels 10-25.

a. Petals less than 3 mm long; fruit less than 4 mm wide. A. subcordatum Raf.
a. Petals more than 3 mm long; fruit 4 mm wide or more; rare in Ohio. A. triviale Pursh

2. Echinodorus Richard

E. rostratus (Nutt.) Engelm. Leaf-blades ovate, base cordate or truncate; inflorescence a panicle; petals white, 5-10 mm long; achenes turgid, ribbed, with an erect beak. Forma lanceolatus
(Engelm.) Fern. has lanceolate blades. Both rare in Ohio.

3. Sagittaria L., (Including Lophotocarpos Th. Durand), Arrow-head

Leaves with sagittate or unlobed blades or without blades; petals white or purple at base, to 3 cm
long; stamens 7-many; achenes winged.

a. Leaves without blades or blades usually not sagittate; filaments hairy.
    b. Stem bent at about base of raceme; lowest carpellate flowers sessile or short-pediceled.
      S. rigida Pursh
    b. Stem straight; flowers with pedicels 1-3 cm long. S. graminea Michx.
a. Leaves usually sagittate.
    b. Sepals of carpellate and perfect flowers appressed, almost circular. S. montevidensis Cham. &
      Schlect. (L. calycinus (Engelm.) J.G. Sm.)
    b. Sepals reflexed, not as above; perfect flowers rarely present; filaments smooth.
      c. Beak of achene horizontal or minute and erect.
        d. Bracts boat-shaped, ovate; beak of achene horizontal. S. latifolia Willd.
        d. Bracts lanceolate; beak of achene erect and minute. S. cuneata Sheld.
      c. Beak of achene broad-based, spreading or recurved, not minute. S. engelmanniana J.G. Smith
        d. Beak of achene strongly recurved; pedicels of carpellate flowers usually shorter than 1 cm
          long. ssp. longirostra (Michele) Bogin (S. australis (J.G. Sm.) Small)
        d. Beak of achene erect or spreading; pedicels of carpellate flowers 1-2 cm long.
          ssp. brevirostra (Mack. & Bush) Bogin (S. brevirostra Mack. & Bush)

BUTOMACEAE, Flowering-rush Family

Butomus L., Flowering-rush

*B. umbellatus L. Tall herb growing in mud or shallow water; leaves linear, basal; flowers hypogynous, regular, bisporangiate, about 2 cm wide, on slender pedicels in a 3-bracted umbel; sepals 3; petals 3, pink; stamens 9; carpels 6, slightly united at base; fruit an aggregate of follicles.

HYDROCHARITACEAE, Frog's-bit Family

Submersed aquatics; our species diecious; fruit indehiscent; seeds few to many.

a. Leaves cauline, whorled or opposite, 1-3 cm long, many. Anacharis
a. Leaves basal, alternate, long and ribbonlike. Vallisneria

Anacharis Rich. (Elodea Michx.), Waterweed

Flowers subtended by sessile or nearly sessile spathes; staminate flowers with 3 sepals, 3 petals, and 9 stamens; carpellate flowers epigynous, with 3 sepals, 3 petals, 1-loculed ovulary with 3 parietal placentae, and 3 stigmas, these flowers, at anthesis, at surface of water as a result of elongation of hypanthium.

a. Leaves usually obtuse, in width from less than 2 mm to about 4 mm, averaging about 2 mm.
   A. canadensis (Michx.) Planch. (E. canadensis Michx.)
a. Leaves acute, in width from less than 1 mm to almost 2 mm, averaging about 1.3 mm. A. nuttallii
   Planch. (E. nuttallii (Planch.) St. John)

Vallisneria L., Tape-grass, Eel-grass

V. americana Michx. Rooted in mud, leaves submersed; sepals 3, petals 3; staminate flowers minute, many in a head borne under water, flowers separating at anthesis and floating, stamens usually 2; carpellate flowers epigynous, white, sessile and solitary in a long-peduncled spathe, stigmas 3, these flowers at surface of water at anthesis, fruit under water as a result of coiling of peduncle.

GRAMINEAE, Grass Family

Herbaceous or rarely woody; rhizomes present or absent; internodes hollow or solid, nodes solid; leaves alternate, 2-ranked, linear to lanceolate, parallel-veined, consisting of sheath and blade, the ligule at junction of blade and sheath; unit of inflorescence a spikelet consisting of an axis (rachilla) bearing 2 rows of scales, the 2 lowest empty, these called the first (lower) and second glumes; scales above the glumes called lemmas; in axil of lemma another scale with back to rachilla, the palea; between lemma and palea, subtended by palea, the flower; lemma, palea, and flower parts constituting a floret; spikelet may contain one to many florets; if rachilla is jointed or articulated below each floret, florets fall at maturity from glumes and pedicel; if joint is in pedicel below glumes, the whole spikelet falls as a unit. Flowers hypogynous, bisporangiate, with sometimes some monosporangiate flowers present, too, or rarely all monosporangiate, the species then monecious or diecious; perianth none or of 2 or 3 small scales (lodicules); stamens usually 3 (sometimes 1, 2, or 6-many); carpels 3, united; ovulary 1-loculed; stigmas 2, usually plumose; seed 1; fruit a grain or rarely a utricle, a nut, or a berry.

Stated lengths of lemmas and glumes do not include awns.

a. Culms woody, 1-2 cm thick, 2-3 m tall; seldom flowering. 1. Arundinaria
a. Culms herbaceous.
  b. Spikelets enclosed in prickly burs. 66. Cenchrus
  b. Spikelets not enclosed in prickly burs.
    c. Florets monosporangiate, the staminate and carpellate in different spikelets on same plant; vestigial florets may be present, also.
      d. The two kinds of spikelets in different parts of same inflorescence.
        e. Carpellate spikelets surrounded by beadlike involucre from top of which staminate part of inflorescence protrudes. 73. Coix
        e. Spikelets not covered with beadlike involucre; upper branches of terminal panicle bearing carpellate spikelets; lower branches bearing staminate spikelets. 59. Zizania

27

      d. The two kinds of spikelets in different inflorescences, the staminate in terminal panicle (tassel), the carpellate in rows on one or more axillary branches. 74. <u>Zea</u>

    c. Florets bisporangiate; some monosporangiate and/or vestigial ones may be present, also.

      d. Spikelets in definite rows on a rachis; inflorescence a single spike or spikelike raceme or a cluster of spikes or spikelike racemes. Key <u>A</u>, p. 28

      d. Spikelets not in definite rows; inflorescence an open or a narrow panicle, sometimes dense and spikelike.

        e. Spikelet with 2 or more bisporangiate florets. Key <u>B</u>, p. 29

        e. Spikelet with 1 bisporangiate floret; 1 or more staminate and/or vestigial florets may be present, also. Key <u>C</u>, p. 31

KEY A. SPIKELETS IN ROWS ON A RACHIS

a. Single spikelets or groups of spikelets in 2 rows on opposite sides of rachis; inflorescence a single spike or spikelike raceme, not one-sided.

  b. Spikelets solitary at each node of rachis.

    c. Each spikelet appressed to and closely fitting into a concavity of rachis; spikelets and spike cylindric. 19. <u>Aegilops</u>

    c. Spikelet and rachis not as above; spikelets flattened laterally; spike not cylindric.

      d. Spikelet with edge to rachis. 24. <u>Lolium</u>

      d. Spikelet with side to rachis.

        e. Glumes ovate; lemma ovate, asymmetric, awned or awnless; spikelets 2-5-flowered. 18. <u>Triticum</u>

        e. Glumes and lemmas lanceolate or narrower.

          f. Spikelet several-flowered; lemma symmetric, awned or awnless; glumes lanceolate. 17. <u>Agropyron</u>

          f. Spikelet 2-flowered; lemma long-awned, ciliate keel on one side of center; glumes linear. 20. <u>Secale</u>

  b. Spikelets 2 or more at each node of rachis (sometimes fewer at base of spike), lateral ones sometimes awnlike.

    c. Spikelets usually 3 at each node of rachis, each with a perfect floret <u>or</u> lateral ones vestigial. 23. <u>Hordeum</u>

    c. Spikelets 2 at each node of rachis, each containing 1 or more perfect florets; <u>or</u> spikelets more than 2 at each node, then each spikelet 2-6-flowered.

      d. Spikelets ascending or erect at maturity; both glumes present, about equal, not much smaller than lemmas. 21. <u>Elymus</u>

      d. Spikelets ascending at first, horizontal at maturity; one or both glumes minute, or awnlike and much shorter than lemma. 22. <u>Hystrix</u>

a. Single spikelets or groups of spikelets in rows, usually on one side of rachis; spikes or spikelike racemes clustered or, if rarely solitary, then one-sided.

  b. Glumes absent or vestigial; spikelet laterally flattened, its edges formed by keel of lemma and of palea; lemma awnless. 58. <u>Leersia</u>

  b. At least one glume well developed.

    c. Spikelets laterally flattened <u>or</u> articulation above glumes (rarely whole spike falling entire); both glumes present; staminate or vestigial florets, when present, above perfect florets.

      d. Branches of inflorescence attached at tip of peduncle, or one or more attached a short distance below tip.

        e. Spikes 5 mm wide; spikelet with 2-several perfect florets. 49. <u>Eleusine</u>

        e. Spikes 2-3 mm wide; spikelet with 1 perfect floret. 50. <u>Cynodon</u>

      d. Branches of inflorescence racemosely attached along an elongate axis.

        e. Branches of inflorescence reflexed, about 2 cm long, falling entire from axis. 54. <u>Bouteloua</u>

        e. Branches of inflorescence appressed or erect <u>or</u>, if spreading, then more than 2 cm long.

          f. Spikelets closely imbricated, with 1 perfect floret and no vestigial florets; articulation below glumes; branches of inflorescence erect or appressed.

            g. Lemma a little longer than the inflated, awnless glumes; spikelet circular; spikes short. 51. <u>Beckmannia</u>

            g. Glumes not inflated, the first shorter than lemma, the second longer than lemma, awned. 52. <u>Spartina</u>

          f. Spikelets not closely imbricated; articulation above glumes; branches of inflorescence slender.

g. Panicle about as wide as long, branches spreading or slightly reflexed at maturity; culm-blades several, 2-7 cm long, about 1 cm wide. 53. Gymnopogon
g. Panicle longer than wide, branches many, spreading, or branches long and ascending; culm-blades fewer, longer, and narrower than in the above. 48. Leptochloa
c. Spikelets not laterally flattened, sometimes dorsally flattened; articulation below glumes; one glume sometimes absent; a staminate or a vestigial floret below perfect floret; sometimes whole spikelet vestigial.
d. Lemma and palea of bisporangiate floret firmer than glumes; below this floret are lower lemma and 1 or 2 glumes, all membranous; if spikelets in pairs, both of pair containing a perfect floret.
e. Both glumes present; lemma of perfect floret pointed or awned, its sides, but not its tip, inrolled around palea. 64. Echinochloa
e. Second glume present; first glume absent or minute.
f. Racemes 1-2 cm long, appressed; upper lemma short-awned. 61a. Eriochloa
f. Racemes not as above; upper lemma acute or obtuse, not awned.
g. Spikelet flattened, apex acute; rachis not wider than rows of spikelets attached to it; lemma of perfect floret not inrolled around palea. 60. Digitaria
g. Spikelet plano-convex, apex blunt, or rachis wider than rows of spikelets; lemma of perfect floret hard, margins inrolled. 62. Paspalum
d. Glumes firmer than lemmas; spikelets in pairs, 1 sessile, containing a perfect floret, the other pediceled and vestigial or containing a staminate floret; rarely both pediceled and/or containing a perfect floret.
e. Inflorescence terminal, of many hairy racemes; each spikelet of pair containing a perfect floret and bearing a tuft of long hairs at base.
f. Panicle fan-shaped, racemes 1-2 dm long; both spikelets of pair pediceled; rachis continuous. 67. Miscanthus
f. Panicle dense, ellipsoid, racemes shorter than axis; 1 spikelet of pair sessile; rachis jointed. 68. Erianthus
e. Inflorescence terminal or lateral or both, of 1-few racemes clustered at or near end of peduncle, or inflorescence a panicle of such clusters.
f. Blades lanceolate, 3-8 cm long; culms weak, reclining or decumbent; pediceled spikelet containing a perfect floret. 69. Eulalia
f. Blades linear, elongate; culms erect; pediceled spikelet vestigial or containing a staminate floret. 70. Andropogon

KEY B. SPIKELETS NOT IN ROWS, EACH CONTAINING TWO OR MORE BISPORANGIATE FLORETS

a. At least 1 glume overtopping body of lemma just above it; or lower lemma awned about middle of back or below; or both characters present.
b. Glumes about 2 cm long, 7-11-nerved, usually exceeding upper lemma; lemma awnless or awned from back; panicle open. 30. Avena
b. Glumes shorter, with fewer nerves.
c. Lower lemma awnless or awned from base of terminal teeth; each glume overtopping floret just above it.
d. Spikelets 2-flowered, 4-5 mm long, many; plant soft-pubescent, gray-green; upper floret staminate, lemma with short hooked awn; lower floret bisporangiate, lemma awnless. 32. Holcus
d. Spikelets several-flowered, 10-14 cm long, few; florets bisporangiate; lemma-awn bent, spiral at base. 33. Danthonia
c. Lower lemma awned from middle of back or below; spikelet 2-flowered; articulation above glumes.
d. Panicle open, spreading, bases of branches naked; both florets of spikelet perfect; lower glume not more than 5 mm long.
e. Glumes sometimes unequal, the shorter usually not overtopping the upper floret; rachilla prolonged behind palea; lemma short-toothed or erose at tip. 28. Deschampsia
e. Glumes equal, ovate, both exceeding body of the upper lemma but not always the awn; rachilla not prolonged; lemma with 2 narrow apical teeth. 29. Aira
d. Panicle narrow, the whorled branches usually bearing spikelets to base; lower floret staminate, upper perfect, both lemmas awned; longer glume 7-9 mm long. 31. Arrhenatherum

a. Neither glume overtopping body of lemma just above it; lemmas awned above middle or awnless.
  b. Articulation below glumes, sometimes in pedicel some distance below.
    c. Rachilla prolonged behind uppermost palea bearing 2-3 empty lemmas enfolded in a club-shaped mass; sheaths closed. 13. Melica
    c. Upper lemmas not in a club-shaped mass; sheaths open; articulation in pedicel below glumes.
      d. Lemmas awnless; second glume much wider than first; upper floret falling early from lower floret and glumes. 26. Sphenopholis
      d. Lower lemma awnless; upper lemma with outwardly curved awn. 27. Trisetum
  b. Articulation above glumes (except in Eragrostis, with continuous rachilla).
    c. Hairs of rachilla equaling or exceeding body of lemma; tall stout plants of wet places. 12. Phragmites
    c. Hairs of rachilla, if present, much shorter than body of lemma.
      d. Spikelets of 2 kinds together in clusters in a spikelike or capitate inflorescence, one kind with a few perfect florets, the other composed of several empty lemmas. 11. Cynosurus
      d. Spikelets uniform, all containing some perfect florets.
        e. Panicle dense and spikelike, axis and branches densely short-hairy; lemma 3-4 mm long, awnless or with short awn from just below tip. 25. Koeleria
        e. Panicle not dense and spikelike or, if rarely so, then axis and branches not densely short-hairy.
          f. Spikelets nearly sessile in dense one-sided clusters at ends of the few panicle-branches. 10. Dactylis
          f. Spikelets not in dense one-sided clusters at ends of panicle-branches.
            g. Lemma 5-many-nerved.
              h. Lemma 2-toothed at apex, usually awned from base of teeth or just below; sheaths closed.
                i. Callus of lemma bearded; tip of ovulary glabrous. 14. Schizachne
                i. Callus of lemma not bearded but body of lemma may be hairy; tip of ovulary pubescent. 2. Bromus
              h. Lemma not 2-toothed at apex but apex may be erose.
                i. Lemma keeled, awnless.
                  j. Lemma folded very flat lengthwise, many-nerved; spikelet 2-4 cm long, laterally flattened. 9. Uniola
                  j. Lemma not folded flat, usually with tuft of long soft hairs at base (web); when, rarely, web is absent, then both keel and marginal nerves pubescent; leaf-blades boat-shaped at tip. 6. Poa
                i. Lemma rounded on back or keeled only at tip; web absent.
                  j. Lemma awned or acute at tip, the 5 nerves obscure; sheaths of awnless species open. 3. Festuca
                  j. Lemma not awned, usually not acute but, if rarely so, then nerves conspicuous and more than 5.
                    k. Lemma 7-9-nerved; sheaths closed; plants of moist, swamp, or aquatic habitats. 5. Glyceria
                    k. Lemma 5-nerved (a few may be 7-nerved); sheaths open.
                      l. Panicle-branches ascending; habitats moist, swampy, or aquatic. 5. Glyceria
                      l. Lower panicle-branches reflexed; habitats dry or moist, sometimes saline. 4. Puccinellia
            g. Lemma 1-3-nerved.
              h. Lateral nerves or midnerve of lemma or both pubescent; or, if rarely all nerves are glabrous, then lemma with tuft of long soft hairs at base (web).
                i. Lemma not toothed at apex, not awned, often with web at base; blades with boat-shaped tip. 6. Poa
                i. Lemma toothed or emarginate at apex; midnerve or all 3 nerves excurrent as short awns.
                  j. Panicle large, open, branches drooping; blades elongate; sheaths not enlarged; all 3 nerves of lemma excurrent. 15. Triodia
                  j. Panicle small, narrow; blades and internodes of culm short and many; sheaths enlarged; midnerve of lemma excurrent. 16. Triplasis
              h. Nerves of lemma glabrous or scabrous; no web at base of lemma.

i. Lemma 6-10 mm long, with sharp point, lemma and palea separated at maturity by turgid beaked grain; tall plants of woodlands. 8. Diarrhena
i. Lemma 3 mm long or less; low to moderately tall plants of various habitats. 7. Eragrostis

KEY C. SPIKELETS NOT IN ROWS, EACH CONTAINING ONE BISPORANGIATE FLORET

a. Staminate and vestigial flowers absent.
  b. Glumes absent or minute.
    c. Glumes absent; spikelet laterally flattened, its edges formed by keel of lemma and of palea; lemma awnless. 58. Leersia
    c. Glumes small, the first sometimes absent.
      d. Culms low and slender, decumbent at base; blades 2-4 mm wide; lemma about 2 mm long. 40. Muhlenbergia
      d. Culms tall, erect; blades about 1 cm or more wide; lemma about 1 cm long. 43. Brachy-elytrum
  b. Glumes present, at least 1/4 as long as lemma.
    c. Hairs of callus of lemma half length of lemma or more; rachilla prolonged behind palea, long-hairy. 34. Calamagrostis
    c. Hairs of callus of lemma, if present, less than half length of lemma.
      d. Inflorescence cylindric or nearly so, dense and spikelike.
        e. Lemma 9 mm long or more; inflorescence 1.5 dm long or more and usually more than 1 cm thick. 35. Ammophila
        e. Lemma much less than 9 mm long; inflorescence usually shorter and narrower.
          f. Glumes awned.
            g. Glumes equal, truncate at tip, keels pectinate-ciliate, body longer than lemma, awn about 1 mm long. 39. Phleum
            g. Glumes tapering to tip, body not longer than lemma, awns variable in length, to 4 mm. 40. Muhlenbergia
          f. Glumes not awned.
            g. Lemma exceeding glumes, 1-nerved, awnless; glumes unequal in width, not con-nate. 42. Heleochloa
            g. Lemma equaling glumes, 5-nerved, awned from back; glumes equal, often connate below. 38. Alopecurus
      d. Inflorescence not dense and spikelike.
        e. Lemma at maturity not firmer than glumes.
          f. Articulation below glumes; lemma with minute straight awn from just below tip. 37. Cinna
          f. Articulation above glumes.
            g. Lemma 1-nerved; palea often longer than lemma. 41. Sporobolus
            g. Lemma 3-5-nerved; palea not longer than lemma.
              h. Lemma usually shorter and thinner than glumes; palea absent or not more than 2/3 as long as lemma. 36. Agrostis
              h. Lemma not shorter than glumes; palea and body of lemma about equal in length.
                i. Rachilla prolonged behind palea as a bristle; panicle open; lemma awned. 36. Agrostis
                i. Rachilla not prolonged; panicle usually narrow, rarely open; lemma pointed at tip or awned. 40. Muhlenbergia
        e. Lemma at maturity obviously firmer than glumes.
          f. Lemma and palea together ellipsoid in shape; callus short and blunt; lemma awnless or with straight awn; glumes ovate or broadly elliptic.
            g. Lemma awnless, edges wrapped around palea, pale at maturity; ligule membranous, to 6 mm long or more. 44. Milium
            g. Lemma awned, awn straight or flexuous, deciduous; lemma dark at maturity. 45. Oryzopsis
          f. Lemma and palea together terete, linear; callus hard and sharp; lemma with bent or branched awn; glumes narrow.
            g. Awn not branched, twice bent, several times length of body of lemma, strongly twisted below. 46. Stipa
            g. Awn 3-parted, lateral branches sometimes short. 47. Aristida

31

a. One or 2 staminate or vestigial florets below or rarely above bisporangiate floret, vestigial floret consisting of lemma and palea, lemma only, or small vestige.
  b. Spikelet with involucre of bristles. 65. <u>Setaria</u>
  b. Spikelet without involucre of bristles.
    c. Spikelets in pairs, one sessile, containing a bisporangiate floret (plus a vestigial one), the other a pedicel, with or without a spikelet; lemma thin and hyaline, less firm than glumes.
      d. Lemma awnless; pedicel bearing a spikelet. 71. <u>Sorghum</u>
      d. Lemma awned; pedicel naked. 72. <u>Sorghastrum</u>
    c. Spikelets not in pairs <u>or</u>, if rarely so, then each spikelet of pair containing a bisporangiate floret; lemma of bisporangiate floret firmer than glumes.
      d. Articulation above glumes.
        e. Glumes equal, as long as or longer than lemma.
          f. Glumes and lemma about equal; spikelet containing 1 perfect floret between 2 staminate florets. 55. <u>Hierochloe</u>
          f. Glumes longer than lemma, keeled or winged; 2 minute scales below perfect floret. 57. <u>Phalaris</u>
        e. Glumes unequal.
          f. Panicle spikelike; spikelet with 1 bisporangiate floret between and exceeded by 2 brown hairy awned scales. 56. <u>Anthoxanthum</u>
          f. Panicle narrow, not spikelike; spikelet with 2 florets, lower staminate, upper bisporangiate; both lemmas awned. 31. <u>Arrhenatherum</u>
      d. Articulation below glumes.
        e. Spikelet with 2 florets, lower bisporangiate, upper staminate; upper lemma with small curved or hooked awn. 32. <u>Holcus</u>
        e. Upper floret bisporangiate, its lemma and palea firmer than glumes; below its palea are lower lemma and first glume, below its lemma is second glume, all membranous.
          f. First glume minute; margin of lemma of perfect floret hyaline, not inrolled; panicle diffuse. 61. <u>Leptoloma</u>
          f. Both glumes obvious; lemma of perfect floret inrolled around palea.
            g. Ligule absent; spikelets often awned, often hispid; tip of lemma not inrolled. 64. <u>Echinochloa</u>
            g. Ligule present; spikelets not awned; lemma inrolled all around palea. 63. <u>Panicum</u>

BAMBUSEAE, Bamboo Tribe

1. <u>Arundinaria</u> Michx., Bamboo, Cane

<u>A. gigantea</u> (Walt.) Chapman Stems very tall, to 2 cm or more in diameter; blades short-petioled, jointed with sheaths, to 4 cm wide; spikelets 8-12-flowered, in racemes or panicles; lemma 1.5-2 cm long. Seldom flowering. Doubtfully native.

FESTUCEAE, Fescue Tribe

2. <u>Bromus</u> L., Brome Grass

Sheaths closed; spikelets 1-4 cm long, of several to many florets, in panicles or rarely racemes; lemma about 1 or 2 cm long, awned from below 2-toothed apex or rarely awnless; stigmas inserted below tip of ovulary.

Actual length of lemma-teeth is hard to ascertain because base of notch between them is covered by awn; <u>in this key</u>, teeth are considered to extend to base of awn.

a. First glume 1-nerved, second glume 3-nerved; length of lemma 2 1/2 times width or more.
  b. Lemma-teeth not more than 0.6 mm long; lemma awnless or awn shorter than body.
    c. Panicle erect, branches whorled, spreading in flower, ascending in fruit; foliage, except sometimes lower sheaths, and lemma usually glabrous; lemma awnless or awn to 3 mm.
      *<u>B</u>. inermis Leyss, Smooth B.
    c. Panicle branches spreading and usually drooping at maturity, spikelets near ends of branches only; sheaths or blades or both usually pubescent; lemma usually pubescent, at least near base and margin.
      d. Nodes 10 or more, all except sometimes the upper, covered by sheaths; flange and auricles at base of blade prominent; flowering Aug. and Sept. <u>B. latiglumis</u> (Shear) Hitchc.

d. Nodes usually fewer than 10, not covered by sheaths except sometimes the lower; flange and auricles absent at base of blade.

  e. Lemma more or less evenly pubescent all over or rarely glabrous; flowering June and July. B. purgans L.

  e. Lemma with band of long hairs on each margin, at least below middle, remainder of surface glabrous; nodes of culm usually pubescent; flowering in July. B. ciliatus L., Fringed B.

b. Lemma-teeth slender and pointed, to 2 mm long or more; lemma narrowly lanceolate, awn equaling or longer than body; weedy grasses of open places.

  c. Inflorescence a panicle, branches of axis or most of them again branched; awn of lemma 1-1.5 cm long. *B. tectorum L., Downy B.

  c. Inflorescence racemelike, branches of axis or most of them not again branched; awn of lemma 2-3 cm long. *B. sterilis L.

a. First glume 3-5-nerved, second glume 5-9-nerved.

  b. Lemma pubescent.

    c. Lemma more than twice as long as wide, teeth minute; panicle drooping, branches slender and flexuous; awn 2-3 mm long. B. kalmii Gray

    c. Lemma not more than twice as long as wide, teeth about 1 mm long; panicle compact, erect, branches shorter than spikelets. *B. mollis L.

  b. Lemma glabrous or scabrous, teeth 0.5-2 mm long, about as long as wide.

    c. Sheaths glabrous or the lower slightly hairy; tip of palea and hairy tip of grain usually projecting slightly beyond tip of lemma at maturity; awn usually short, sometimes crimped; anthers 2 mm long or less. *B. secalinus L., Cheat, Chess.

    c. Sheaths, except sometimes the upper, hairy; palea not exceeding lemma.

      d. Lemma almost as broad as long, inflated, obtuse; awn very short or absent; spikelets flattened, about 1 cm wide. *B. brizaeformis Fisch & Mey., Rattlesnake Chess.

      d. Lemma obviously longer than wide, awn well developed.

        e. Lower sheaths soft-villous, hairs somewhat entangled; awn at maturity usually divaricate; palea 1-2 mm shorter than lemma; anther about 1 mm long; panicle diffuse, branches divergent. *B. japonicus Thunb.

        e. Hairs of lower sheaths less soft, usually shorter, less dense, not entangled, or lower sheaths subglabrous; awn not divaricate at maturity; palea little shorter than lemma.

          f. Teeth of lemma longer than wide, acute; anther 3-4 mm long; awn about equaling body of lemma; panicle usually diffuse; lower sheaths short-downy to subglabrous; spikelets usually somewhat purple at maturity. *B. arvensis L.

          f. Teeth of lemma broad and obtuse; anther 2 mm long or less; awn usually shorter than body of lemma; panicle not diffuse; lower sheaths pubescent, sometimes sparsely.

            g. Lower lemmas of spikelet 9-11 mm long; common. *B. commutatus Schrad.

            g. Lower lemmas of spikelet 6.5-8.5 mm long; apparently rare. *B. racemosus L.

3. Festuca L., Fescue

Culms short to moderately tall; blades flat or rolled; spikelets few- to several-flowered; glumes 1- and 3-nerved; lemma rounded on back or keeled at tip, awned or awnless, 5-nerved, nerves often obscure.

a. Blades flat, more than 3 mm wide; lemma awnless or with awn to 2 mm long; perennial.

  b. Panicle-branches spikelet-bearing only near tips; spikelets about 5 mm long; blades without auricles. F. obtusa Spreng., Nodding F.

  b. Panicle-branches spikelet-bearing almost to base; spikelets about 1 cm long; blades with auricles. *F. elatior L., Meadow F.

    c. Auricles with smooth margin. var. *elatior.

    c. Auricles with ciliate margin. var. *arundinacea (Schreb.) Wimmer

a. Blades rolled, folded, setaceous, or flat and less than 3 mm wide.

  b. Slender annuals without a basal mass of old leaf-bases.

    c. First glume at most half as long as second glume. *F. myuros L.

    c. First glume longer, more than half as long as second.

      d. Spikelet usually 6-10-flowered; lemma 2.5-5 mm long, its awn usually 1-5 mm long; sheaths and/or culms sometimes puberulent. F. octoflora Walt.

      d. Spikelets usually 4-6-flowered; lemma 7-8 mm long; awn to 14 mm long; sheaths and culms glabrous. *F. dertonensis (All.) Aschers. and Graebn.

33

b. Perennials with a dense basal mass of old leaf-bases; lemma awnless or awn shorter than body.
  c. Culms usually decumbent at base; sheaths red or brown, soon disintegrating into fibers.
    F. rubra L., Red F.
  c. Culms erect at base; sheaths drab or light in color, not readily disintegrating into fibers.
    *F. ovina L., Sheep F.

4. Puccinellia Parl., Alkali Grass

*P. distans (L.) Parl. Blades narrow; panicle open, to 1.5 dm long, lower branches usually reflexed; spikelets much like those of Glyceria; lemma 2 mm long, obtuse.

5. Glyceria R. Br., Manna Grass

Usually of aquatic or wet-land habitats; sheaths usually closed or partly closed; spikelets of few to many florets, in terminal panicles, glumes 1-3-nerved; lemma rounded on back, nerves usually prominent, apex usually obtuse and hyaline.

a. Sheaths flattened; spikelets linear, 1-4 cm long; panicle narrow, to 5 dm long, branches erect.
  b. Palea with 2 apical acuminate teeth, extending 1.5-3 mm beyond acute tip of lemma. G. acutiflora Torr.
  b. Palea not toothed or only short-toothed at tip, extending less than 1 mm or not at all beyond obtuse tip of lemma.
    c. Lemma 4-5 mm long, scabrous between nerves, firm, usually shorter than palea, hyaline tip strongly nerved. G. septentrionalis Hitchc.
    c. Lemma 3-4 mm long, almost glabrous between nerves, thin, slightly longer than palea, hyaline tip without nerves. G. borealis (Nash) Batchelder
a. Sheaths not or little flattened; spikelets not linear, less than 1 cm long.
  b. Sheaths open; second glume with 3 (rarely 5) nerves; lemma with 5 (rarely 7) nerves; panicle open. G. pallida (Torr.) Trin.
  b. Sheaths closed, at least below; second glume 1-nerved; lemma 7-nerved.
    c. Panicle narrow, long, nodding at tip, branches erect or ascending; ligule less than 1 mm long; lemma 2-2.5 mm long. B. melicaria (Michx.) Hubb.
    c. Panicle open, lower branches spreading; ligule 2-6 mm long.
      d. Spikelets more than 2.5 mm wide, broadly ovate; lemma round-ovate, 3-4 mm long; palea less than twice as long as wide, shorter and wider than lemma. G. canadensis (Michx.) Trin., Rattlesnake Grass.
      d. Spikelet not more than 2.5 mm wide; lemma 2-2.5 mm long; palea more than twice as long as wide.
        e. Culms slender; panicle 1-2 dm long; glumes usually 0.5 and 1 mm long (to 1 and 1.4 mm); lemma about 2 mm long. G. striata (Lam.) Hitchc.
        e. Culms stout; panicle 2-4 dm long; glumes about 1.5 and 2 mm long; lemma about 2.5 mm long; spikelets often purple. G. grandis S. Wats.

6. Poa L., Bluegrass, Meadow Grass

Culms short to moderately tall; tip of blades boat-shaped; spikelets 2-several-flowered, in terminal panicles; first glume with 1 nerve (rarely 3), second with 3 nerves (rarely 5); lemma keeled, 2-5 mm long, usually with tuft of cobwebby hairs (web) at base, 5-nerved, intermediate nerves sometimes faint.

a. All or most of the spikelets consisting not of florets but of bulblets. *P. bulbosa L.
a. Spikelets consisting of florets of usual structure.
  b. Annuals (or winter annuals) of open ground; culms usually 2-3 dm high; panicle ovoid, usually 2-5 cm long.
    c. Lemma without web, the 5 distinct nerves usually pubescent, at least toward base; anther 0.5-1 mm long. *P. annua L., Annual B.
    c. Lemma with web, the 3 distinct nerves pubescent; anthers about 0.2 mm long. P. chapmaniana Scribn.
  b. Perennials; culms usually taller; panicles usually longer.
    c. Lower panicle-branches 1-3 at a node.

34

    d. Panicle narrow, branches short, usually 2 at a node, bearing spikelets to base; culms flattened, 2-edged, not tufted, with rhizomes; distinct nerves of lemma 3; at least lowest lemma of spikelet usually sparsely webbed. *P. compressa L., Canada B.
    d. Panicle open, the branches long, bearing spikelets only toward tips; culms not 2-edged.
      e. Lemma without basal web, pubescent on keel and marginal nerves and between nerves, without rhizomes. P. autumnalis Muhl.
      e. Lemma, at least the lowest in the spikelet, with basal web.
        f. Keel and marginal nerves of lemma glabrous; lemma 2-3 mm long, obtuse at tip; without rhizomes. P. languida Hitchc.
        f. Keel and marginal nerves of lemma pubescent.
          g. Lemma 2.5-3.5 mm long, with 3 distinct nerves; without rhizomes. P. paludigena Fern & Wieg.
          g. Lemma 3.5-5 mm long, with 5 distinct nerves.
            h. Panicle nodding, branches ascending; anthers 1.5 mm long or less; without rhizomes. P. wolfii Scribn.
            h. Panicle usually erect, lower branches spreading or reflexed; anthers 2-3 mm long; web sparse; surface of uppermost lemma sometimes pubescent; blades abruptly sharp-pointed, basal ones very long, those of the culm short; rhizomes slender. P. cuspidata Nutt.
  c. Lower panicle-branches 4 or more at a node; lemma with basal web and pubescent keel.
    d. Marginal nerves of lemma glabrous (sometimes pubescent in P. trivialis).
      e. Lemma with 5 distinct nerves; tips of glumes strongly incurved; ligule of upper leaves 4-10 mm long; panicle long-exsert, some branches spikelet-bearing nearly to base. *P. trivialis L., Rough B.
      e. Lemma with 3 distinct nerves; tips of glumes slightly incurved; ligule of upper leaves 1-2 mm long; base of panicle included, at least at first; spikelets few, borne only at tips of branches. P. alsodes Gray
    d. Marginal nerves of lemma pubescent.
      e. Distinct nerves of lemma 3; glumes acuminate; without rhizomes.
        f. Ligule 1 mm long or less; panicle 4-10 cm long; stem slender. *P. nemoralis L., Wood B.
        f. Ligule 2-5 mm long; panicle 1-3 dm long, often bronze or purple. P. palustris L., Fowl M.
      e. Distinct nerves of lemma 5; ligule of upper leaves 1-2 mm long.
        f. Panicle ovoid, exserted, at least some branches with spikelets to base, lower usually in whorl of 5; lemma pubescent on lower 1/2 to 2/3 of keel; with rhizomes; usually of open places. *P. pratensis L., Kentucky B.
        f. Panicle oblong; branches bearing a few spikelets on upper half; lemma pubescent on whole length of keel, somewhat pubescent between keel and marginal nerves; without rhizomes; of woodlands. P. sylvestris Gray

7. Eragrostis Beauv., Love Grass

Ligule a ring of hairs; spikelets flattened, of few to many often closely imbricated florets, in panicles; glumes 1-nerved or the second 3-nerved; lemmas small, 3-nerved; rachilla continuous in Ohio species, lemma and grain falling from persistent palea.

a. Stems prostrate, rooting at nodes; flowering branches short, erect or ascending; of mud flats and stream banks. E. hypnoides (Lam.) BSP., Creeping L.
a. Stems erect, sometimes bent at base, not prostrate.
  b. Glands present on some or all of the following parts: pedicels, glumes, lemmas, sheaths, blades; florets of spikelet many, rarely few.
    c. Spikelet 2.5 mm wide or more; lemma more than 2 mm long. *E. megastachya (Koel.) Link, Strong-scented L.
    c. Spikelet 2 mm wide or less, lemma less than 2 mm long; glands often more abundant on sheaths and blades than elsewhere. *E. poaeoides Beauv.
  b. Without glands.
    c. Perennial; panicle ovoid, 2/3 height of entire plant, little or not longer than wide, branches slender, rigid, spreading, spikelets red- or pink-purple, florets 6-12, rarely few. E. spectabilis (Pursh) Steud., Tumble L.

c. Annual; branches of panicle slender or capillary, not rigid; spikelets usually dark-gray or -purple.
   d. Florets of spikelet 5 or fewer; panicle cylindric or ellipsoid.
      e. Panicle to 4 dm long, diffuse, 2/3 as long as entire plant; pedicels much longer than spikelets; grain with lengthwise furrow; plants of dry ground. E. capillaris (L.) Nees, Capillary L.
      e. Panicle to 15 cm long, about half as long as entire plant; pedicels not much longer than spikelets; grain without lengthwise furrow; culms much branched; plants of moist ground. E. frankii C. A. Meyer
   d. Florets of spikelet more than 5.
      e. Spikelets 1 mm wide or less, not appressed to branches; panicle delicate and diffuse, branches capillary; lateral nerves of lemma obscure. *E. pilosa (L.) Beauv.
      e. Spikelets about 1.5 mm wide, appressed to panicle-branches; panicle open but not diffuse; lateral nerves of lemma conspicuous. E. pectinacea (Michx.) Nees

8. Diarrhena Beauv.

D. americana Beauv. Tall, with rhizomes; blades 1-2 cm wide, elongate; spikelets few-flowered, few, in a long narrow drooping panicle; first glume 1-nerved, second 3-5-nerved; lemma broad, coriaceous, shining, 3-nerved, pointed; palea firm, obtuse, shorter than lemma; lemma and palea at maturity widely separated by the large ovoid beaked grain.

a. Lemma 7-10 mm long, ovate, gradually tapering to cusp. var. americana
a. Lemma shorter, obovate, abruptly rounded to cusp. var. obovata Gl.

9. Uniola L.

U. latifolia Michx. Blades 1-2 cm wide; panicle drooping; spikelets very flat, large, florets several to many, closely imbricated; lemmas about 1 cm long, striate-nerved, keeled, lowest empty; palea shorter than lemma, keel winged; stamen 1; flowers cleistogamous.

10. Dactylis L.

*D. glomerata L., Orchard Grass. Growing in large tufts; blades glaucous; spikelets 2-several-flowered, nearly sessile in dense 1-sided clusters at ends of the few panicle-branches; lemma keeled, awn-pointed, 5-nerved, 5-8 mm long.

11. Cynosurus L., Dog's-tail

Spikelets of two kinds, one kind consisting of 2 glumes and a few perfect florets, the other kind fanlike, of 2 glumes and several lemmas but no flowers, the two kinds nearly sessile, in a spikelike or capitate cluster.

a. Inflorescence spikelike, linear; lemma-awns short and inconspicuous. *C. cristata L., Crested D.
a. Inflorescence ovoid or ellipsoid, lemma-awns conspicuous. *C. echinatus L.

12. Phragmites Trin., Reed Grass

P. communis Trin. Culms stout, 2-4 m tall; with rhizomes; growing in colonies; blades 1-5 cm wide; panicle large, purple when young, rachilla-hairs conspicuous; spikelets 1-1.5 cm long, several-flowered; glumes lanceolate, acute; lemma narrow, 3-nerved, long-acuminate, much longer than palea, lowest lemma of spikelet empty or subtending a staminate flower.

13. Melica L., Melic Grass

M. nitens Nutt. Sheaths closed; spikelets usually 3-flowered, falling entire, pendent, few in a panicle, rachilla projecting beyond uppermost floret bearing a small club-shaped mass of enfolded empty lemmas; glumes large, membranous, with scarious margins; lemma 7-9 mm long, several-nerved, scarious-margined.

14. Schizachne Hack.

S. purpurascens (Torr.) Swallen   Sheaths closed; spikelets several-flowered, in slender panicles; glumes unequal; lemma 2-toothed at apex, about 1 cm long, long-awned, prominently several-nerved, pilose on callus; keels of palea pubescent.

15. Triodia R. Br.

T. flava (L.) Smyth   Purpletop, Grease Grass.   Sheaths bearded at summit; panicle open, axis, branches, and pedicels viscid, the axils pubescent; spikelets purple or sometimes yellow or green, about 6-8-flowered, 5-8 mm long; lemma broad, obtuse, about 4 mm long, 3-nerved, nerves pubescent below, excurrent.

16. Triplasis Beauv., Sand Grass

T. purpurea (Walt.) Chapm., Purple S.   Annual; blades short and rigid; nodes many, pubescent; sheaths enlarged, enclosing axillary panicles or sometimes single spikelets; spikelets few-flowered, purple, in small panicles; lemma 3-nerved, 2-lobed at apex, nerves pubescent, midnerve excurrent; palea divergent from lemma, upper half of keels densely villous.

HORDEAE, Barley Tribe

17. Agropyron Gaertn., Quack Grass, Wheat Grass

Spikelets several-flowered, usually solitary at nodes of rachis, side to rachis; glumes firm, usually several-nerved, acute or awned; lemma firm, 5-7-nerved, acute or awned.

a. Spikelets crowded on rachis, spreading, each reaching almost to tip of one next above on same side.
 *A. desertorum (Fisch.) Schult.
a. Spikelets ascending or appressed, more remote, each sometimes overlapping base of one next above on same side but not reaching almost to its tip.
 b. Without rhizomes; anthers 1-2.5 mm long.   A. trachycaulum (Linke) Malte
 b. With rhizomes; anthers 3-7 mm long.
  c. Glumes tapering from near base to a sharp point; spikelets 6-12-flowered; blades mostly 2-4 mm wide; bands of upper nodes shorter than wide. *A. smithii Rydb., Western W.
  c. Glumes tapering from above middle; spikelets 4-8-flowered; blades mostly 5-10 mm wide; bands of upper nodes about as long as wide. *A. repens (L.) Beauv., Quack Grass.

18. Triticum L., Wheat

*T. aestivum L., Common Wheat.   Annual; blades 1-2 cm wide; spikelets 2-5-flowered, solitary at nodes of rachis, side to rachis; glumes broadly ovate, mucronate; lemmas broad, awned or awnless, keeled at one side of center.

19. Aegilops L., Goat Grass

*A. cylindrica Host   Annual; spikelet about 1 cm long, few-flowered, solitary at node of rachis, side closely appressed to concavity of rachis, whole spike thus cylindric; spike breaking between spikelets; glumes several-nerved, keeled on one side of center, keel projecting as an awn; lemmas several-nerved, upper with awn 4-5 mm long, lower with shorter awn or awnless.

20. Secale L., Rye

*S. cereale L.   Tall annual; somewhat blue-green; leaves about 1 cm wide; spikelets usually with 2 perfect flowers, solitary at nodes or rachis, side to rachis; glumes narrow, shorter than lemma; lemma long-awned.

21. Elymus L., Wild-rye

Ours perennials; spikelets 1-several-flowered, 2 or more at each node of rachis, side to rachis; glumes narrow or setaceous, asymmetric, usually awned; lemma rounded on back, nerveless or faintly nerved, usually with long awn from tip.

a. Awns divergently curved at maturity; spikelets usually 3-6-flowered, often more than 2 at a node; palea (from lowest floret of spikelet from middle of spike) usually 10 mm long or more; base of glume not terete. E. canadensis L., Canada W.

a. Awns straight at maturity, rarely absent; spikelets 2-4-flowered, usually 2 at a node; palea less than 9 mm long; base of glume terete.
   b. Florets readily falling from glumes; glumes scarcely widened above, base little or not bowed out; base of spike exserted.
      c. Spikelet 1-2-flowered; spike to 12 cm long; palea about as long as body of lemma; upper surface of blades usually short-villous. E. villosus Muhl. In the typical form, lemma and glumes are villous; in rarer forma arkansanus (Scribn. & Ball) Fern., lemma and glumes are glabrous or scabrous.
      c. Spikelet 2-4-flowered; spike to 2 dm long; palea of lowest floret of spikelet 2 mm or more shorter than body of lemma; sheaths and blades glabrous or scabrous; lemma and glumes hispidulous to glabrous. E. riparius Wieg.
   b. Florets not falling from glumes; glumes widened upward, somewhat twisted, base yellowish, hard, bowed out. E. virginicus L.
      c. Glumes and lemmas awnless or nearly so. var. submuticus Hook.
      c. Glumes and lemmas awned.
         d. Lower part of spike usually included in inflated upper sheath.
            e. Lemma glabrous or scabrous. var. virginicus.
            e. Lemma hirsute. var. virginicus forma hirsutiglumis (Scribn.) Fern.
         d. Spike well exserted; upper sheath not inflated; lemma glabrous to hirsute. var. glabriflorus (Vasey) Bush

22. Hystrix Moench

H. patula Moench, Bottle-brush Grass. Spikelets 2-4-flowered, usually 2 at each node of rachis, horizontally spreading at maturity, falling early; glumes none or setaceous; lemma narrow, rigid, with slender straight awn.

23. Hordeum L., Barley

Rachis of inflorescence breaking at maturity at base of each segment (except in H. vulgare), segment and attached spikelets falling together; spikelets of 1, rarely 2, florets, usually 3 at each node of rachis, central one sessile, lateral ones sometimes pediceled and often vestigial; glumes narrow or setaceous, in front of spikelet; lemma rounded on back, awned.

a. Awn of lemma (at least of central spikelet) several times as long as body, or lemma awnless and 3-lobed at apex.
   b. Inflorescence narrow, awns erect; lemmas of all spikelets with awn much longer and coarser than awns of glumes (in Beardless Barley, lemma awnless and 3-lobed at summit). *H. vulgare L., Common B.
   b. Inflorescence as wide as long, awns long and spreading; lemma of central spikelet and glumes similarly awned. *H. jubatum L., Squirreltail B.
a. Awn of lemma of central spikelet equaling or longer than body but not several times as long; inflorescence slender, awns erect. H. pusillum Nutt., Little B.

24. Lolium L., Rye Grass

Spikelets with several to many florets, solitary at nodes of rachis, edge to rachis; first glume (on side of spikelet next to rachis) usually absent except in terminal spikelet; second glume 3-5-nerved; lemma convex on back, 5-7-nerved.

a. Lemma awnless or nearly so; spikelet usually 6-10-flowered; blades folded in bud. *L. perenne L., Perennial R.
a. Lemma usually awned, at least in upper spikelets; spikelet usually 10-20-flowered; blades rolled in bud. *L. multiflorum Lam., Italian R.

38

25. Koeleria Pers.

K. cristata (L.) Pers. Tufted; leaves 1-3 mm wide; panicle silvery-green, narrow; spikelets usually 2-flowered; glumes acute, first 1-nerved, second 3-5-nerved; lemma acute, awnless or rarely short-awned from below apex.

26. Sphenopholis Scribn.

Tufted; panicles rather narrow, shining; spikelets usually 2-flowered; first glume narrow, second obovate; lemma almost nerveless, awnless in our species; upper floret falling early from remainder of spikelet; joint in pedicel some distance below glumes, lower floret, glumes, and segment of pedicel later falling as a unit.

a. Panicle dense, erect, spikelets crowded; second glume about as wide as long, tip rounded, somewhat cucullate, with broad papery margin. S. obtusata (Michx.) Scribn.
a. Panicle lax but narrow, often nodding; second glume 2/5-2/3 as wide as long, not cucullate.
  b. First glume (folded) subulate, usually shorter than second glume; lemma glabrous; lower sheaths and blades usually glabrous. S. intermedia Rydb.
  b. First glume (folded) linear-lanceolate, about equaling second glume; at least upper lemma scabrous-papillose; lower sheaths and blades usually hairy. S. nitida (Biehler) Scribn.

27. Trisetum L.

T. pensylvanicum (L.) Beauv. Spikelets 2-flowered, rarely 3-flowered, in a narrow lax panicle; first glume 3-nerved, the second wider, 3-5-nerved; lemmas acuminate or acute, the lower usually awnless, the second with an outwardly-curving awn about 5 mm long from below the 2-toothed apex; pedicel jointed, spikelet and a length of pedicel falling together.

28. Deschampsia Beauv., Hair Grass

Our species with dense tufts of narrow basal leaves; spikelets 2-flowered, shining, in open panicles; rachilla extending beyond upper floret; lemma toothed or erose at tip, faintly 5-nerved, midnerve becoming the awn; callus bearded.

a. Blades filiform, flexuous; awn twisted below, extending beyond tip of lemma; rachilla-extension inconspicuous. D. flexuosa (L.) Trin.
a. Blades narrow but not filiform; awn nearly straight, usually not exserted, rachilla-extension conspicuous. D. caespitosa (L.) Beauv.

29. Aira L.

*A. caryophyllea L., Silver Hair Grass. Small delicate annual; blades narrow; panicle open; spikelets 2-flowered, about 3 mm long, shining; glumes ovate, equal, exceeding body of uppermost lemma; lemmas awned on back, awn bent and twisted.

30. Avena L., Oats

Our species annual; panicles loose and open; spikelets large; rachilla hairy between florets; glumes membranous, several-nerved, usually exceeding uppermost floret; lemma indurate, 2-toothed at apex, awned or awnless.

a. Lemma glabrous; awn straight or absent; spikelet usually 2-flowered. *A. sativa L., Common O.
a. Lemma hirsute at base and sometimes on back; awn bent and twisted; spikelet usually 3-flowered. *A. fatua L., Wild O.

31. Arrhenatherum Beauv., Oat Grass

*A. elatius (L.) Mert. & Koch Tufted, tall; panicle narrow; spikelets shining, 2-flowered, lower floret staminate, upper floret perfect; glumes unequal, broad, first 1-nerved, second about equaling upper floret, 3-nerved; lemmas 5-7-nerved, the lower with a twisted bent awn from below middle, awn of upper short, straight, or rarely longer, twisted and bent; callus hairy.

32. Holcus L.

*H. lanatus L., Velvet Grass. Gray-green, velvety-pubescent, tufted; panicle dense; spikelets 2-flowered, upper flower staminate, lower perfect; glumes pubescent, exceeding florets, 1- and 3-nerved; lemma smooth and shining, obscurely nerved, of upper floret with a short hooked awn from below apex, of lower floret awnless.

33. Danthonia DC.

Blades narrow, often curved or twisted; panicles of few several-flowered spikelets; glumes acute, usually exceeding uppermost floret; lemma rounded on back, faintly several-nerved, 2-toothed a flat twisted awn borne between the teeth.

a. Upper leaf often not reaching inflorescence; lemma-teeth less than 2 mm long, acute to lance-attenuate; inflorescence often a raceme. D. spicata (L.) Beauv., Poverty Oat Grass.
a. Upper leaf often reaching inflorescence; lemma teeth aristate, 2-3 mm long; inflorescence often somewhat paniculate. D. compressa Aust.

AGROSTIDEAE, Bent Tribe

34. Calamagrostis Adans., Reed Grass

Panicle open or narrow; rachilla prolonged behind the single floret into a usually hairy bristle; glumes about equal, usually longer than lemma; lemma faintly 5-nerved, with delicate dorsal awn, callus with abundant white hairs.

a. Awn bent, twisted below, attached about halfway between middle and base of lemma; rachilla-extension hairy throughout, its hairs and those of callus reaching halfway or more to tip of lemma.
   C. insperata Swallen
a. Awn straight, attached near middle of lemma or above middle.
   b. Rachilla-extension with hairs only at tip, hairs reaching almost to tip of lemma; lemma-awn attached about halfway between middle and apex. C. cinnoides (Muhl.) Bart.
   b. Rachilla-extension hairy throughout; awn attached at about middle of lemma.
      c. Panicle open; tip of lemma translucent; callus hairs abundant, about equaling lemma; awn delicate. C. canadensis (Michx.) Nutt., Bluejoint.
      c. Panicle narrow; branches ascending or erect; lemma scabrous, hardly translucent; callus hairs 1/2-3/4 as long as lemma. C. inexpansa Gray

35. Ammophila Host, Beach Grass

A. breviligulata Fern. Coarse erect perennial of sand dunes; rhizomes extensive; blades elongate, rolled, curved; panicle dense, spikelike, nearly cylindric, base often enclosed in upper sheath; rachilla-extension hairy; glumes 1- and 3-nerved; lemma 5-nerved; callus hairy.

36. Agrostis L., Bent Grass

Culms often slender; ligule membranous; panicle usually open, often delicate; spikelet 1-flowered; glumes membranous; lemma awned or awnless, rarely 2-toothed at apex, usually shorter and thinner than glumes; palea often minute or wanting.

a. Palea, lemma, and glumes about equal; rachilla prolonged behind palea; lemma firm, long-awned from below tip. *A. spica-venti L.
a. Palea wanting or not more than 2/3 as long as lemma; rachilla not prolonged; lemma thin, shorter than glumes.
   b. Palea 1/2-2/3 as long as lemma, 2-nerved.
      c. Ligule 1-2 mm long; panicle open, not densely-flowered, branches naked at base, spreading; lemma sometimes with bent awn from near base. *A. tenuis Sibth., Rhode Island B.
      c. Ligule 2-6 mm long; some of the several panicle-branches at each node spikelet-bearing to base; panicle often red. *A. alba L., Redtop.
   b. Palea wanting or minute and nerveless.
      c. Annual; lemma 1-1.5 mm long, with delicate flexuous awn to 10 mm long from just below tip.
         A. elliottiana Schult.
      c. Perennial; lemma awnless or rarely short-awned.

40

d. Panicle diffuse, the capillary branches, very rough when rubbed downward, forking above middle, bearing spikelets only near tips.
    e. Spikelet to 1.7 mm long; lemma to 1.2 mm long, hardly longer than grain; spikelets in small clusters; flowering May and June. A. hyemalis (Walt.) BSP.
    e. Spikelets 2-2.7 mm long, not clustered; lemma 1.5 mm long or more, longer than grain; flowering June and July. A. scabra Willd.
d. Panicle open, longer than wide, branches slender, slightly rough when rubbed downward, forking at or below middle; spikelet 2-3 mm long; flowering Aug. and Sept. A. perennans (Walt.) Tuckerm.

37. Cinna L., Wood Reed

Ligule hyaline, conspicuous; spikelet 1-flowered; floret stipitate; glumes keeled; lemma with short straight awn from below apex; palea 1-keeled; articulation below glumes.

a. Second glume 3-nerved, hyaline margins narrow; panicle dense, silvery-green, branches ascending. C. arundinacea L.
a. Second glume 1-nerved or with 2 faint additional nerves at base; hyaline margins wide; panicle open, branches spreading; rare. C. latifolia (Trev.) Griseb.

38. Alopecurus L., Foxtail

Panicle dense, spikelike; spikelet 1-flowered, flattened laterally; glumes equal, keeled, keels usually pubescent; lemma wide, obtuse or truncate, 5-nerved, dorsally awned, margins united at base; palea none; articulation below glumes.

a. Spikelet 4-7 mm long; anthers 2-3 mm long; awn bent.
    b. Inflorescence usually more than 5 mm thick; glumes long-villous on keel and pubescent on nerves. *A. pratensis L.
    b. Inflorescence usually not more than 5 mm thick; glumes short-pubescent or scabrous on keel and nerves. *A. myosuroides Huds.
a. Spikelet 2-3 mm long; anthers 1 mm long or less.
    b. Awn exserted beyond glumes not more than 1 mm, or included, straight, attached near middle of lemma; perennial. A. aequalis Sobol
    b. Awn exserted beyond glumes 2 mm or more as short bristle, bent, attached near base of lemma; annual. A. carolinianus Walt.

39. Phleum L., Timothy

*P. pratense L. Stem unbranched, the base bulbous; panicle dense and spikelike; spikelet 1-flowered; glumes membranous, truncate, keeled, the keel pectinate-ciliate, awn about 1 mm long; lemma shorter than glumes, hyaline, broad, truncate, unawned.

40. Muhlenbergia Schreb.

Erect or decumbent; scaly rhizomes usually present; panicles usually narrow, rarely open; spikelets 1-flowered; glumes membranous or hyaline, acute or awned, rarely minute or absent; lemma narrow, often pilose at base, usually 3-nerved, acute, awnless or awned from tip or just below.

a. Panicle open or diffuse; spikelets on capillary pedicels.
    b. Lemma 1.5-2 mm long, glabrous, awnless; spikelets sometimes 2-flowered. M. asperifolia (Nees & Meyen) Parodi, Scratch Grass.
    b. Lemma 3-4 mm long, puberulent at base, awned; spikelets purple-red. M. capillaris (Lam.) Trin.
a. Panicle narrow, no more than 4 cm wide, branches ascending, or rarely dense and spikelike.
    b. Lemma not pilose at base, the surface minutely pubescent, the tip cuspidate; blades 1-2 mm wide, erect or ascending; tufted, with hard bulblike base, without rhizomes. M. cuspidata (Torr.) Rydb.
    b. Lemma pilose at base.
        c. One or both glumes small, less than half length of lemma, the first sometimes wanting; stem weak, decumbent at base; lemma awned.

d. Second glume not more than 0.5 mm long, first glume minute or wanting; lemma about 2 mm long, awn equaling to twice as long as body. M. schreberi Gmel., Nimblewill.

d. Glumes variable, minute to almost half as long as lemma; lemma 2.5-3 mm long, awn shorter than body. M. curtisetosa (Scribn.) Bush

c. Both glumes well developed, half length of lemma or more, with pointed or awned tips; culms erect or reclining; rhizomes with overlapping scales.

d. Panicle dense and spikelike; glumes lanceolate, awned, awns exceeding lemma; anthers 1-1.5 mm long. M. glomerata (Willd.) Trin.

d. Panicle not dense and spikelike but branches sometimes short and dense.

e. Glumes lanceolate, tapering from base to tip, at least one nearly as long as to longer than lemma; scales of rhizome ovate, base cucullate-arched; anthers 0.3-0.6 mm long; lemma of each species awned or awnless.

f. Plants much branched, sprawling; internodes glabrous; lemma typically awnless (awned in one form), short-pilose at base; panicles many. M. frondosa (Poir.) Fern.

f. Plants usually not sprawling; internodes puberulent; lemma long-pilose at base.

g. Panicle rather stiff, spikelets short-pediceled or subsessile, densely imbricated on short branches; lemma typically awnless (awned in one form). M. mexicana (L.) Trin.

g. Panicle slender and flexuous, branches erect, slender, often elongate, not densely flowered; lemma usually awned. M. sylvatica Torr.

e. Glumes ovate, abruptly pointed, 1/2-3/4 as long as lemma; scales of rhizome narrowly ovate, closely appressed; anther 0.8-1.5 mm long; panicle slender and elongate; culms erect, blades spreading.

f. Nodes and internodes glabrous or scabrous; lemma 2-2.5 mm long, blunt or appearing 3-toothed, awnless or awn shorter than body. M. sobolifera (Muhl.) Trin.

f. Nodes and at least upper part of internodes pubescent; lemma 3-4 mm long, awn equaling or longer than body. M. tenuiflora (Willd.) BSP.

41. Sporobolus R. Br., Dropseed

Blades flat or rolled; panicles narrow or open; spikelet 1-flowered; glumes membranous, 1-nerved, shorter than to longer than lemma; lemma 1-nerved, awnless; palea often longer than lemma; fruit readily falling from spikelet; pericarp free from seed.

a. A conspicuous tuft of white hairs at top of sheath; base of panicle often included in sheath. S. cryptandrus (Torr.) Gray

a. No tuft of hairs at top of sheath.

b. Panicle ellipsoid, open; spikelets gray; glumes unequal, the first much shorter than lemma; fruit globose, splitting palea as it enlarges. S. heterolepis Gray, Prairie D.

b. Panicle narrow, with appressed branches, usually included in sheath.

c. Plants moderately tall, perennial; blades long, often rolled, tapering to narrow tips; upper sheath often inflated.

d. Lemma and palea glabrous, blunt at tip; upper sheath usually inflated, enclosing or partly enclosing terminal panicle. S. asper (Michx.) Kunth

d. Lemma and palea pubescent, tapering to narrow point at tip; uppermost sheath not or scarcely inflated. S. clandestinus (Bieler) Hitchc.

c. Plants short, annual; cauline blades short; all sheaths inflated, panicles usually wholly or partly included.

d. Lemma pubescent; spikelet 3 mm long or more. S. vaginiflorus (Torr.) Wood

d. Lemma glabrous; spikelet 2-3 mm long. S. neglectus Nash

42. Heleochloa Host

*H. schoenoides (L.) Host   Low tufted annual; blades short; panicle spikelike, to 4 cm long, 8-10 cm thick, base included in inflated sheath; spikelet 3 mm long, 1-flowered.

43. Brachyelytrum Beauv.

B. erectum (Schreb.) Beauv.   Blades to 16 mm wide, lanceolate, tapering to both ends; panicle narrow, of few spikelets; spikelet 1-flowered; rachilla prolonged behind floret as a bristle; glumes minute, the first often wanting, the second sometimes awned; lemma about 1 cm long, narrow, firm, with long terminal awn.

44. Milium L.

M. effusum L. Smooth; blades to 1.5 cm wide; spikelets 1-flowered, 3-4 mm long, on upper half of branches of open terminal panicle; glumes thin, a little longer than lemma; lemma firmer than glumes, smooth and shining, margins inrolled around palea; ligule to 6 mm long, membranous, toothed.

45. Oryzopsis Michx., Rice Grass

O. racemosa (Sm.) Ricker Blades 5-15 mm wide; spikelets 1-flowered, 7-9 mm long, rather few, near ends of the few panicle-branches; glumes about 7-nerved; lemma a little shorter than glumes, indurate, convolute, dark in fruit, awn deciduous.

46. Stipa L., Needle Grass

Blades often rolled; spikelets 1-flowered, in narrow panicles; glumes thin, narrow, acuminate; lemma indurate, convolute, narrow, terete, with terminal bent twisted awn; callus sharp-pointed, hairy.

a. Body of lemma about 1 cm long; glumes 1.5 cm long; awn to 6 cm long. S. avenacea L., Black Oat Grass.
a. Body of lemma 1.6-2.5 cm long; glumes 3-4 cm long; awn to 20 cm long. S. spartea Trin., Porcupine Grass.

47. Aristida L., Triple-awn Grass

Usually tufted; blades narrow, often rolled; spikelets 1-flowered, in narrow panicles; glumes narrow, acute to awned; lemma linear, indurate, convolute, terete, with 3-parted awn; callus sharp.

a. Middle awn of lemma at least twice as long as lateral awns; annual.
  b. Middle awn up to 10 mm long, coiled 1/2 to 1 turn, horizontally divergent; glumes 6-8 mm long. A. dichotoma Michx.
  b. Middle awn up to 2 cm long, bent at about a right angle but not spiral; glumes about 5 mm long. A. longespica Poir.
a. Middle awn not much longer than lateral awns.
  b. Lemma about 2 cm long; awns 3-7 cm long; annual. A. oligantha Michx.
  b. Lemma about 7 mm long; awns 1.5-3 cm long; perennial. A. purpurascens Poir.

CHLORIDEAE, Grama Tribe

48. Leptochloa Beauv.

Ohio species annual; spikelets 2-several-flowered, in slender, usually many, spikelike racemes which are racemosely attached to main axis of inflorescence; glumes 1-nerved, longer or shorter than lower lemma; lemma 3-nerved.

a. Lemma 1-1.5 mm long; spikelet about 2 mm long, sheaths papillose-pilose. L. filiformis (Lam.) Beauv.
a. Lemma 4-5 mm long; spikelet 7-12 mm long; sheaths glabrous. L. fascicularis (Lam.) A. Gray

49. Eleusine Gaertn.

*E. indica (L.) Gaertn. Smooth tufted annual; spikelets few- to several-flowered, sessile, imbricated, in 2 rows in a one-sided spike, the spikes digitate at tip of main axis or 1 or 2 below tip; glumes 1-nerved; pericarp free from seed.

50. Cynodon Richard

*C. dactylon (L.) Pers., Bermuda Grass. Perennial, with extensive rhizomes or stolons or both; ligule a ring of hairs; spikelets 1-flowered, in one-sided spike, the spikes, usually 4-5, digitate at tip of main axis; rachilla prolonged, sometimes bearing a rudiment; lemma 2 mm long, boat-shaped, keel pubescent.

51. Beckmannia Host, Slough Grass

B. syzigachne (Steud.) Fern. Spikelets flat, orbicular, 1-flowered, in short, erect or ascending, one-sided spikes racemosely attached to main axis or branches; glumes equal, transversely wrinkled, inflated, slightly shorter than acuminate lemma.

52. Spartina Schreb., Cord Grass

S. pectinata Link, Slough Grass, Prairie C. Tall stout perennial with horizontal rhizomes; blades long, tapering to slender tip, involute when dry; spikelets laterally flattened, 1-flowered, closely imbricated in one-sided spikes, spikes racemose on main axis; first glume acuminate, shorter than lemma, second glume awned, longer than lemma; lemma keeled, with 2 blunt apical teeth.

53. Gymnopogon Beauv.

G. ambiguus (Michx.) BSP. Blades crowded, stiffly spreading, short; sheaths overlapping; spikelet with 1 perfect flower, rachilla prolonged bearing a rudiment that is sometimes awned; spikelets inconspicuous, in one-sided spikes racemosely arranged on main axis, inflorescence sometimes half height of entire plant; glumes narrow, acuminate; lemma narrow, awned from between 2 apical teeth.

54. Bouteloua Lag., Grama

B. curtipendula (Michx.) Torr., Side-oats G. Tufted; blades narrow; spikelet with 1 perfect floret, rachilla prolonged, bearing 1 or more rudiments or a staminate flower; spikelets in short one-sided spikes which are racemose and reflexed on elongate main axis; glumes narrow; lemma narrow, 3-toothed; rudiment 3-awned.

PHALARIDEAE, Canary Grass Tribe

55. Hierochloe R. Br., Vanilla Grass

H. odorata (L.) Beauv. Fragrant perennial; blades of flowering culm short; inflorescence an open panicle; spikelet about 5 mm long, tawny, shining, with 1 perfect floret above and 2 staminate florets below; glumes broad, thin; lemma awnless, of staminate flowers about as long as glumes, of perfect flower shorter.

56. Anthoxanthum L., Vernal Grass

Plants fragrant; inflorescence a spikelike panicle; spikelets 7-10 mm long, with 1 perfect floret above and 2 rudimentary lemmas below; glumes unequal, with wide hyaline margins; lemma of perfect floret indurate, awnless, brown, shorter than the brown-hairy awned rudiments.

a. Perennial; culms erect, unbranched. *A. odoratum L., Sweet V.
a. Annual; culms often bent at base, branched. *A. aristatum Boiss. (A. puelii).

57. Phalaris L., Canary Grass

Spikelet laterally flattened, with 1 perfect floret, at base of which are small scalelike rudiments of 2 lemmas; glumes equal, boat-shaped, keeled, keel sometimes winged; lemma of perfect flower coriaceous, shining, shorter than glumes.

a. Panicle dense and spikelike, usually 5 cm long or less; annual.
  b. Spike ovoid or ovoid-oblong; glumes 7-8 mm long, keels broadly winged, a prominent green stripe on each side of keel. *P. canariensis L., Canary G.
  b. Spike lance-ovoid or narrowly ellipsoid; glumes 5-6 mm long, keel narrowly winged. P. caroliniana Walt.
a. Panicle 7 cm long or more, narrow and dense but with obvious branches; keels of glumes wingless or narrowly winged; perennial, with rhizomes. P. arundinacea L., Reed C.

ORYZEAE, Rice Tribe

58. Leersia Swartz

Stems weak, often decumbent; spikelets 1-flowered, sessile or nearly so on slender panicle-branches; glumes none; lemma boat-shaped, folded, awnless, 5-nerved; palea narrower, 3-nerved, keeled, margins clasped by margins of lemma.

a. Edges of blades very harsh and cutting; lower branches of panicle more than 1 at a node; rhizome slender. L. oryzoides (L.) Swartz, Rice Cutgrass.
a. Edges of blades not cutting; panicle-branches one at a node, few, naked below; rhizome stout, scaly. L. virginica Willd., White Grass.

ZIZANIEAE, Indian Rice Tribe

59. Zizania L., Wild Rice

Z. aquatica L. Very tall aquatic annual; blades 1-4 cm wide; spikelets 1-flowered, monosporangiate; glumes none or vestigial; lemma of staminate floret narrow, membranous, about 1 cm long; lemma of carpellate floret 2 cm long, firm, long-awned; panicle large, lower branches spreading, bearing staminate spikelets, upper branches ascending, bearing early-deciduous carpellate spikelets.

PANICEAE, Panic Grass Tribe

60. Digitaria Heist., Crab Grass

Annual; spikelets sessile or short-pediceled, in 2's or 3's or rarely solitary, in 2 rows in one-sided spikelike racemes, these racemes in 1 or more whorls or somewhat racemose at summit of peduncle; first glume minute or absent; second glume and lower lemma of like texture, membranous.

a. Second glume about half as long as greenish lemma of perfect floret; rachis winged; blades and lower sheaths usually hairy. *D. sanguinalis (L.) Scop. Common C.
a. Second glume 3/4 to nearly as long as dark purple lemma of perfect floret.
　b. Rachis obviously winged; spikelets single or in 2's; sheaths usually glabrous; usually spreading and decumbent. *D. ischaemum (Schreb.) Muhl., Small C.
　b. Rachis triangular, scarcely winged; spikelets sometimes in 3's; lower sheaths pubescent; usually erect. D. filiformis (L.) Koel., Slender C.

61. Leptoloma Chase

L. cognatum (Schult.) Chase, Fall Witch Grass. Perennial; culms branching, decumbent at base; panicle diffuse, breaking away at maturity; spikelets solitary on capillary pedicels, elliptic, acuminate, about 3 mm long, with 1 perfect floret, lemma and palea firm, lemma-margins hyaline; first glume minute or absent; second glume resembling lower lemma, both several-nerved, pubescent between nerves.

61a. Eriochloa HBK., Cupgrass

*E. contracta Hitchc. Blades hairy; spikelets 4-5 mm long, in panicled spikelike racemes; first glume absent; second glume and lower lemma lanceolate, villous, awn-tipped; upper lemma awn-tipped.

62. Paspalum L.

Inflorescence of 2-several spikelike racemes digitate or racemose on the main axis, or rarely of 1 raceme; spikelets usually plano-convex, subsessile or petioled, solitary or in pairs in 2 rows on one side of rachis; spikelet with a perfect floret above, lemma and palea firm; first glume usually absent; second glume and lower lemma similar, membranous, with appearance of 2 glumes.

a. Rachis wider than the 2 rows of spikelets, leaflike; panicle-branches usually many; spikelet flattened, acute; of water or wet soil. P. fluitans (Ell.) Kunth
a. Rachis not broader than the rows of spikelets attached to it; spikelets plano-convex, blunt or rounded at apex.
　b. Spikelets borne singly, not in pairs, on rachis, 2.5-3.2 mm long, orbicular or broadly oval. P. laeve Michx.
　b. Spikelets borne in pairs on rachis (one spikelet of pair sometimes vestigial, but pedicel present).
　　c. Spikelets more than 2.5 mm long; racemes usually 4-8.
　　　d. Spikelets about 3 mm long, obovoid, usually lying in 4 rows; culms decumbent at base, rooting at lower nodes. P. pubiflorum Rupr. var. glabrum Vasey
　　　d. Spikelets about 4 mm long, ovoid, not lying in 4 rows; culms erect. P. floridanum Michx.
　　c. Spikelets less than 2.5 mm long; racemes 1-3, slender.

    d. Spikelet 1. 7 mm long or less, elliptic-obovate. P. setaceum Michx.
       e. Blades densely villous. var. setaceum.
       e. Blades glabrous to sparsely strigose. var. longipedunculatum (Le Conte) Wood
    d. Spikelet 1. 8-2. 3 mm long, as wide or nearly as wide as long. P. ciliatifolium Michx.
       e. Upper surface of blade long-pilose. var. muhlenbergii (Nash) Fern.
       e. Upper surface of blade puberulent, sometimes with long hairs intermixed.
          var. stramineum (Nash) Fern.

63. Panicum L. , Panic Grass

    Inflorescence a panicle; spikelet with 1 perfect floret above, lemma and palea firm to indurate, margins of lemma inrolled around palea; below perfect floret a lemma, which sometimes subtends a palea with or without a staminate flower, and 2 glumes, first glume smaller, second glume about as long as lower lemma, and similar to it in appearance.

a. Basal and cauline leaves similar; without winter rosettes; without an autumnal phase different from vernal phase; culms sometimes branched; flowering once a year; spikelets, in ours, glabrous, acuminate to subacute.
  b. Lower lemma and second glume covered with tubercles. P. verrucosum Muhl.
  b. Lower lemma and second glume without tubercles; spikelets acute or acuminate at apex.
    c. First glume broad, obtuse or rounded at apex, not more than 1/4 length of spikelet; spikelet lance-ovoid, 2-3. 5 mm long; anthers orange; glabrous or blades sparsely hairy; annual. P. dichotomiflorum Michx.
    c. First glume acuminate or acute at apex, 1/3 to more than 1/2 length of spikelet or sometimes short and blunt when plants are hairy.
      d. Lower part of culms compressed, sheaths often keeled; spikelets short-pediceled, crowded on branches, subsecund; perennial.
        e. Spikelet 3-4 mm long, bent above first glume at angle of about 30 degrees; stout scaly rhizomes present. P. anceps Michx.
        e. Spikelet less than 3 mm long, not bent above first glume; plants tufted, with short caudex.
          f. Panicle (in ours) long and narrow, branches erect or ascending; ligule 2-3 mm long, with fimbriate summit; blades erect. P. longifolium Torr.
          f. Panicle ellipsoid or ovoid, branches densely-flowered; ligule about 1 mm long, membranous.
            g. Panicle ellipsoid, lower branches ascending; spikelet 2. 5 mm long or more; perfect floret with stipe 0. 2-0. 4 mm long above remainder of spikelet. P. stipitatum Nash
            g. Panicle ovoid, lower branches divaricate, remote; spikelet less than 2. 5 mm long; perfect floret sessile or nearly so. P. agrostoides Spreng.
      d. Culms rounded, sheaths not keeled; panicles open or diffuse, spikelets rarely crowded.
        e. Tall perennial; sheaths glabrous; spikelet 3. 5-5. 5 mm long, first glume 1/2-3/4 length of spikelet; anthers purple. P. virgatum L. , Switchgrass.
        e. Annuals; sheaths pubescent.
          f. Spikelet ovoid, 4. 5-5. 5 mm long; blades to 2 cm wide. *P. miliaceum L. , Broomcorn Millet.
          f. Spikelet narrower and shorter, and blades narrower, than above.
            g. Spikelets acuminate, 3 mm long or more or, if shorter, then panicle half height of plant or more.
              h. Spikelets 3-3. 5 mm long, narrow; panicle narrowly ellipsoid, 2-3 times as long as wide, branches ascending; blades erect. P. flexile (Gatt.) Scribn.
              h. Spikelets mostly less than 3 mm long, rarely to 4 mm; panicle diffuse, as wide as long, terminal one 1/2-2/3 height of plant. P. capillare L. , Witch Grass.
            g. Spikelets less than 3 mm long, acute.
              h. Culm slender; blades 3-8 mm wide; terminal panicle exserted, ovoid or rhombic; spikelets few, usually 2 mm long or less, usually in 2's on short pedicels near end of each branchlet. P. philadelphicum Bernh. and P. tuckermani Fern.
              h. Culms stout; blades 6-10 mm wide; panicles many, terminal and axillary, ellipsoid, base usually included; spikelets 2-2. 5 mm long, often racemose along upper part of branchlets. P. gattingeri Nash

a. Cauline blades 20 times as long as wide or more, erect, 5 mm wide or less; without winter rosettes; in autumnal phase, branching from base, small panicles partly hidden among basal leaves. (See third a.)
   b. Spikelet more than 3 mm long; second glume and lower lemma forming a slightly twisted beak above upper lemma. P. depauperatum Muhl.
   b. Spikelet less than 3 mm long; second glume and lower lemma not forming a beak above upper lemma. P. linearifolium Scribn. Var. linearifolium has pilose sheaths; var. werneri Scribn. has glabrous sheaths.
a. Basal leaves usually shorter and wider than cauline; with winter rosettes; vernal phase with unbranched culms and terminal panicles; autumnal phase with branches and panicles from some or all nodes; cauline blades rarely as much as 20 times as long as wide; spikelets usually rounded at apex.
   b. Principal cauline blades 15 mm wide or more.
      c. Nodes retrorsely bearded; spikelets 4-4.5 mm long. P. boscii Poir.
      c. Nodes sometimes pubescent but not retrorsely bearded; spikelets less than 4 mm long.
         d. Vernal sheaths papillose-hispid, or at least papillose, rarely glabrous; autumnal sheaths overlapping, bristly-hairy, inflated, enclosing panicles; spikelets 2.7-3 mm long. P. clandestinum L.
         d. Vernal sheaths not or slightly papillose-hispid; autumnal sheaths not bristly-hairy.
            e. Spikelet 3.4-3.7 mm long, rarely shorter; blades 1.5-4 cm wide. P. latifolium L.
            e. Spikelet usually 3 mm long or less, rarely to 3.2 mm long.
               f. Spikelet 2.9-3 mm long; blades 9-15 mm wide; primary nerves differentiated from secondary on lower surface. P. calliphyllum Ashe
               f. Spikelet 2.2-3.2 mm long; blades 5-25 mm wide, primary nerves scarcely differentiated from secondary on lower surface. P. commutatum Schultes (See third f.)
               f. Spikelet 1.8 mm long or less, ellipsoid when young, spherical-obovoid at maturity; rosette blades wide, with white cartilaginous margins.
                  g. Length of panicle 1-2 times width; blades of culm 7-15 mm wide, usually reduced upward. P. sphaerocarpum Ell.
                  g. Length of panicle 2-4 times width; blades of culm 15-25 mm wide, upper ones, usually 3, widest. P. polyanthes Schultes
   b. Principal cauline blades less than 15 mm wide.
      c. Some or all blades 10-20 times as long as wide, erect or ascending, uppermost usually longest, glabrous except for ciliate base, 3-8 mm wide; spikelet 2.3-2.8 mm long, long-pediceled. P. bicknellii Nash
      c. Blades usually not more than 10 times as long as wide.
         d. Spikelet more than 2.5 mm long.
            e. Spikelet 3.4-4 mm long, long-villous, hairs to 1 mm long or more; first glume narrow, 1/2 length of spikelet or more; blades papillose-hispid. P. leibergii (Vasey) Scribn.
            e. Spikelet shorter; hairs of spikelet, if present, shorter.
               f. Vernal sheaths papillose-hispid, or at least papillose, rarely glabrous; autumnal sheaths overlapping, inflated, enclosing panicles, with bristlelike hairs; spikelets 2.7-3 mm long. P. clandestinum L.
               f. Without the above set of characters.
                  g. Culms spreading-pilose, nodes retrorsely villous; blades velvety pubescent; spikelets pilose, 2.9-3 mm long. P. malacophyllum Nash
                  g. Culms ascending- or appressed-pilose, puberulent, or glabrous; nodes not retrorsely villous; blades not velvety.
                     h. Spikelet obovoid, turgid, blunt, 3-3.5 mm long; second glume and lower lemma strongly nerved; culms and sheaths more or less pubescent. P. oligosanthes Schultes var. scribnerianum (Nash) Fern.
                     h. Spikelet ellipsoid; culms and sheaths glabrous or puberulent, nodes sometimes pubescent.
                       i. Spikelets 2.9-3 mm long; blades 9-15 mm wide; culms, sheaths, and blades glabrous, nodes sometimes pubescent; primary nerves differentiated from secondary on lower surface. P. calliphyllum Ashe
                       i. Spikelets 2.2-3.2 mm long; blades 5-25 mm wide, primary nerves scarcely differentiated from secondary on lower surface; culms and sheaths usually puberulent; blades puberulent or glabrous. P. commutatum Schultes
         d. Spikelet 2.5 mm long or less.

e. Sheaths, at least the upper, glabrous (sometimes ciliate at margin, sometimes pubescent at nodes); culms usually glabrous.
 f. Ligule of hairs 3-5 mm long, clearly visible at base of blade; spikelets 1. 3-2. 1 mm long.   Varieties of P. lanuginosum Ell.
 f. Ligule not more than 1 mm long.
  g. Nodes densely retrorsely bearded; spikelet 1. 5-1. 8 mm long, usually glabrous; blades 8-15 mm wide.   P. microcarpon Muhl.
  g. Nodes not retrorsely bearded, sometimes pubescent, or the lower slightly bearded.
   h. Spikelet 1. 8 mm long or less, pubescent, spherical-obovoid, ellipsoid when young; rosette blades wide, with white cartilaginous margins; panicle-branches viscid-spotted.   P. sphaerocarpon Ell.
   h. Spikelet 2 mm long or more, ellipsoid at maturity.
    i. Spikelet glabrous (rarely pubescent in P. dichotomum).
     j. Sheaths pale-spotted; second glume and lower lemma extended beyond upper lemma in a short, slightly twisted, point; spikelet 2. 3-2. 5 mm long. P. yadkinense Ashe
     j. Sheaths not pale-spotted; second glume not longer than, often shorter than, upper lemma; spikelet 2-2. 2 mm long.   P. dichotomum L.
    i. Spikelet pubescent.
     j. Spikelet 2-2. 2 mm long, 1 mm wide; nodes usually glabrous; upper blades erect or nearly so.   P. boreale Nash
     j. Spikelet longer and wider; nodes puberulent; blades ascending or spreading.   P. commutatum Schult.
e. Sheaths or culms, usually both, pubescent or puberulent; spikelet pubescent.
 f. Ligule of dense hairs 2-5 mm long; culms and sheaths villous or papillose-pilose.
  g. Spikelet 2-2. 5 mm long; hairs of culms and sheaths long and abundant, spreading or appressed.   P. villosissimum Nash
  g. Spikelet 1. 3-2. 1 mm long; hairs of culms and sheaths varying in length and abundance.   Varieties of P. lanuginosum Ell.
 f. Ligule less than 2 mm long.
  g. Sheaths pilose with long hairs; nodes retrorsely bearded; spikelet 1. 9-2. 3 mm long.   P. laxiflorum Lam.
  g. Sheaths puberulent, sometimes with long hairs intermixed; nodes not retrorsely bearded.
   h. Spikelet 1. 4-2 mm long; sheaths and culms puberulent, sometimes with long hairs intermixed; blades 3-7 mm wide, glabrous, puberulent, or appressed pubescent.   P. columbianum Scribn.
   h. Spikelet 2. 2-3. 2 mm long; sheaths and culms puberulent, without long hairs; blades 5-25 mm wide.   P. commutatum Schult.

64. Echinochloa Beauv.

Coarse annuals, our species without ligules; inflorescence a panicle of one-sided racemes; spikelets subsessile, often spiny-hispid, glumes mucronate or awned; upper floret bisporangiate, lemma and palea firm, lemma inrolled around palea except at tip; lower floret consisting of a lemma which sometimes subtends a palea and sometimes also a staminate flower.

a. Second glume and lower lemma both awned; lower sheaths usually papillose-hispid. E. walteri (Pursh) Heller
a. Second glume awnless, pointed; sheaths glabrous.
 b. Lower lemma awnless, pointed at apex; branches of panicle appressed; spikelets short-pubescent to glabrous, in 4 rows. *E. colonum (L.) Link
 b. Lower lemma awned or awnless; branches of panicle ascending or spreading; spikelets usually not in rows, hispid on nerves, usually some or all of the hairs papillose-based.   Considered by several authors as not one species but the two below. E. crusgalli (L.) Beauv., Barnyard Grass.
  c. Papillose-based hairs of spikelet wanting or confined to margin; tip of upper lemma differing in texture from lustrous body, withering. E. crusgalli (L.) Beauv.
  c. Papillose-based hairs of spikelet usually present and abundant; tip of upper lemma merging gradually into lustrous body, firm and sharp. E. muricata (Beauv.) Fern. (E. pungens (Poir.) Rydb.)

65. Setaria Beauv. Foxtail Grass

Our species annuals; inflorescence a cylindric bristly spikelike panicle; spikelet with upper bisporangiate floret and lower lemma enclosing a hyaline palea and sometimes a staminate flower; one or more slender bristles below each spikelet; abortive spikelets sometimes present; lemma and palea of perfect floret chartaceous.

a. Bristles downwardly barbed, short, 1 per spikelet; spikelet about 2 mm long; second glume about 3/4 as long as upper lemma. *S. verticillata (L.) Beauv.
a. Bristles upwardly barbed; bristles 1-many below each spikelet.
  b. Second glume 1/2-2/3 as long as rugose upper lemma; spikelet about 3 mm long and 2 mm wide; bristles 5-20 below each spikelet; panicle yellow at maturity. *S. glauca (L.) Beauv.
  b. Second glume 3/4 to as long as slightly rugose upper lemma; 1-several bristles below each spikelet.
    c. Panicle somewhat lobed (obviously branched), often more than 1 cm thick, excluding bristles, variously colored, often red or purple; at maturity, upper lemma, palea, and grain falling away from remainder of spikelet. *S. italica (L.) Beauv., Italian or Hungarian Millet.
    c. Panicle usually not lobed, usually 1 cm or less thick, excluding bristles, usually green; at maturity, whole spikelet deciduous from pedicel.
      d. Panicle usually 7 cm long or less; spikelet 2-2.5 mm long; bristles 1-3 below each spikelet; blades glabrous. *S. viridis (L.) Beauv., Green F.
      d. Panicle longer, usually conspicuously nodding; spikelet 2.5-3 mm long; bristles often 3-6 below each spikelet; blades usually pubescent, at least above. *S. faberii Herrm.

66. Cenchrus L., Sandbur Grass

C. pauciflorus Benth. (C. longispinus (Hack.) Fern.). Inflorescence a spikelike raceme of burs, the bur with subglobose body 4-5 mm wide, usually enclosing 2 spikelets, falling from axis without breaking open; spikelet with upper bisporangiate floret, lemma and palea firm, and a lower staminate or vestigial floret.

ANDROPOGONEAE, Beard Grass Tribe

67. Miscanthus Anderss.

*M. sinensis Anderss. "Eulalia." Tall coarse perennial; inflorescence a large terminal hairy panicle of elongate racemes; spikelets of pair pediceled, each containing a perfect floret, each with tuft of hair at base; lemma awned; rachis not jointed.

68. Erianthus Michx., Plume Grass

E. alopecuroides (L.) Ell., Silver P. Tall, with large terminal hairy panicle; one spikelet of pair pediceled, one sessile, each containing a perfect floret, each with tuft of hair at base; lemma awned.

69. Eulalia Kunth

*E. viminea (Trin.) Kuntze Culms weak, branching, rooting at nodes; blades lanceolate, slightly petioled; inflorescence of 1-few racemes clustered at or near tip of peduncle; each spikelet of pair containing a perfect floret; lemma awned or awnless.

70. Andropogon L., Beard Grass

Culms solid; spikelets in racemes, the racemes solitary, paired, or digitate on a terminal peduncle or one arising from axil of a usually spathelike leaf; if axillary peduncles many, the whole then a panicle or corymb; spikelets paired at nodes, one sessile with a perfect floret and a lower hyaline lemma, one pediceled with a staminate or a rudimentary floret or without a floret; rachis and pedicel often villous.

a. Tall to very tall; racemes 2-6 at tip of peduncle; axillary peduncles few; pediceled spikelet staminate, awnless. A. gerardi Vitman, Big Bluestem.
a. Moderately tall; racemes 1-few on each peduncle, delicate; axillary peduncles several to many.
  b. Each peduncle bearing 1 raceme; sessile spikelet 6-8 mm long; pediceled spikelet vestigial or staminate. A. scoparius Michx., Little Bluestem.

b. Each peduncle bearing usually 2, rarely 3-4, racemes; rachis-segments and pedicels bearing long white hairs; pediceled spikelet vestigial or only a pedicel.
   c. Sheaths that subtend peduncles much overlapping and inflated, red-brown; awn loosely twisted; sessile spikelet 4-5 mm long. A. elliotii Chapm.
   c. Sheaths little or not inflated, scattered, whole inflorescence elongate; awn straight; sessile spikelet 3-4 mm long. A. virginicus L., Broom-sedge.

71. Sorghum Moench

Racemes short, inflorescence appearing a usual panicle; sessile spikelet of pair ovoid, consisting of a perfect floret and a lower lemma with a staminate or a vestigial flower, glumes indurate, lemma hyaline, with bent spiral awn, palea of similar texture or absent; pediceled spikelet narrow, with a staminate or a vestigial floret.

a. Perennial with extensive rhizomes; blade usually less than 2 cm wide, white midrib prominent.
   *S. halepense (L.) Pers., Johnson Grass.
a. Annual; blades often more than 2 cm wide. *S. vulgare Pers., Sorghum.

72. Sorghastrum Nash, Indian Grass

S. nutans (L.) Nash   Tall perennial with narrow yellow or brown terminal panicle; racemes of panicle short; at each node of rachis a pedicel usually without a spikelet and a sessile spikelet with 1 perfect floret and a lower lemma, glumes coriaceous, lemma hyaline; lemma of perfect floret awned, awn bent; spikelets and pedicels hairy.

TRIPSACEAE (MAYDEAE), Maize Tribe

73. Coix L.

*C. lachryma-jobi L., Job's-tears. Blades to 4 cm wide; peduncles in leaf-axils, each bearing a beadlike involucre within which are 2 vestiges and a spikelet of 1 carpellate flower and 1 lemma; staminate inflorescence protruding from orifice of involucre; staminate spikelets 2-flowered.

74. Zea L., Corn

*Z. mays L. Tall coarse annual; stem solid; staminate spikelets 2-flowered, in pairs, in a one-sided spikelike raceme, the racemes in terminal panicle (tassel); carpellate spikelets usually 2-flowered (1 flower vestigial), in double rows on 1 or more lateral branches (cob), spikelets and cob forming the ear, which is covered with husks; glumes of carpellate spikelet firm, lemma and palea hyaline, style a long silk.

CYPERACEAE, Sedge Family

Herbs, usually perennial; stem often triangular; leaves alternate, sometimes basal, cauline often 3-ranked, blades linear, sheaths usually closed; flowers hypogynous, each subtended by a bract (scale) and sometimes enclosed in a saclike bract (perigynium); flowers solitary or clustered in spikes or spikelets, the cluster often subtended by leaflike bracts; perianth none or of bristles or scales; stamens 1-6, usually 3; carpels 2-3, united; ovulary 1-loculed; style 1 with 2-3 branches; fruit an achene.

a. Each carpellate flower enclosed in a perigynium; flowers all monosporangiate. 12. Carex
a. Flowers not enclosed in perigynia.
   b. Flowers and subtending scales 2-ranked on axis of spikelet.
      c. Leaves basal; inflorescence terminal, capitate or umbelliform, subtended by 1-several leaf-like bracts. 1. Cyperus
      c. Spikelets 2-ranked on peduncles axillary to the upper of numerous cauline leaves. 2. Dulichium
   b. Flowers and subtending scales not 2-ranked or, if rarely so, then achene with a tubercle.
      c. Base of style persistent on achene as a tubercle or a conspicuous beak.
         d. Culms without leaf-blades; spikelets solitary, terminal, without a foliaceous bract.
            3. Eleocharis
         d. Leaf-blades present, cauline, basal, or both; spikelets clustered.
            e. Tubercle at tip of achene minute; blades filiform; plants tufted, to 2 dm high.
               4. Bulbostylis

      e. Tubercle conspicuous, pointed; plants taller; spikelets in fascicles or glomerules; blades not filiform. 9. <u>Rhynchospora</u>

c. Achene without a tubercle but sometimes with a small pointed tip.

    d. Perianth bristles soft, hairlike, white or tawny, becoming much longer than scales; heads cottony or silky. 7. <u>Eriophorum</u>

    d. Perianth bristles not as above; if rarely spikelets have a woolly appearance, then bristles brown and not more than 8 per achene.

      e. Spikelet with not more than one perfect or carpellate flower; 1 or 2 staminate flowers and/or some empty scales may be present in spikelet, also; some spikelets may be wholly staminate.

        f. Spikelets in a single sessile subglobose head or 2-3 confluent heads, subtended by 2-3 long bracts; tufted annuals 1 or 2 dm high; spikelet 1-flowered, flower perfect. 1. <u>Cyperus</u>

        f. Spikelets not in subglobose heads; plants taller.

          g. Achene hard, whitish, often rough, rounded at base to short disk narrower than achene; flowers monosporangiate; some spikelets staminate. 11. <u>Scleria</u>

          g. Achene brown, ovoid, not rough, pointed at tip, base truncate and as wide as achene or wider; uppermost flower of spikelet perfect, others staminate or vestigial. 10. <u>Cladium</u>

      e. Spikelet with 2 or more flowers; flowers all perfect.

        f. Style dilated at base, deciduous below dilation; perianth bristles none; spikelets ovoid to linear, solitary or in groups of 2-3 in cymose clusters; tufted annual with slender stems and leaves 1-2 mm wide. 5. <u>Fimbristylis</u>

        f. Style not dilated at base.

          g. Perianth none; achene cylindric or oblong-obovoid; tufted annual 1-2 dm high with filiform stem and leaves; spikelets ovoid, 2-5 mm long, sessile, few in terminal cluster. 8. <u>Hemicarpha</u>

          g. Perianth bristles 1-several, rarely none; achene lenticular, plano-convex, or 3-angled; spikelets sessile or peduncled, solitary or in capitate, spicate, or umbelliform clusters. 6. <u>Scirpus</u>

1. Cyperus L.

    Stem unbranched below inflorescence; leaves basal; flowers perfect, without perianth, 2-ranked in spikelets, spikelets in sessile or peduncled spikes; spikes solitary or in heads or umbels, subtended by 1 to several leaflike bracts; stamens 1-3; style 2-3-branched; achene 3-angled or lenticular.

a. Tips of scales outwardly curved or recurved; stigmas 3; achenes 3-angled; spikes 6-many-flowered, flattened; heads sessile or peduncled; annuals.

    b. Tips of scales strongly recurved; spikelets linear. <u>C. aristatus</u> Rottb. (<u>C. inflexus</u> Muhl.).

    b. Tips of scales not strongly recurved; spikelets wider. <u>C. acuminatus</u> Torr. & Hook.

a. Tips of scales ascending or erect.

    b. Spikelets 1-3-flowered, crowded in a dense globular or slightly oblong head.

      c. Heads sessile, solitary or 2-3 confluent; spikelet 1-flowered; stigmas 2; slender annual with culms to 2 dm. <u>C. tenuifolius</u> (Steud.) Dandy

      c. Heads sessile or peduncled or both, usually more than 1; spikelet 1-3-flowered; stigmas 3.

        d. Heads globose; scales elliptic, 3 mm long or more. <u>C. ovularis</u> (Michx.) Torr.

        d. Heads short-cylindric; scales ovate, less than 3 mm long. <u>C. retrorsus</u> Chapm.

    b. Spikelets 3-many-flowered, in loose spherical or hemispherical heads. (See third b.)

      c. Stigmas 2; scales nerved at center but not in margin; spikelet flattened; achene not 3-angled; stamens 2-3; annual.

        d. Achene black, white-encrusted, rough, nearly as wide as long; scales green or yellow. <u>C. flavescens</u> L. var. <u>poaeformis</u> (Pursh) Fern.

        d. Achene brownish, longer than wide; scales with some markings of purple or brown on sides.

          e. Style cleft to base, exserted 2-4 mm beyond scale; scale with band of brown or purple along margin and at tip. <u>C. diandrus</u> Torr.

          e. Style cleft to middle, often deciduous, exserted 1-1.5 mm beyond scale; whole side of scale red-purple. <u>C. rivularis</u> Kunth

      c. Stigmas 3; achene 3-angled; scales nerved in center and in margin.

        d. Involucral bracts and spikelets ascending.

          e. Scales ovate, 3 mm long or more, strongly mucronate. <u>C. schweinitzii</u> Torr.

          e. Scales subcircular, less than 3 mm long, minutely mucronate. <u>C. houghtonii</u> Torr.

        d. Involucral leaves and lateral spikelets spreading; scales minutely mucronate. C. filiculmis Vahl.
  b. Spikelets in an obconic or a loose cylindric spike, the spikes in an umbelliform cluster; stigmas 3.
      c. Spikelets all ascending; scales conspicuously mucronate; spikelets flattened. C. schweinitzii Torr.
      c. Some spikelets horizontal or reflexed; scales not conspicuously mucronate.
         d. Scales more than 3 mm long, narrow, appressed; spikelet terete; achene linear.
            e. Spikelet 1-3-flowered, subulate, rigid, all but uppermost becoming reflexed; spike obconic. C. retrofractus (L.) Torr. var. dipsaciformis (Fern.) Kukenth.
            e. Spikelet usually with more than 3 flowers; spike oblong or short-cylindric, none but lower spikelets strongly reflexed.
               f. Spikelets many, their attachments closely crowded; tip of scale obviously overlapping one next above on same side. C. lancastriensis Porter
               f. Spikelets in loose clusters, attachments not crowded; tip of scale not or little overlapping base of one next above on same side. C. refractus Engelm.
         d. Scales 3 mm long or less, or spikelet flattened, or both characters present.
            e. Scales 4-6 mm long, 7-nerved; spikelet flattened, golden-yellow. C. strigosus L.
            e. Scales 3 mm long or less.
               f. Tip of scale not overlapping one next above on same side. C. engelmanni Steud.
               f. Tip of scale overlapping one next above on same side.
                  g. Scales 1-1.5 mm long, nerved in center, not in margin, closely imbricated, divergent; achene less than 1 mm long, little longer than wide. C. erythrorhizos Muhl.
                  g. Scales 1.5 mm long or more, nerved in center and in margin; achene linear or oblong, more than 1 mm long.
                     h. Annual with stout base; without rhizomes and tubers; rachilla breaking into short segments. C. odoratus L.
                     h. Perennial with slender rhizomes ending in tubers; rachilla not breaking into segments. C. esculentus L.

2. Dulichium Pers.

    D. arundinaceum (L.) Britt. Stems tall, hollow; leaves many, short, the lower bladeless; spikelets in axillary spikes; flowers 2-ranked, perfect; perianth-bristles 6 or more; stamens 3; style 2-cleft; achene linear-oblong, flattened, style persistent as a beak.

3. Eleocharis R. Br., Spike-rush

    In wet soil or water; culms unbranched; leaf-sheaths bladeless or lower rarely with blades; spikelets terminal, solitary, erect, without involucre; flowers perfect, spirally arranged, lower 1-3 scales usually empty; perianth none or of 1-12 bristles; stamens 2-3; styles 2-3-branched, base dilated, persisting as a tubercle on the 3-angled or lenticular achene.

a. Culm quadrangular, about as wide as the spikelet. E. quadrangulata (Michx.) R. & S.
a. Culm usually not quadrangular, narrower than the spikelet.
  b. Tubercle little or not at all differentiated from achene-body.
    c. Culm capillary; spikelet 7 mm long or less, 2-7-flowered. E. pauciflora (Lightf.) Link
    c. Culm 1-2 mm wide; spikelet to 2 cm long, 12-20-flowered. E. rostellata Torr.
  b. Tubercle clearly differentiated from achene-body, often constricted at base.
    c. Tubercle flattened dorsi-ventrally, sharp-edged, as wide to half as wide as achene-body.
      d. Tubercle about half as wide and half as long as achene-body. E. ovata (Roth.) R. & S.
      d. Tubercle about 3/4 as wide and 1/3 - 1/2 as long as achene-body; bristles exceeding achene; common. E. obtusa (Willd.) Schult. (See third d)
      d. Tubercle about as wide and 1/6 - 1/4 as long as achene-body; bristles short or absent; rare, if present in Ohio. E. engelmanni Steud.
    c. Tubercle not flattened as above, not more than half as wide as achene-body.
      d. Stigmas 2; achene biconvex or lenticular.
        e. Sheaths loose, tips white, scarious; achene olive to brown. E. olivacea Torr.
        e. Sheaths tight, tips not white and scarious.
          f. Spikes lance-ovoid, acute; achene yellow to brown; usually with rhizomes.

g. Basal empty scales 2-3; scales with conspicuous hyaline tips, the median acute; sheath blackish at top; tubercle ovoid-conic. E. smallii Britt.

g. Basal empty scale 1, circular, base encircling spikelet-base; scales obtuse; tubercle conic, longer than wide. E. erythropoda Steud. (E. calva Torr.)

f. Spikes subglobose to ovoid, obtuse; scales subcircular; achene black; tubercle small, spongy, pale; plant tufted. E. geniculata (L.) R. & S.

d. Stigmas 3; achene 3-angled or terete.

e. Achene with lengthwise ridges and crosswise lines, terete, ellipsoid or obovoid.

f. Culms filiform; bristles sometimes present. E. acicularis (L.) R. & S.

f. Culms 1-2 mm wide, flattened; bristles none; rare. E. wolfii Gray

e. Achene pitted, pits sometimes in lines, or roughened, obovoid, width 2/3 length.

f. Culms tufted, unequal in length; achene yellow to olive-brown, pitted; tubercle conic, pointed, longer than wide. E. intermedia (Muhl.) Schult.

f. Culms from horizontal rhizomes; achene rough; scales with wide pale margins.

g. Culms flattened; scales wide, white apex often split; achene yellow to brown, rough-granular; tubercle low to globose-conic. E. compressa Sulliv.

g. Culms angled; tubercle pyramidal or low. E. tenuis (Willd.) Schult. Var. tenuis has culms 4-5-angled and achenes olive; var. borealis (Svenson) Gl. (E. elliptica Kunth) has culms 6-8-angled and achenes yellow or orange.

4. Bulbostylis (Kunth) Clarke

B. capillaris (L.) Clarke Tufted annual; stem erect, slender; leaves very narrow, basal or sub-basal; sheaths pubescent; flowers perfect, spirally imbricated; spikelets in bracted capitate or umbelliform clusters; perianth none; style 3-cleft, base persistent as minute tubercle; achene cross-wrinkled.

5. Fimbristylis Vahl

F. autumnalis (L.) R. & S. Tufted annual; stem and leaves slender, leaves basal or subbasal; flowers perfect, spirally arranged in spikelet; spikelets in simple or compound cymose clusters subtended by 2-3 bracts; perianth none; style 3-cleft, its dilated base deciduous; achene 3-angled.

6. Scirpus L., Bulrush

Growing in wet soil or in water; stems leafy or only basal sheaths present; few to many flowers spirally arranged in spikelets, each flower in axil of a scale, spikelets sessile or peduncled, solitary or in clusters; involucre absent or of 1 to several sometimes leaflike bracts; perianth absent or of 1 to several bristles; stamens 2-3; style 2-3-cleft; achene 3-angled, plano-convex, or lenticular.

a. Involucre none; outermost scale of solitary spikelet elongate, appearing as an involucre; tufted; blades elongate. S. verecundus Fern.

a. Involucre of only 1 primary bract; leaves, if present, only near base. (See third a.)

b. Spikelets sessile or subsessile, single or 2-12 in a glomerule.

c. Stem sharply 3-angled; perennial, with rhizomes.

d. Stigmas 2; achene plano-convex, short-pointed; scale notched, mucronate. S. americanus Pers.

d. Stigmas 3; achene 3-angled, long-pointed; scale acute, not notched. S. torreyi Olney

c. Stem bluntly 3-angled or terete; tufted annuals.

d. Stigmas 2, rarely 3; involucral bract often deflexed at maturity; achene biconvex. S. purshianus Fern.

d. Stigmas 2; involucre erect or arching; achene plano-convex; rare. S. smithii Gray

d. Stigmas 3; achene 3-angled, cross-wrinkled; stem terete; rare. S. saximontanus Fern.

b. At least some spikelets or glomerules of spikelets stalked in an umbelliform inflorescence.

c. Stem soft, easily compressed; scale about equaling achene; spikelet ovoid. S. validus Vahl

c. Stem firm; scale longer than achene; spikelet lance-ovoid, acute. S. acutus Muhl.

a. Involucre of 2 or more leaflike bracts; stem leafy; spikelets or groups of spikelets in umbelliform cluster.

b. Stem sharply 3-angled, tall, stout; spikelet 5 mm thick or more, 1-2.5 cm long. S. fluviatilis (Torr.) Gray, Great River B.

b. Stem terete or slightly 3-angled; spikelet narrower and shorter.

c. Lower sheaths red-tinged, strongly septate-nodulose; stem about 1 cm thick at base; blades 1-2.5 cm wide; stamens and stigmas 3. <u>S</u>. <u>expansus</u> Fern.
c. Sheaths not red-tinged; stem more slender.
  d. Leaves 10-20, 2-ranked; spikelets red-brown, in small glomerules; bristles 6, contorted, about twice as long as pale achene. <u>S</u>. <u>polyphyllus</u> Vahl
  d. Leaves 10 or fewer, not 2-ranked.
    e. Spikelets dark-green or dark-gray, becoming blackish, in dense glomerules; bristles straight, equaling or shorter than the pale achene or wanting. <u>S</u>. <u>atrovirens</u> Willd.
    e. Spikelets brown or red-brown, in clusters of few to several or solitary; bristles flexuous, much longer than pale achene, at maturity long exserted, inflorescence then a mass of brown wool. <u>S</u>. <u>cyperinus</u> (L.) Kunth   (See third e.)
    e. Spikelets oblong or cylindric, brown, mostly solitary on pedicels; scale with excurrent green midrib; involucral leaves mostly shorter than rays; bristles curling, longer than brown achene. <u>S</u>. <u>lineatus</u> Michx.

7. Eriophorum L., Cotton-sedge

Bog plants; flowers perfect; spikelets many-flowered, solitary or in clusters; foliaceous bracts present in Ohio species; stamens 1-3; style 3-cleft; achene surrounded by the many persistent elongate white to brown cottony branches of bristles.

a. Involucral leaf 1; leaves channeled. <u>E</u>. <u>gracile</u> W. D. J. Koch
a. Involucral leaves more than 1; leaves flat below the middle.
  b. Stamens 3; scale with 1 prominent midnerve. <u>E</u>. <u>viridi-carinatum</u> (Engelm.) Fern.
  b. Stamen 1; scale with 3-5 nerves. <u>E</u>. <u>virginicum</u> L.

8. Hemicarpha Nees

<u>H</u>. <u>micrantha</u> (Vahl) Britt. Small tufted annual; stem-leaves usually 2, the lower only a sheath; flowers perfect; scales with short recurved tips; spikelets sessile, solitary or 2-3 together, subtended by 2-3 foliaceous bracts, lowest of which looks like continuation of culm; perianth none; stamen 1; style 2-cleft; achene obovoid-oblong.

9. Rhynchospora Vahl, Beak-rush

Stem usually leafy; spikelets 1-several-flowered, in terminal and often axillary cymose clusters; usually lowest scales empty; flowers perfect or the terminal often staminate; perianth of bristles, rarely absent; stamens 2-3; style 2-cleft or 2-toothed, base persisting as a beak on the flat or lenticular achene.

a. Tubercle conic, about as wide as long; achene subglobose, with transverse ridges. <u>R</u>. <u>globularis</u> (Chapm.) Small var. <u>recognita</u> Gale
a. Tubercle longer than wide; achene obovoid.
  b. Stamens usually 2; scales white or pinkish-tan; bristles 8-14, usually pubescent at base; blades setaceous to narrowly linear. <u>R</u>. <u>alba</u> (L.) Vahl
  b. Stamens 3; scales brown; bristles usually 6, not pubescent.
    c. Blades filiform; width of achene about half the length, exclusive of beak. <u>R</u>. <u>capillacea</u> Torr.
    c. Blades 2.5-7 mm wide; width of achene about 2/3 the length, exclusive of beak. <u>R</u>. <u>capitellata</u> (Michx.) Vahl

10. Cladium P. Br., Twig-rush

<u>C</u>. <u>mariscoides</u> (Muhl.) Torr. Culm leafy; blades narrow; spikelet few-flowered, lowest scales empty, middle flowers staminate or rudimentary, uppermost flower perfect; spikelets in capitate clusters, the clusters in terminal and axillary umbelliform cymes, the terminal subtended by a leaflike bract; perianth none; stamens 2; style deciduous: achene ovoid.

11. Scleria Berg., Nut-rush

Stem 3-angled, leafy; spikelets in small terminal and axillary clusters; flowers monosporangiate, the 2 kinds in separate spikelets; carpellate spikelets 1-flowered, lower scales empty; staminate spikelets few-flowered; perianth none; stamens 1-3; stigmas 3; achenes hard, white in ours, globose or ovoid, usually with smooth or rough disk at base.

a. Achene smooth; disk covered with white crust; blades 4-9 mm wide. S. triglomerata Michx.
a. Achene rough; blades narrowly linear.
    b. Perennial, with rhizomes; disk with about 6 tubercles; spikelets in terminal and sometimes peduncled axillary clusters. S. pauciflora Muhl.
    b. Annual; disk very small, without tubercles; spikelets in sessile clusters that form an interrupted spike. S. verticillata Muhl.

12. Carex L.

Stems mostly triangular; leaves 3-ranked; flowers monosporangiate, in spikes, the spikes solitary or clustered; staminate flower of 3 stamens in axil of a scale; carpellate flower of 2 or 3 united carpels, enclosed in a sac (perigynium) which is in axil of a scale, stigmas protruding from tip of perigynium at anthesis; ovulary 1-loculed; fruit an achene.

a. Stigmas 2; achene lenticular.
    b. Spikes similar in appearance, sessile, usually both staminate and carpellate flowers in each.
        c. At least the terminal spike with staminate flowers at tip and carpellate flowers at base (androgynous).
            d. Culms solitary or few together, from long rhizomes; perigynia with thin or winged margins; blades 2-4 mm wide. 1. ARENARIAE
            d. Culms tufted, from usually short thick rhizomes.
                e. Culms soft, easily compressed; inner band of leaf-sheath usually cross-wrinkled or red-dotted; perigynium-base spongy or corky; spikes all on main axis or some on lateral branches. 5. VULPINAE
                e. Culms firm, not easily compressed; perigynium-base sometimes spongy or corky.
                    f. Spikes usually fewer than 10, all on main axis of inflorescence, sometimes not separated; perigynium usually plano-convex, usually green. 2. BRACTEOSAE
                    f. Spikes many, usually some attached to lateral branches of main axis; inner band of leaf-sheath cross-wrinkled or red-dotted or red-brown at mouth.
                        g. Perigynia flattened, pale; inner band of leaf-sheath cross-wrinkled; bracts of inflorescence usually several, setaceous. 3. MULTIFLORAE
                        g. Perigynia biconvex, dark at maturity; inner band of sheath red-dotted or with red-brown mouth, not wrinkled; bracts usually short and few. 4. PANICULATAE
        c. At least the terminal spike with staminate flowers at base (gynecandrous) or scattered among the carpellate flowers.
            d. Perigynium very flat, scalelike, margin winged, base not thick or corky. 9. OVALES
            d. Perigynium not so flat, margin sometimes thin, not winged, base thickened or corky.
                e. Perigynium narrowly lanceolate, 4 mm long or more, closely appressed. 7. DEWEYANAE
                e. Perigynium shorter than above.
                    f. Perigynium ellipsoid, not bent outward, length about twice width, finely dotted, abruptly narrowed to short, not thin-edged, beak. 6. HELEONASTES
                    f. Perigynium lance- to round-ovate, strongly bent outward at maturity, firm, not dotted, length less than twice width or beak thin-edged. 8. STELLULATAE
    b. Spikes not all similar in appearance, the lowest usually carpellate, sometimes peduncled.
        c. Scales with rough awns; achenes indented on one side. C. crinita Lam. Lower sheaths of var. crinita are smooth, of var. gynandra (Schwein.) Schwein. & Torr., are scabrous.
        c. Scales obtuse to acuminate; achenes not indented on one side.
            d. Perigynium plump, almost terete in cross section, beakless, obovoid, orange or white-powdery; lowest bract sheathing; scales dark or pale. 13. BICOLORES
            d. Perigynium biconvex, plano-convex, or flattened, usually with short beak; bracts sheathless or nearly so; scales dark. 14. ACUTAE
a. Stigmas 3; achene 3-angled or terete.
    b. Style jointed with the achene, deciduous from mature achene.
        c. Margins of staminate scales overlapping around axis of spike or somewhat united; spike 1 per culm, or additional ones on slender almost basal peduncles.
            d. Perigynium beakless, with blunt, usually notched, apex. C. leptalea Wahl.
            d. Perigynium beaked; lowest carpellate scale leaflike. 10. PHYLLOSTACHYAE

c. Margins of staminate scales not overlapping, not united; spikes **2** or more per culm, uppermost usually wholly or partly staminate, lowest usually carpellate.
  d. Bract at base of inflorescence without, or with very short, sheath.
    e. Perigynia or blades or sheaths pubescent or puberulent, or perigynia scabrous.
      f. Terminal spike carpellate above or, if wholly staminate, then perigynium beakless.
        17. VIRESCENTES
      f. Terminal spike wholly staminate; perigynium beaked.
        g. Perigynium scabrous, beak nearly as long as body, curved outward; blades of sterile shoots 6-18 mm wide. <u>C. scabrata</u> Schwein.
        g. Perigynium pubescent, sometimes minutely so; beak straight.
          h. Culms, blades, and sheaths pubescent. <u>C. hirtifolia</u> Mackenz.
          h. Culms and blades glabrous; sheaths usually glabrous.
            i. Carpellate spikes globose or oblong, usually 1 cm long or less, not more than 3 cm apart; plants low; perigynium stipitate. 11. MONTANAE
            i. Carpellate spikes 1.5-5 cm long, cylindric, far apart. 16. HIRTAE
    e. Perigynia, blades, and sheaths glabrous.
      f. Perigynium with evident beak.
        g. Carpellate spikes linear-cylindric, the lower long-peduncled and drooping; perigynium 2-edged, later bulging with 3-angled achene. 18. GRACILLIMAE
        g. Carpellate spikes oblong to globose, sessile or subsessile. 24. EXTENSAE
      f. Perigynium beakless or almost beakless.
        g. Terminal spike staminate; roots covered with felt. 15. LIMOSAE
        g. Terminal spike carpellate above, staminate below.
          h. Perigynium as wide as long, obovoid, cross-wrinkled. <u>C. shortiana</u> Dew.
          h. Perigynium ellipsoid, green or whitish, papillate, not cross-wrinkled.
            <u>C. buxbaumii</u> Wahl.
  d. Bract at base of inflorescence with long closed sheath.
    e. Bract at base of inflorescence bladeless or with rudimentary blade.
      f. Blades 10-25 mm wide; sheaths red-tinged. 23. LAXIFLORAE
      f. Blades much less than 10 mm wide.
        g. Blades filiform; carpellate scales whitish. <u>C. eburnea</u> Boott
        g. Blades 2-3 mm wide; carpellate scales purple-brown. 12. DIGITATAE
    e. Bract at base of inflorescence with well-developed blade.
      f. Perigynia densely hairy; lateral spikes far apart, subsessile. 16. HIRTAE
      f. Perigynia glabrous or puberulent or with a few hairs near beak.
        g. Perigynia spreading or reflexed at maturity, beaked; carpellate spikes subglobose or oblong, sessile or short-peduncled. 24. EXTENSAE
        g. Perigynia ascending; carpellate spikes oblong- to linear-cylindric.
          h. Terminal spike carpellate above, staminate below; perigynia beakless or short-beaked, lateral spikes usually drooping. 18. GRACILLIMAE
          h. Terminal spike staminate or, if rarely not, then perigynia obviously beaked.
            i. Carpellate spikes linear-cylindric, the lower slender-stalked and drooping; perigynia beaked, not 3-angled, nerves usually few or weak.
              j. Perigynium-body subglobose, abruptly narrowed to beak as long as body. <u>C. sprengelii</u> Dew.
              j. Perigynium-body ellipsoid or fusiform, gradually narrowed to beak; perigynium or achene stipitate. 19. SYLVATICAE
            i. Carpellate spikes short, linear or oblong, ascending or erect; if rarely drooping, then perigynia sharply 3-angled.
              j. Nerves of perigynia impressed, many, fine. 21. OLIGOCARPAE
              j. Nerves of perigynia raised.
                k. Perigynia rounded at base, subterete, symmetric. 20. GRANULARES
                k. Perigynia tapering to base, 3-angled, often asymmetric.
                  l. Carpellate scales usually white or pale with green centers; leaf-blades flat, 2-40 mm wide. 23. LAXIFLORAE
                  l. Carpellate scales red or purple; leaf-blades sometimes plicate or rolled, usually 4 mm wide or less. 22. PANICEAE
b. Style not jointed with achene, persistent.
  c. Body of perigynium obovoid, truncate at top; spike sometimes 1. 25. SQUARROSAE
  c. Body of perigynium not obovoid and truncate.

d. Perigynia firm, sometimes with long out-curved teeth, sometimes pubescent; basal sheaths usually red.   26. PALUDOSAE
d. Perigynia papery, inflated, usually glabrous.
    e. Perigynia lance-ovoid, 1-1.5 cm long; spikes short-cylindric or subglobose; scales awned, almost equaling perigynia. C. folliculata L.
    e. Perigynia shorter or, if 1 cm long, then ovoid.
        f. Awns of carpellate scales rough, as long as or longer than blades of scales.
            27. PSEUDO-CYPEREAE
        f. Awns of carpellate scales absent or shorter than blades of scales or not rough.
            g. Perigynia usually less than 10 mm long; nerves 12 or fewer. 29. VESICARIAE
            g. Perigynia 10 mm long or more; nerves more than 12. 28. LUPULINAE

1. ARENARIAE

a. Perigynium ovate, beak 1/3 as long as body or less; inner band of leaf-sheath green-nerved.
    C. sartwellii Dew.
a. Perigynium lanceolate, beak 1/2 as long as body or more; inner band of leaf-sheath without green nerves. C. foenea Willd.

2. BRACTEOSAE

a. Leaf-sheaths tight, usually neither green and white mottled nor septate-nodulose.
  b. Perigynium spongy at base.
    c. Perigynium-beak smooth-margined; scales acuminate to cuspidate; blades 1-3 mm wide.
      d. Perigynium biconvex, base of inner face striate, distended. C. retroflexa Muhl.
      d. Perigynium plano-convex, nerveless. C. texensis (Torr.) Bailey
    c. Perigynium-beak serrate-margined.
      d. Perigynium 4-6 mm long; scales red-tinged, acuminate to awned; blades 1.5-3 mm wide; spikes contiguous to somewhat separated. *C. spicata Huds.
      d. Perigynium usually shorter; scales white with green midvein, obtuse; spikes separate.
        e. Stems capillary, spreading or reclining; blades about 1 mm wide; stigmas coiled.
          C. radiata (Wahl.) Dew.
        e. Stems erect or ascending, slender.
          f. Blades 1.5-3 mm wide; stigmas coiled. C. convoluta Mackenz.
          f. Blades 1-2 mm wide; stigmas recurved. C. rosea Schkuhr
  b. Perigynium not spongy at base; upper, or all, spikes contiguous.
    c. Perigynium widest near truncate base, 2-3.5 mm long; blades 1-3 mm wide; inflorescence oblong to ovoid, spikes contiguous. C. leavenworthii Dew.
    c. Perigynium widest at about middle, base rounded or tapering; blades 1-4 mm wide.
      d. Inflorescence short-cylindric, lower spikes usually separate; perigynium 3-3.5 mm long, strongly ribbed, or rarely upper face scarcely ribbed. C. muhlenbergii Schkuhr
      d. Inflorescence ovoid or oblong, short, spikes contiguous.
        e. Perigynium 2.5 mm long. C. cephalophora Muhl.
        e. Perigynium 3-3.5 mm long. C. mesochorea Mackenz.
a. Leaf-sheaths loose, green and white mottled, septate-nodulose.
  b. Perigynia drab or brownish, almost equalled by the acuminate or awned scales; spikes mostly contiguous; sutures and teeth of perigynia prominent. C. gravida Bailey
  b. Perigynia green, longer than the scales.
    c. Top of inner band of leaf-sheath firm; stigmas long. C. aggregata Mackenz.
    c. Top of inner band of leaf-sheath fragile; stigmas short.
      d. Spikes usually contiguous; inflorescence to 4 cm long. C. cephaloidea Dew.
      d. At least lower spikes separated; inflorescence to 15 cm long. C. sparganioides Muhl.

3. MULTIFLORAE

a. Leaves usually equaling or overtopping culm; beak of perigynium 1/2 as long as body or more.
    C. vulpinoidea Michx.
a. Leaves usually shorter than culm; beak of perigynium less than half as long as body. C. annectens Bickn.

4. PANICULATAE

a. Perigynium obovoid, biconvex, abruptly narrowed to beak. C. decomposita Muhl.
a. Perigynium ovoid or lance-ovoid, gradually tapering to beak.
  b. Upper edge of inner band of leaf-sheath brown. C. prairea Dew.
  b. Upper edge of inner band of leaf-sheath pale. C. diandra Schrank

5. VULPINAE

a. Corky base of perigynium an enlarged cushion, wider than portion above; beak of uniform width,
   2-3 times as long as body. C. crus-corvi Shuttlew.
a. Perigynium tapering from base to apex.
  b. Beak of perigynium not longer than body. C. conjuncta Boott
  b. Beak of perigynium longer than body.
    c. Thin band of leaf-sheath fragile, prolonged at top, usually cross-wrinkled. C. stipata Muhl.
    c. Thin band of leaf-sheath firm, concave or truncate at top, usually not cross-wrinkled.
      C. laevivaginata (Kukenth.) Mackenz.

6. HELEONASTES

a. Culms filiform; spikes 1-3, widely separated, 1-5-flowered. C. trisperma Dew.
a. Culms slender; spikes more than 3, not widely separated.
  b. Spikes usually with more than 10 flowers; leaves glaucous. C. canescens L.
  b. Spikes usually with fewer than 10 flowers; leaves green. C. brunnescens (Pers.) Poir.

7. DEWEYANAE

a. Perigynia strongly nerved. C. bromoides Schkuhr
a. Perigynia nerveless or faintly nerved near base. C. deweyana Schwein.

8. STELLULATAE

a. Perigynium widest at middle, strongly nerved, body circular, beak short, smooth. C. seorsa Howe
a. Perigynium widest near base; beak serrulate.
  b. Perigynium-teeth obscure, about 0.25 mm long; scales shorter than perigynium-body.
    c. Culms weak, very slender; blades 1 mm wide or less; scales pale. C. howei Mackenz.
    c. Culms wiry; blades 1-3 mm wide; scales brownish. C. interior Bailey
  b. Perigynium-teeth obvious, to 0.5 mm long.
    c. Perigynium broadly ovate, width 2/3 - 3/4 length, base truncate or cordate, lower half spongy;
      both faces strongly nerved, on lower face the lateral nerves strongly bowed outward at base;
      scales 2/3 length of body or more. C. incomperta Bickn.
    c. Perigynium ovate or lance-ovate, about equaling upper scales; upper face lightly nerved.
      d. Margin of perigynium-beak sharp, rough, ciliate-serrate; spikes wholly staminate, wholly
        carpellate, or of both kinds of flowers. C. sterilis Willd.
      d. Margin of perigynium-beak serrulate; spikes staminate at base, carpellate at tip.
        C. cephalantha (Bailey) Bickn.

9. OVALES

a. Perigynium lanceolate, gradually tapering from near base to tip of beak, at least 3 times as long as
  wide.
  b. Spikes acute at both ends, 1.5-2.5 cm long; perigynium 7-10 mm long. C. muskingumensis
    Schwein.
  b. Spikes and perigynium shorter; perigynium not more than 2 mm wide.
    c. Spikes usually acute at tip; blades 1-3 mm wide; wing of perigynium extending to base but
      often narrow near base. C. scoparia Schk.
    c. Spikes usually not acute at tip; blades wider; wing of perigynium abruptly ending at about mid-
      dle of achene.
      d. Inflorescence not moniliform; sheaths green-striped on ventral side. C. tribuloides Wahl.
      d. Inflorescence moniliform; sheaths white-hyaline on ventral side, fragile at summit, green
        and white mottled and septate-nodulose; rare. C. projecta Mackenz.

a. Perigynium lanceolate to broadly ovate, less than 3 times as long as wide, usually, but not always, abruptly narrowed to beak.
  b. Tips of perigynia at maturity strongly spreading, beak usually twisted at base; wing ending abruptly at about middle of achene; sheaths loose; blades 3-7 mm wide. C. cristatella Britt.
  b. Tips of perigynia not strongly spreading, beak not twisted.
    c. Either scales as long as perigynia and covering them or scales awn-tipped, or both.
      d. Scales as long as perigynia, covering them, except sometimes at edges; inflorescence moniliform. C. argyrantha Tuckerm.
      d. Scales awn-tipped.
        e. Body of perigynium widest above middle; beak less than half as long as body. C. alata Torr. & Gray
        e. Body of perigynium widest at middle; beak half as long as body or longer; spikes with long staminate bases. C. straminea Willd.
    c. Scales not awn-tipped, not reaching to tip of perigynium-beak.
      d. Inflorescence compact, spikes densely crowded; blades narrow.
        e. Perigynium-body 2-4 mm long, about half as wide. C. bebbii Olney
        e. Perigynium-body 2-3.4 mm long and wide, beak very short. C. cumulata (Bailey) Mackenz.
      d. Spikes not densely crowded in a compact inflorescence.
        e. Perigynium not more than 2 mm wide; sheaths prolonged ventrally.
          f. Blades mostly 4-6 mm wide; sheaths green and white mottled; perigynium about 7-nerved on upper face; inflorescence rarely moniliform, if so, straight. C. normalis Mackenz.
          f. Blades mostly less than 4 mm wide; inflorescence moniliform, arching or flexuous, or bent.
            g. Inflorescence flexuous or bent; spikes without long staminate bases. C. tenera Dewey
            g. Inflorescence arching; spikes with long staminate bases; sheaths green and white mottled; style bent just above achene. C. festucacea Schk.
        e. Perigynium more than 2 mm wide, abruptly narrowed to beak.
          f. Spikes subacute at tip and base; blades 2-3 mm wide; inner side of sheath green. C. suberecta (Olney) Britt.
          f. Spikes usually rounded at tip; inner side of sheath partly hyaline. C. brevior (Dewey) Mackenz. and C. molesta Mackenz.

10. PHYLLOSTACHYAE

a. Body of perigynium and of achene oblong; staminate scales 6-12, tapering or rounded at tip; carpellate flowers 3-10. C. willdenowii Schkuhr
a. Body of perigynium and of achene globose; staminate scales 8-20, truncate and erose at tip; carpellate flowers 2-4. C. jamesii Schwein.

11. MONTANAE

a. Body of perigynium about as long as wide, subglobose.
  b. With long rhizomes; lower sheaths strongly fibrillose; achene-tip not bent. C. pensylvanica Lam.
  b. Without long rhizomes; lower sheaths not or little fibrillose; achene-tip bent. C. communis Bailey
a. Body of perigynium longer than wide.
  b. Perigynia almost hidden by red-purple scales. C. nigromarginata Schwein.
  b. Perigynia not hidden.
    c. Spikes closely aggregated; stem drooping or reclining; rare. C. emmonsii Dew.
    c. Spikes separated; stem erect to ascending. C. artitecta Mackenz.

12. DIGITATAE

a. Terminal spike staminate above, carpellate below; basal peduncles present; carpellate scales cuspidate. C. pedunculata Muhl.
a. Terminal spike wholly staminate; basal peduncles absent; carpellate scales obtuse or acute. C. richardsonii R. Br.

13. BICOLORES

a. Perigynia, at maturity, fleshy, orange or brown; carpellate scales acute, spreading. C. aurea Nutt.
a. Perigynia, at maturity, not as above, granular; carpellate scales obtuse, appressed. C. garberi Fern.

14. ACUTAE

a. Perigynium-tip twisted or bent at maturity; carpellate spikes often curved. C. torta Boott
a. Perigynium-tip not twisted; carpellate spikes erect; perigynium biconvex or flattened.
   b. Carpellate spikes usually 1-3 cm long, rarely staminate at tip, scales spreading at maturity; lower sheaths bladeless, rarely fibrillose; perigynium somewhat inflated. C. haydenii Dew.
   b. Carpellate spikes to 6 cm long or more, sometimes staminate at tip; scales not spreading.
      c. Perigynium flattened, obovate or elliptic; lower sheaths with blades; blades rough toward apex. C. aquatilis Wahl.
      c. Perigynium biconvex; lower sheaths bladeless.
         d. Basal sheaths often fibrillose; lowest carpellate spike usually remote. C. stricta Lam.
         d. Basal sheaths not fibrillose; carpellate spikes usually overlapping. C. emoryi Dew.

15. LIMOSAE

a. Carpellate scales lanceolate, long-acuminate, narrower than perigynia. C. paupercula Michx.
a. Carpellate scales ovate, about as wide as perigynia. C. limosa L.

16. HIRTAE

a. Blade 2 mm wide or less, rolled, smooth; tip of achene bent. C. lasiocarpa Ehrh.
a. Blade 2-5 mm wide, flat, scabrous; tip of achene not bent. C. lanuginosa Michx.

17. VIRESCENTES

a. Perigynia glabrous.
   b. Terminal spike wholly staminate. C. pallescens L.
   b. Terminal spike staminate only at base.
      c. Blades glabrate; sheaths pubescent; perigynia plump. C. caroliniana Schwein.
      c. Blades pubescent; perigynia flattened on upper side. C. hirsutella Mackenz.
a. Perigynia pubescent.
   b. Carpellate spikes subglobose to short-cylindric; achene-tip bent. C. swanii (Fern.) Mackenz.
   b. Carpellate spikes linear-cylindric; achene-tip not bent. C. virescens Muhl.

18. GRACILLIMAE

a. Blades pubescent.
   b. Lateral spikes short-stalked; perigynia not exceeding scales. C. davisii Schwein & Torr.
   b. Lateral spikes long-stalked, gynecandrous; perigynia exceeding scales. C. formosa Dew.
a. Blades glabrous.
   b. Perigynium ovate, beak prominent, curving. C. prasina Wahlenb.
   b. Perigynium ellipsoid, beakless. C. gracillima Schwein.

19. SYLVATICAE

a. Perigynium stipitate; achene almost sessile, beak short, abrupt; basal leaves 6-10 mm wide. C. arctata Boott
a. Perigynium almost sessile, slender, fusiform; achene stipitate; basal leaves 6 mm wide or less. C. debilis Michx. Var. rudgei Bailey has shorter wider perigynia and brown-tinged keeled carpellate scales.

20. GRANULARES

a. Staminate spike sessile or short-stalked; culms tufted. C. granularis Muhl.
a. Staminate spike long-stalked; culms solitary. C. crawei Dew.

21. OLIGOCARPAE

a. Carpellate spikes dense; perigynium-beak short.
  b. Carpellate peduncles and margins of bract-sheaths rough. C. conoidea Schkuhr
  b. Carpellate peduncles and margins of bract-sheaths smooth..
    c. Leaves usually green; carpellate spikes 3-20-flowered. C. amphibola Steud.
    c. Leaves glaucous; carpellate spikes 7-60-flowered. C. flaccosperma Dew. var. glaucodea
      (Tuckerm.) Kukenth.
a. Perigynia few or remote in spikes, with definite beaks.
  b. Bract-sheath glabrous; achene-beak not bent. O. oligocarpa Schkuhr
  b. Bract-sheath hispid; achene-beak bent. C. hitchcockiana Dew.

22. PANICEAE

a. Blades gray-green or glaucous, stiff; carpellate spikes dense, 5-10 mm thick. C. meadii Dew.
a. Blades green, thin; carpellate spikes 3-5 mm thick, lower perigynia remote. C. tetanica Schkuhr
a. Blades green, not thick; lower sheaths bladeless, red; perigynia 2-3-ranked. C. woodii Dew.

23. LAXIFLORAE

a. Blades of flowering stem absent or rudimentary; sheaths, culm-bases, and staminate scales red-
    purple. C. plantaginea Lam.
a. Blades of flowering stem well developed.
  b. Perigynia acutely angled above, tapering to short base, many-nerved.
    c. Blade of lowest bract-sheath not more than 3 times as long as sheath.
      d. Sheaths red-purple; perigynium 5-6.5 mm long, beak conic. C. careyana Torr.
      d. Sheaths green; perigynium 2.5-4.5 mm long, beak short. C. platyphylla Carey
    c. Blade of lowest bract-sheath more than 3 times as long as sheath; peduncles of carpellate
        spikes slender, drooping.
      d. Carpellate spike with 1 or 2 staminate flowers at base; rosette blades 6-12 mm wide, often
          glaucous. C. laxiculmis Schw.
      d. Carpellate spike with no staminate flowers at base, perigynia alternate; rosette blades usu-
          ally 5 mm wide or less. C. digitalis Willd.
  b. Perigynia long-tapering to base, its 3 angles obtuse.
    c. Culm winged; rosette leaves to 3.5 cm wide; carpellate scales fan-shaped, truncate, whitish
        with green midvein, shorter than perigynia, awnless. C. albursina Sheldon
    c. Culm not winged; carpellate scales with long or short awn.
      d. Flowerless shoots with culms; perigynia strongly asymmetric.
        e. Staminate spike short and sessile, often hidden; carpellate spikes short and densely-
            flowered; perigynium-beak short, abruptly bent. C. blanda Dew.
        e. Staminate spike usually long-peduncled; carpellate spike not crowded, perigynia grad-
            ually tapering to conspicuous outwardly curved beak. C. gracilescens Steud.
      d. Flowerless shoots without culms; perigynia not strongly asymmetric, beak not much
          curved, whitish. C. laxiflora Lam.

24. EXTENSAE

a. Perigynium 3 mm long or less, beak 1/3 to 1/2 as long as body. C. viridula Michx.
a. Perigynium more than 3 mm long, beak 1/2 to as long as body, strongly reflexed.
  b. Beak serrulate. C. flava L.
  b. Beak smooth. C. cryptolepis Mackenz.

25. SQUARROSAE

a. Carpellate scales long-awned, exceeding perigynia; terminal spikes usually wholly staminate; bracts
    as long as entire culm. C. frankii Kunth
a. Carpellate scales awnless or short-awned, shorter than perigynia; terminal spike carpellate above;
    bracts shorter than culm.
  b. Spike usually solitary; perigynia-beaks spreading or reflexed; style bent at base.
      C. squarrosa L.
  b. Spikes one or more; perigynia-beaks ascending; style straight or bent above base.
      C. typhina Michx.

26. PALUDOSAE

a. Perigynia or sheaths pubescent; beaks and teeth prominent.
  b. Perigynia pubescent; sheaths glabrous. C. trichocarpa Muhl.
  b. Perigynia glabrous; sheaths pubescent; teeth 2-3 mm long. C. atherodes Spreng.
a. Perigynia and sheaths glabrous; teeth less than 1 mm long.
  b. Nerves of perigynium conspicuous, raised; ligule twice as long as wide; lower sheaths fibrillose.
    C. lacustris Willd.
  b. Nerves of perigynium faint, impressed; ligule not twice as long as wide; lower sheaths not fibrillose. C. hyalinolepis Steud.

27. PSEUDO-CYPEREAE

a. Perigynia about 10-nerved.
  b. Perigynia 40-100 per spike, beak shorter than body; blades 2-7 mm wide. C. lurida Wahl.
  b. Perigynia 20-40 per spike, beak longer than body; blades 2-4 mm wide. C. baileyi Britt.
a. Perigynium 12-20-nerved.
  b. Teeth of perigynium outwardly curved, 1-2 mm long. C. comosa Boott
  b. Teeth of perigynium erect, 1 mm long or less.
    c. Perigynium triangular in cross section, coriaceous, stipitate, rare. C. pseudo-cyperus L.
    c. Perigynium rounded in cross section, membranous, scarcely stipitate. C. hystricina Muhl.

28. LUPULINAE

a. Carpellate spikes spherical.
  b. Perigynia smooth, shining, usually 12 or fewer to a spike; achene angled; angles blunt.
    C. intumescens Rudge
  b. Perigynia usually hairy below middle, dull, usually more than 12 to a spike; achene obscurely angled. C. grayii Carey
a. Carpellate spikes oblong or cylindric.
  b. Achenes as long as wide, sides concave, angles prominently knobbed. C. lupuliformis Sartwell.
  b. Achenes longer than wide, sides flat or slightly concave.
    c. Carpellate spike 20-35 mm thick, perigynia 20-75; blades 4-15 mm wide. C. lupulina Muhl.
    c. Carpellate spike narrower, perigynia 10-30; blades 2-6 mm wide. C. louisianica Bailey

29. VESICARIAE

a. Blades filiform, rolled; carpellate spike globose or short-oblong. C. oligosperma Michx.
a. Blades wider, flat; carpellate spikes oblong to cylindric.
  b. Achenes constricted on one side; perigynia 5 mm wide or more. C. tuckermani Boott
  b. Achenes not constricted on one side; perigynia less than 4 mm wide.
    c. Perigynia horizontally spreading or reflexed; bracts more than 3 times as long as whole inflorescence. C. retrorsa Schwein.
    c. Perigynia ascending or spreading; bracts 1-3 times inflorescence.
      d. Perigynia ascending; lower sheaths fibrillose; stem sharply angled above; leaves not conspicuously septate-nodulose. C. vesicaria L.
      d. Perigynia spreading to horizontal; lower sheaths not fibrillose; stems bluntly angled above; leaves conspicuously septate-nodulose, spongy at base. C. rostrata Stokes

ARACEAE, Arum Family

Perennial herbs; leaves cauline and alternate or all basal; inflorescence a spadix subtended by a spathe; flowers hypogynous, monosporangiate or bisporangiate; perianth of 4-6 segments sometimes present in bisporangiate flowers, absent in monosporangiate ones; stamens 2-8; ovulary 1-many-loculed; style present or stigma sessile.

a. Leaves linear; spadix cylindric, diverging from and not surrounded by the linear leaflike spathe.
  5. Acorus.
a. Leaves broader; spathe broad, surrounding at least base of spadix.
  b. Leaves compound; flowers on only base of spadix. 1. Arisaema.
  b. Leaves simple; spadix covered by flowers except sometimes at very tip.

c. Spathe open, ovate, white, acuminate; lateral veins of ovate-cordate leaf-blades parallel.
   3. Calla.
c. Spathe ovoid, fleshy, mottled with purple, brown, yellow, or green; leaf-blades ovate-cordate, lateral veins anastomosing. 4. Symplocarpus.
c. Spathe green with pale margin, covering the elongate slender spadix; leaf-blades sagittate.
   2. Peltandra.

1. Arisaema Mart. , Indian-turnip

With corms; monecious or diecious; flowers monosporangiate; staminate flower a cluster of almost sessile anthers, carpellate, an ovulary with broad stigma, when on one spadix, the staminate above; fruit a red 1-3-seeded berry.

a. Spathe shorter than long-pointed spadix; leaflets more than 3. A. dracontium (L.) Schott, Green Dragon
a. Spathe longer than club-shaped spadix, arching over it; leaflets 3.
  b. Tube of spathe shallowly furrowed.
    c. Lateral leaflets strongly asymmetric; flange of spathe 2-8 mm wide, hood 3-6 cm wide; common. A. atrorubens (Ait.) Blume, Jack-in-the-pulpit
    c. Lateral leaflets not strongly asymmetric; flange of spathe 0. 5-2 mm wide, hood 2-3 cm wide; rare. A. triphyllum (L.) Schott
  b. Tube of spathe deeply furrowed, ridges white; rare. A. stewartsonii Britt.

2. Peltandra Raf. , Arrow Arum

P. virginica (L.) Schott & Endl. Monecious; of swamps and shallow water; leaves basal, long-petioled; flowers monosporangiate, covering spadix except sometimes at tip, staminate above; staminate flower a cluster of sessile anthers, carpellate a carpel surrounded by scalelike staminodes; fruit a 1-3-seeded brown berry.

3. Calla L. , Calla

C. palustris L. , Wild C. Of bogs and pond margins; leaves basal; spadix short-cylindric, 2. 5 cm long or less, covered by flowers; flowers perfect or those at the tip staminate; spathe longer than spadix, persistent; perianth none; stamens 6; ovulary 1-loculed; fruit a berry.

4. Symplocarpus Salisb. , Skunk-cabbage

S. foetidus (L.) Nutt. Leaves large, ovate, cordate at base; spathe almost sessile, fleshy, margins inrolled; spadix globose, covered with the bisporangiate flowers; perianth segments 4; stamens 4; ovulary embedded in spadix, 1-loculed, 1-ovuled.

5. Acorus L. , Sweet-flag, Calamus

A. calamus L. Of swamps, shallow water; rhizome aromatic; midvein of leaf usually at one side of center; flowers yellow, bisporangiate; perianth of 6 segments; stamens 6; ovulary 2-3-loculed; fruit 1- to few-seeded.

LEMNACEAE, Duckweed Family

Small floating herbs; plant body a globose, convex, or flat thallus, with or without roots from lower side; plants solitary or, when they do not separate after budding, in colonies; flowers (rarely observed) at edge of or on upper surface of thallus, perianth none; staminate flower of 1 stamen, carpellate of an ovulary with 1 to few ovules.

a. Plant with roots.
  b. Several roots on each thallus. 1. Spirodela.
  b. Only 1 root on each thallus. 2. Lemna.
a. Plant without roots.
  b. Thallus globose, ovoid, or ellipsoid, usually solitary. 3. Wolffia.
  b. Thallus linear, solitary or in starlike clusters. 4. Wolffiella.

1. Spirodela Schleid., Duckweed

S. polyrhiza (L.) Schleid., Greater D. Thallus flat or somewhat convex below, oval or ovate, to about 1 cm long, green above, purple below, with 5 or more nerves; roots few to 15 or more.

2. Lemna L., Duckweed

Thallus flat or convex beneath, 1-5-nerved, with 1 root.

a. Thallus oblong, up to 1 cm long, with a stalk at one end to 1.5 cm long. L. trisulca L.
a. Thallus oval or oblong, without a stalk, 5 mm long or less.
  b. Thallus symmetric or nearly so.
    c. Thallus convex on both surfaces, 3-nerved. L. minor L.
    c. Thallus convex above, flat below, 1-nerved. L. minima Phillipi
  b. Thallus asymmetric.
    c. Thallus obovoid, 3-nerved. L. perpusilla Torr.
    c. Thallus oblong, 1-nerved. L. valdiviana Philippi

3. Wolffia Horkel

Thallus minute, to 1.5 mm long, globose or ellipsoid, rootless, nerveless, the individuals separating soon after budding.

a. Both upper and lower surface rounded, usually 3 papillae along upper median line. W. columbiana Karst.
a. Lower surface rounded, upper surface flat; brown-dotted; with 1 papilla.
  b. Symmetric; papilla at one end. W. punctata Griseb.
  b. Asymmetric; papilla at center. W. papulifera Thompson

4. Wolffiella Hegelm.

W. floridana (J. D. Sm.) Thompson Thallus flat, linear, curved, to 1.5 mm long, rootless, brown-dotted; solitary or in star-shaped groups.

XYRIDACEAE, Yellow-eyed-grass Family

Xyris L., Yellow-eyed-grass

Leaves basal or nearly so, linear or linear-lanceolate; flowers perfect; hypogynous, bracted, in ovoid heads or spikes 5-15 mm long; sepals 3, one larger; petals 3, yellow; stamens 3; staminodes 3; capsule 1-loculed; placentae 3, parietal.

a. Leaves and scapes often conspicuously twisted; lateral sepals with ciliate keel; base of plant hard, bulbous. X. torta Sm.
a. Leaves and scape not conspicuously twisted; lateral sepals with erose keel; base of plant not as above. X. caroliniana Walt.

ERIOCAULACEAE, Pipewort Family

Eriocaulon L., Pipewort

E. septangulare With. Herbs at edge of or in shallow water; stem often 7-angled, its base enclosed by a tubular sheath; leaves basal, narrow, much shorter than stem; flowers hypogynous, monosporangiate, both kinds in a solitary head 1 cm wide or less, each flower in axil of a scarious bractlet, head subtended by an involucre; perianth white-pubescent; sepals 2; petals 2; stamens 4; carpels united; fruit a capsule.

COMMELINACEAE, Spiderwort Family

Leaves sheathing at base, entire; flowers hypogynous, perfect, clustered, ephemeral; sepals 3, green; petals 3, sometimes unequal; stamens 6, some of them sometimes sterile; carpels 3, united; ovulary 3-loculed; ovules 1-few in each locule or 1 locule empty; style 1, stigma 1; fruit a capsule.

a. Fertile stamens 6; corolla regular. Tradescantia.
a. Fertile stamens 3; corolla zygomorphic. Commelina.

Commelina L. , Dayflower

Inflorescence subtended by a cordate-ovate folded spathe; sepals of unequal size, slightly united; anthers of fertile stamens unlike, 3 stamens sterile; filaments glabrous; one locule of ovulary sometimes without an ovule.

a. Larger petals blue, ovate; smaller petal paler; anthers 6. C. communis L.
a. Petals all blue, the larger reniform; anthers 5. C. diffusa Burm. f.

Tradescantia L. , Spiderwort

Inflorescence subtended by a leaflike bract; sepals often somewhat tinged with rose-purple; petals 1-2 cm long, blue, rose, or white; filaments bearded.

a. Blade much wider than circumference of sheath, a short petiolelike base between blade and sheath.
    T. subaspera Ker
a. Blade little, if any, wider than circumference of sheath; petiolelike base absent.
    b. Leaves glaucous; pedicels glabrous; sepals glabrous or hairy at apex. T. ohiensis Raf.
    b. Leaves not glaucous; pedicels and sepals hairy, not viscid. T. virginiana L. (See third b)
    b. Leaves not glaucous; pedicels and sepals viscid-hairy; rare. T. bracteata Small

PONTEDERIACEAE, Pickerel-weed Family

Aquatic or marsh herbs; leaves or their petioles usually sheathing; flowers hypogynous, perfect; perianth segments 6, united in ours; stamens on perianth-tube; ovulary 3- or 1-loculed.

a. Cauline leaf solitary, ovate or narrower; flowers in spikes; perianth blue. Pontederia
a. Leaves alternate on stem; flowers solitary; perianth yellow. Heteranthera

Pontederia L. , Pickerel-weed

P. cordata L. Cauline leaf solitary, similar to basal ones; blades cordate to tapering at base, ovate to linear-lanceolate; spike dense, subtended by a spathelike bract; perianth about 1 cm long, blue, 2-lipped, the limb spreading, as long as or longer than the tube; stamens 6, 3 of them sometimes not perfect; filaments pubescent; ovulary 3-loculed, 2 locules empty; fruit 1-seeded.

Heteranthera R. & P.

H. dubia (Jacq.) MacM. , Water Stargrass. Submersed or growing on mud; leaves alternate, linear, sessile; flower solitary, subtended by a spathe; perianth yellow, regular, tube 2-3 cm long, slender, limb spreading; stamens 3; ovulary 1-loculed; placentae 3, parietal.

JUNCACEAE, Rush Family

Annual or perennial grasslike or sedgelike herbs; leaves linear, sometimes auricles at summit of sheath; flowers small, hypogynous, perfect; individual flowers or glomerules of flowers in cymes, spikes, or umbels; sepals and petals 3 each, separate, regular, glumelike, green, brown, or red; stamens 6 or 3; style 1; stigmas 3; carpels 3, united; ovulary 3- or 1-loculed; fruit a capsule.

a. Plants glabrous; leaf-sheaths open; capsule many-seeded. Juncus
a. Plants usually hairy; leaf-sheaths closed; capsule 3-seeded. Luzula

Juncus L. , Rush

Blades flat or terete, sometimes absent; stamens 6, opposite petals and sepals, or 3, opposite sepals; ovulary 3-loculed, placentae axile, or 1-loculed, placentae parietal; seeds many, sometimes with tail-like appendages.

a. Blades flat or terete, not septate, sometimes absent below involucre of inflorescence.
    b. Each flower with 2 small bracts (prophylls) immediately below perianth; flowers not in heads.
        c. Inflorescence apparently lateral, a single involucral leaf appearing to be a continuation of
            stem; leaves below involucre only bladeless sheaths.
            d. Perianth green or pale brown; culms tufted; anthers equaling filaments. J. effusus L.
            d. Perianth-parts with prominent chestnut stripe each side of midvein; culms usually in a row
              from elongate rhizome; anthers much longer than filaments. J. balticus Willd.

c. Inflorescence terminal; blades present below involucre.
 d. Usually less than 2 dm high, annual; inflorescence half the height of plant. J. bufonius L. , Toad R.
 d. Usually more than 2 dm high, perennial; inflorescence less than half height of plant.
  e. Perianth-parts with brownish stripes, tips incurved, about equaling stamens; leaf-sheaths extending halfway up the culm or more; rare. J. gerardi Loisel. , Black-grass.
  e. Tips of perianth-parts erect or spreading, acute; leaf-sheaths on only lower portion of culm.
   f. Capsule longer than perianth, truncate at tip; blades filiform, terete or narrowly channeled; rare. J. greenei Oakes & Tuckerm.
   f. Capsule not longer than perianth.
    g. Auricles delicate, hyaline, much longer than wide, to 5 mm long; bract usually exceeding inflorescence. J. tenuis Willd.
    g. Auricles rounded at summit, usually not delicate, about 1 mm long, not longer than wide.
     h. Inflorescence exceeding involucral leaf; flowers conspicuously secund; auricles pale. J. secundus Beauv.
     h. Inflorescence exceeded by involucral leaf; flowers not conspicuously secund.
      i. Auricles coriaceous, yellow or brown; prophylls with rounded or subacute apex. J. dudleyi Wieg.
      i. Auricles not coriaceous, pale; prophylls with acute apex.
       j. Auricles firm; inner basal sheaths purple-tinged; rare. J. platyphyllus (Wieg.) Fern.
       j. Auricles membranous; sheaths drab or pale brown; rare. J. interior Wieg.
 b. No bracts (prophylls) immediately below perianth; flowers in glomerules of 2 or more; stamens almost as long as perianth.
  c. Stems tufted, slender; blades 3 mm wide or less; heads fewer than 25; stamens usually not persistent in fruit. J. marginatus Rostk.
  c. Stems stout, usually arising singly from rhizome, 3 mm thick or more at base; blades 4 mm wide or more; heads usually more than 25; stamens usually evident in fruit. J. biflorus Ell.
a. Leaf blades terete, septate; flowers in heads; no bracts (prophylls) immediately below perianth.
 b. Seeds with white tips or tail-like appendages.
  c. Combined length of tails equaling or exceeding length of seed-body; perianth equaling or a little shorter than capsule. J. canadensis J. Gay
  c. Combined length of tails or tips much less than length of seed-body.
   d. Glomerules small, 2-5-flowered; capsule much exceeding the acute perianth-segments. J. brachycephalus (Engelm.) Buch.
   d. Glomerules mostly 5-many-flowered; capsule a little longer than the acute perianth-segments. J. subcaudatus (Engelm.) Cov. & Blake
 b. Seeds with minute dark tips, not with tail-like appendages.
  c. Stamens 3.
   d. Heads densely spherical.
    e. Capsule about as long as perianth; rare or absent in Ohio. J. scirpoides Lam.
    e. Capsule 1/2 to 2/3 as long as perianth. J. brachycarpus Engelm.
   d. Heads hemispherical.
    e. Capsule about twice as long as perianth; inflorescence very diffuse. J. diffusissimus Buckl.
    e. Capsule equaling or slightly longer than perianth; inflorescence about half as wide as long, heads relatively few. J. acuminatus Michx.
  c. Stamens 6.
   d. Capsule slender, subulate; sepals and petals narrowly lanceolate, tapering to slender pointed apex; heads dense, spherical.
    e. Stem stout; blades abruptly divergent from stem; auricles 2 mm long or more; heads greenish-brown, to 1.5 cm wide; sepals longer than petals. J. torreyi Cov.
    e. Stem slender; blades erect or ascending; auricles 1 mm long; heads reddish-brown, to 12 mm wide; sepals and petals equal or petals longer. J. nodosus L.
   d. Capsule ovoid or oblong, obtuse or acute at apex, longer than the obtuse to acute sepals and petals; heads hemispheric or few-flowered.
    e. Inflorescence-branches spreading; capsule red-brown, conspicuously longer than perianth. J. articulatus L.

e. Inflorescence-branches ascending; capsule often pale brown, a little longer than perianth. J. alpinus Vill.

Luzula DC., Wood-rush

Leaf-blades flat and soft; perianth green or brown, often scarious; stamens 6; ovulary 1-loculed; ovules and seeds 3, seeds often with caruncles.

a. Flowers usually solitary at ends of spreading inflorescence-branches. L. acuminata Raf. In var.
  acuminata, the primary inflorescence-branches are mostly unbranched; in var. carolinae (Wats.)
  Fern., they are mostly branched.
a. Flowers in heads at ends of inflorescence-branches.
  b. Plants with white coral-like tubers at base. L. bulbosa (Wood) Rydb.
  b. Plants without white coral-like tubers.
    c. Heads short-cylindric on erect or ascending branches. L. multiflora (Retz.) Lejeune
    c. Heads subglobose on slender, often divergent, branches. L. echinata (Small) F. J. Herm.

LILIACEAE, Lily Family

Mostly perennial herbs with corms, bulbs, or rhizomes; leaves linear to broad, alternate, opposite, or whorled, often all basal, usually parallel-veined; flowers hypogynous or partly epigynous;
perianth usually regular, of 6 or 4 parts, petals and sepals usually similar, separate or united; stamens 6 or 4, free or adnate to perianth; carpels usually 3, united; locules 3; styles 1 or 3 or absent;
fruit a berry or capsule.

a. Flowers solitary and axillary or flowers in axillary clusters.
  b. Leaves bractlike, with filiform branches in the axils. 17. Asparagus
  b. Leaves ordinary; filiform branches not present.
    c. Perianth cylindric, lobed; flowers 1-few at a node. 23. Polygonatum
    c. Perianth parts separate or united slightly at base.
      d. Leaves petioled; flowers in umbels; stems usually climbing by tendrils. 28. Smilax
      d. Leaves not petioled; flowers single or in 2's; stems not climbing.
        e. Perianth yellow, 1.5-5 cm long; leaves perfoliate or sessile. 7. Uvularia
        e. Perianth rose-color or greenish-white, 1 cm long; leaves sessile or clasping.
        22. Streptopus
a. Flowers solitary and terminal or flowers in terminal clusters.
  b. Flowers solitary.
    c. Leaves in a single whorl of 3; sepals green, petals white or colored. 26. Trillium
    c. Leaves opposite, alternate, whorled, or all basal, but not in a single whorl of 3.
      d. Cauline leaves several to many; perianth parts separate.
        e. Perianth more than 5 cm long, red, orange, or yellow, frequently spotted; anthers
          versatile. 11. Lilium
        e. Perianth smaller.
          f. Filaments shorter than anthers; fruit a capsule; leaves sometimes perfoliate.
            7. Uvularia
          f. Filaments longer than anthers; fruit a berry; leaves not perfoliate. 21. Disporum
      d. Cauline leaves not more than 2 or 3, or none.
        e. Flowers nodding; perianth parts spreading or reflexed. 12. Erythronium
        e. Flowers erect; perianth parts erect. Tulipa
  b. Flowers in clusters.
    c. Perianth parts united below, separate above, 4-12 cm long; leaves basal; scape sometimes
      with leaflike bracts.
      d. Perianth yellow to red-orange; leaves linear. 10. Hemerocallis
      d. Perianth white or bluish; leaves broader. Hosta
    c. Perianth parts separate, or shorter than 4 cm, or both.
      d. Perianth 5 cm long or more, red, orange, or yellow, frequently spotted; anthers versa-
        tile. 11. Lilium
      d. Perianth not as above.
        e. Flowers in umbels.
          f. Leaves in 2 whorls on the stem; perianth greenish-yellow. 25. Medeola
          f. Stem-leaves, if present, not whorled.
            g. Leaves linear or absent at flowering time.

h. Plants with odor of onion. 8. <u>Allium</u>

h. Plants without odor of onion. 9. <u>Nothoscordum</u>

g. Leaves broader than linear, present at flowering time.

h. Leaves 2-4, large, basal or near-basal. 18. <u>Clintonia</u>

h. Leaves several to many, alternate on the stem.

i. Filaments shorter than anthers; fruit a capsule; leaves sometimes perfoliate. 7. <u>Uvularia</u>

i. Filaments longer than anthers; fruit a berry; leaves not perfoliate. 21. <u>Disporum</u>

e. Flowers in spikes, racemes, corymbs, or panicles.

f. Perianth segments united half their length or more.

g. Ovulary partly inferior; perianth rough on outside. 27. <u>Aletris</u>

g. Ovulary superior; perianth smooth.

h. Leaves linear; perianth blue or purple. 15. <u>Muscari</u>

h. Leaves elliptic; perianth white. 24. <u>Convallaria</u>

f. Perianth segments separate or united only a little at base.

g. Leaves present on flowering stem.

h. Leaves linear, grasslike, elongate.

i. Ovulary partly inferior; perianth segments without glands or claws. 3. <u>Stenanthium</u>

i. Ovulary superior; perianth segments clawed, each with 2 glands at base. 5. <u>Melanthium</u>

h. At least some of the leaves broader, not linear, not grasslike.

i. Perianth 4-parted. 20. <u>Maianthemum</u>

i. Perianth 6-parted.

j. Diecious; inflorescence a spikelike raceme. 1. <u>Chamaelirium</u>

j. Not diecious; flowers monosporangiate or perfect.

k. Flowers in a panicle; perianth yellow-green or maroon, 6 mm long or more, longer than or equaling the stamens. 6. <u>Veratrum</u>

k. Flowers in a raceme or panicle; if in a panicle, then perianth 3 mm long or less, shorter than stamens. 19. <u>Smilacina</u>

g. Leaves all basal or nearly basal; scape sometimes with bracts.

h. Woody plants; leaves rigid, sword-shaped; flowers white, 3.5-7 cm long, in large panicle. 16. <u>Yucca</u>

h. Herbs; leaves not as above.

i. Ovulary partly inferior; perianth segments with 1 or 2 glands at base. 4. <u>Zygadenus</u>

i. Ovulary superior; perianth without glands.

j. Scape glutinous with dark glands; styles 3. 2. <u>Tofieldia</u>

j. Scape not glutinous; style 1.

k. Filaments slender; perianth blue. 13. <u>Camassia</u>

k. Filaments broad; perianth white. 14. <u>Ornithogalum</u>

1. Chamaelirium Willd.

<u>C</u>. <u>luteum</u> (L.) Gray, Devil's Bit. Basal leaves oblanceolate, petioled; stem-leaves smaller; diecious; flowers in racemes; perianth segments separate, white, about 3 mm long; stamens about as long as perianth, filaments flat, anthers white; carpellate flowers with vestigial stamens; stigmas 3; fruit a capsule.

2. <u>Tofieldia</u> Huds., False Asphodel

<u>T</u>. <u>glutinosa</u> (Michx.) Pers. Leaves linear, 2-ranked, basal or subbasal; inflorescence a raceme; scape glutinous with dark glands, usually with 1 bractlike leaf; perianth segments separate, white, about 4 mm long; seed appendaged at each end.

3. Stenanthium Gray

<u>S</u>. <u>gramineum</u> (Ker) Morong, Featherbells. Stems tall; leaves elongate, linear, many, chiefly cauline; flowers in a long panicle; perianth segments separate, white or greenish, 1 cm long or less; ovulary partly inferior; fruit a capsule.

4. Zygadenus Michx.

Z. glaucus Nutt., White Camass. Leaves linear, elongate, chiefly basal; flowers bisporangiate or monosporangiate, in a raceme or panicle; perianth segments separate or united at base, about 1 cm long, greenish-white, sometimes bronze or purplish beneath, each with 2-lobed gland at base; ovulary partly inferior; fruit a capsule.

5. Melanthium L., Bunch-flower

M. virginicum L. Stem tall and leafy; leaves broadly linear; flowers in a large terminal panicle with pubescent axis, some bisporangiate, some monosporangiate; perianth segments separate, 1.5 cm long or less, spreading, clawed, each with 2 glands at base; stamens adnate to the claws; fruit a capsule.

6. Veratrum L., False Hellebore

Stems tall, leafy; leaves oval, elliptic, or oblanceolate; flowers bisporangiate or monosporangiate, in panicle; perianth segments separate or united at base, about 6-12 cm long; fruit a capsule.

a. Perianth yellow-green; leaves of about same width to top of stem. V. viride Ait.
a. Perianth dark maroon; upper leaves much narrowed. V. woodii Robbins

7. Uvularia L., Bellwort

With rhizomes; stem scaly below, leafy above; leaves oval, oblong, or lanceolate; flowers nodding, solitary or rarely in 2's, terminal, later appearing lateral as result of prolongation of branches; perianth segments separate, greenish-yellow, 3-4 cm long; styles separate or united to middle; fruit a capsule.

a. Leaves not perfoliate. U. sessilifolia L., Sessile-leaved B.
a. Leaves perfoliate.
    b. Perianth rough-glandular on inner surface; leaves usually glabrous beneath, 1-4 below fork of stem. U. perfoliata L., Perfoliate B.
    b. Perianth smooth on inner surface; leaves usually pubescent beneath, 1-2 below fork of stem. U. grandiflora Sm., Large-flowered B.

8. Allium L., Onion, Garlic, Leek

Stems from tunicate bulbs; leaves narrowly linear or elliptic; flowers in umbels subtended by 1-3 bracts; perianth segments separate or nearly so, about 0.5-1.5 cm long, white, pink, purple, or green; filaments sometimes flat, sometimes toothed or lobed at apex; style filiform or 3-lobed; capsule lobed; seeds black, 1-2 in each locule.

a. Leaves elliptic-lanceolate, appearing early and shriveling before anthesis; perianth white.
    A. tricoccum Ait., Wild L., Ramp.
a. Leaves linear, present at anthesis.
    b. Leaves terete or nearly so, at least the lower portion hollow.
        c. Umbel usually bearing bulblets; inner filaments with 2 long-appendaged teeth at apex; stem leafy to about the middle; bract 1. *A. vineale, Field G.
        c. Umbel usually without bulblets; inner filaments without teeth or the teeth not appendaged.
            d. Stem inflated; perianth white or green; leaves basal; bract 1. *A. cepa L., Common O.
            d. Stem slender; perianth pink or purple; bracts 2-3. *A. schoenoprasum L., Chives.
    b. Leaves flat, solid.
        c. Stem leafy to about middle; inner filaments with 2 long-appendaged teeth at apex.
            d. Umbel bearing bulblets, flowers usually few or none; bract 1, green, long-beaked.
                *A. sativum L., Garlic.
            d. Umbel without bulblets; flowers many; bract 1; stem stout; blades 1-3 cm broad.
                *A. ampeloprasum L., Wild L.
        c. Leaves basal or subbasal; inner filaments without teeth or teeth unappendaged; bracts 2-3.
            d. Umbel erect, bearing bulblets; bulb coat reticulate, the areas diamond-shaped; ovulary not crested. A. canadense L., Meadow G.
            d. Umbel nodding, without bulblets; bulb coat longitudinally nerved, not reticulate; ovulary crested. A. cernuum Roth, Nodding O.

9. Nothoscordum Kunth, False Garlic

N. bivalve (L.) Britt. Bulb tunicate; leaves linear, basal; flowers in terminal umbel; perianth white or greenish, about 1 cm long, segments separate or united at base; capsule subglobose.

10. Hemerocallis L., Day-lily

Leaves basal, linear, 2-ranked; flowers large, lasting but a day, in terminal clusters; perianth funnelform, narrow, limb spreading; stamens inserted at top of perianth-tube; style elongate; stigma capitate; fruit a capsule.

a. Perianth orange, about 1 dm long. *H. fulva L., Common D.
a. Perianth yellow, somewhat smaller. *H. flava L., Yellow D.

11. Lilium L., Lily

Stems tall, from scaly bulbs; leaves lanceolate, sessile, all alternate, or some alternate and some whorled on same stem; flowers large and showy, in clusters or solitary; perianth segments separate, yellow, orange, or red (white or pink in some cultivated species), often purple- or brown-spotted; anthers versatile; style 1; stigma 3-lobed.

a. Perianth segments with claws; flowers erect. P. philadelphicum L.
a. Perianth segments without claws; flowers drooping, buds and fruits erect.
   b. Stem scabrous; leaves all alternate, rounded at base, bulblets in upper axils. *L. tigrinum L., Tiger L.
   b. Stem glabrous; some leaves whorled; no bulblets in axils.
      c. Perianth segments spreading. L. canadense L., Canada L.
      c. Perianth segments recurved, the tips at or near perianth-base.
         d. Midvein of sepals low and rounded; anthers mostly less than 2 mm long; some blades more or less scabrous on veins beneath. L. michiganense Farw.
         d. Midvein of sepals raised and 2-ridged; anthers mostly 2-2.5 mm long; blades smooth beneath. L. superbum L., Turk's-cap L.

12. Erythronium L., Dog-tooth Lily, Fawn Lily

Stem from a deep corm; leaves 2, smooth, somewhat fleshy, mottled, elliptic, appearing basal; flowers solitary, nodding; perianth segments separate, 2-4 cm long; filaments flattened below; fruit a capsule.

a. Perianth white tinged with blue or pink; stigmas spreading. E. albidum Nutt., White F.
a. Perianth yellow; stigmas erect.
   b. Mature capsule rounded at tip, style not persistent. E. americanum Ker, Yellow F.
   b. Mature capsule beaked by persistent style; rare. E. rostratum Wolf

13. Camassia Lindl.

C. scilloides (Raf.) Cory, Wild Hyacinth. Stem scapose; leaves linear, basal; flowers in a raceme, bracts about as long as pedicels; perianth segments separate, blue, about 1-1.5 cm long; filaments elongate; style slender; fruit a capsule.

14. Ornithogalum L., Star-of-Bethlehem

Leaves basal, linear, fleshy; stem scapose, bracted; perianth segments separate, white with median green dorsal band; filaments flat; style 3-angled; stigma 3-lobed or -ridged; fruit a capsule.

a. Flowers 1-2 cm long, in a corymb; filaments tapering to apex. *O. umbellatum L.
a. Flowers about 3 cm long, in a raceme; filaments 2-toothed at summit, same width throughout. *O. nutans L.

15. Muscari Mill., Grape-hyacinth

Leaves basal, linear, fleshy; stem scapose; perianth dark blue or purple, 6-toothed, 4-6 mm long; stamens short, on perianth tube, style short; stigma 1; fruit a capsule.

70

a. Perianth oblong; leaves 2-3 mm wide, tips recurved. *M. racemosum (L.) Mill.
a. Perianth globular; leaves 6-12 mm wide, tips not recurved. *M. botryoides (L.) Mill.

16. Yucca L.

*Y. filamentosa L. , Adam's-needle. Stem short and woody; leaves stiff, linear, spine-tipped; flowers in a large panicle; perianth segments separate, white, 5 cm long or more; stigmas 3; capsule cylindric, somewhat narrowed near the middle.

17. Asparagus L. , Asparagus

*A. officinalis L. , Garden A. Stem at first fleshy and scaly, later tall and much branched; leaves minute, with filiform branchlets in their axils; flowers bisporangiate or monosporangiate, solitary or in 2's, pedicels jointed; perianth greenish-white, campanulate, 5 mm long or less; style short; berry red.

18. Clintonia Raf.

Leaves large, oval, 2-4 near base of stem; flowers in terminal umbel; perianth segments separate; style long and slender; ovulary 3-loculed; fruit a berry.

a. Perianth yellow, 1.5-2 cm long; berry blue, each locule several-seeded. C. borealis (Ait.) Raf.
a. Perianth white, speckled with purple-brown, about 1 cm long; berry black, each locule 2-seeded.
   C. umbellulata (Michx.) Morong

19. Smilacina Desf. , False Solomon's-seal

Rhizome long; stem scaly below, leafy above; leaves oblong to lanceolate, sessile or nearly so, 2-ranked; flowers small, white or greenish; perianth segments separate; style short; berry eventually red, 1-2-seeded.

a. Inflorescence a many-flowered panicle. S. racemosa (L.) Desf.
a. Inflorescence a few-flowered raceme.
   b. Leaves many. S. stellata (L.) Desf. , Starry F.
   b. Leaves 2-4. S. trifolia (L.) Desf. , Three-leaved F.

20. Maianthemum Weber, False Lily-of-the-valley

M. canadense Desf. Stem short; leaves ovate, usually a single cordate leaf accompanying the 2-leaved flowering stem; flowers fragrant, in a raceme; perianth-parts 4, white, separate, about 2 mm long; stamens 4; ovulary 2-loculed; berry pale red, 1-2-seeded.

21. Disporum Salisb.

Rhizome present; stems branched; leaves many, sessile or clasping, ovate or oblong; flowers few in an umbel or solitary; perianth segments white or yellow-green, separate, 1.5-2.5 cm long; style 1; stigmas 3 or 1; berry red or yellow, few-seeded.

a. Perianth yellow-green, unspotted, longer than stamens; fruit glabrous, not wrinkled. D. lanuginosum (Michx.) Nicholson, Yellow Mandarin.
a. Perianth white, purple-spotted, shorter than or equaling stamens; fruit pubescent, wrinkled.
   D. maculatum (Buckl.) Britt.

22. Streptopus Michx. Twisted Stalk

S. roseus Michx. With rhizome; stem sometimes branched; leaves ovate-lanceolate, acuminate; flowers solitary, axillary, flower-stalks jointed; perianth-segments separate, about 1 cm long, rose-color; filaments widened at base; anther 2-pointed at apex; style 3-lobed; berry red, many-seeded; rare.

23. Polygonatum Mill., Solomon's Seal

Stem unbranched, scaly below, leafy above, from a rhizome; leaves oval to narrowly elliptic; flowers drooping, pedicels jointed; perianth cylindric, 6-lobed, greenish; stamens inserted on perianth-tube; style slender; fruit a dark-blue berry, 3-6-seeded.

a. Leaves minutely hairy on veins beneath; peduncles usually 1-2-flowered. P. pubescens (Willd.) Pursh
a. Leaves glabrous beneath.
  b. Stem stout; leaves not flat, margin somewhat ruffled; flowers in groups of 2-5 or more, peduncles flattened. P. canaliculatum (Muhl.) Pursh
  b. Stem slender; leaves flat; flowers in groups of 1-3, peduncles terete or nearly so. P. biflorum (Walt.) Ell.

24. Convallaria L., Lily-of-the-Valley

*C. majalis L. With rhizome; leaves elliptic, basal; flowers bell-shaped, fragrant, in 1-sided racemes, bracted; perianth white, less than 1 cm long, parts united, lobes short and recurved; filaments short; style short; stigma almost unlobed; fruit a red berry.

25. Medeola L., Indian Cucumber-root

M. virginiana L. Rhizome short; leaves ovate to oblanceolate, in 2 whorls, the upper usually of 3, the lower of 5-9; perianth greenish-yellow, the segments separate, about 8 mm long, recurved; styles 3, recurved; berry dark-purple.

26. Trillium L., Trillium, Wake Robin

Stem with a whorl of 3 oval to broadly ovate leaves, from a rhizome; perianth of 3 green sepals and 3 white or colored petals; petals usually 2-5 cm long; anthers linear; ovulary and capsule lobed or angled, sometimes winged; ovules many.

a. Flowers sessile; petals maroon to yellow-green or green.
  b. Leaves sessile; sepals not recurved. T. sessile L., Sessile T.
  b. Leaves short-petioled; sepals recurved; rare. T. recurvatum Beck
a. Flowers peduncled.
  b. Leaves evidently petioled.
    c. Leaves and petals obtuse at tip; petals white. T. nivale Riddell, Snowy T.
    c. Leaves acuminate at tip; petals white marked with purple-red, wavy-margined, acute; rare. T. undulatum Willd., Painted T.
  b. Leaves sessile.
    c. Stigmas slender throughout, erect or spreading, more than half as long as ovulary; petals obovate or oblanceolate, white, becoming pink with age, longer than sepals. T. grandiflorum (Michx.) Salisb.
    c. Stigmas stout, tapering, recurved, about half as long as ovulary.
      d. Ovulary dark red-purple; petals red-purple, green, or white; stamens exceeding stigmas; flowers erect or deflexed. T. erectum L., Ill-scented T.
      d. Ovulary white or only tinged with color.
        e. Filaments a little longer than or equaling anthers; petals white or pink; peduncles deflexed. T. cernuum L., Nodding T.
        e. Filaments half as long as anthers or less; petals white or maroon; peduncles erect to deflexed. T. flexipes Raf.

27. Aletris L., Colic-root

A. farinosa L. Leaves lanceolate, in a basal rosette; flowers small, bracted, in a spikelike raceme; scape bracted; perianth tubular, white, about 1 cm long, rough; ovulary partly inferior; style 3-cleft at apex, each stigma 2-lobed.

28. Smilax L., Greenbrier

Herbaceous or woody, twining or climbing by stipule-tendrils; blades longitudinally ribbed, ovate, alternate; flowers small, monosporangiate; perianth yellow-green, segments separate, about 6 mm long; carpellate flowers with 1-6 vestigial stamens; stigmas 3 or 1, sessile or nearly so; fruit a berry, black in ours.

a. Stems herbaceous; flowers carrion-scented.
  b. Mostly without tendrils; stems erect; peduncles mostly from axils of bracts on leafless lower part of stem; umbels with fewer than 25 flowers. <u>S.</u> <u>ecirrhata</u> (Engelm.) Wats.
  b. Tendrils present; stems twining; peduncles from axils of foliage leaves; umbels with usually more than 25 flowers. <u>S. herbacea</u> L., Carrion-flower.
    c. Leaves glabrous beneath; common. var. <u>herbacea</u>.
    c. Leaves pubescent beneath. var. <u>lasioneuron</u> (Small) Rydb.
a. Stem woody, often prickly.
  b. Blades glaucous beneath; prickles slender-based. <u>S. glauca</u> Walt.
  b. Blades not glaucous beneath.
    c. Stems terete; prickles weak and slender-based; leaf margins minutely serrulate. <u>S. tamnoides</u> L. var. <u>hispida</u> (Muhl.) Fern.
    c. Stems angled or terete; prickles broad-based, rigid; leaf margins not serrulate. <u>S. rotundifolia</u> L. Round-leaf G.

DIOSCOREACEAE, Yam Family

<u>Dioscorea</u> L., Yam

Twining vines; leaves with broad palmately-veined blades usually cordate at base; diecious; flowers small, greenish; carpellate flowers epigynous, in spikes; staminate flowers in panicles; perianth regular, 6-parted; fertile stamens 6 or 3; carpels 3, united; styles 3; ovulary 3-loculed; fruit a 3-winged capsule; seeds flat, winged.

a. Blades usually halberd-shaped; tubers usually in the axils. *<u>D.</u> <u>batatas</u> Dcne., Air-potato.
a. Blades ovate; tubers absent from axils.
  b. Leaves all alternate or the lower in whorls of not more than 3; petioles glabrous or nearly so. <u>D. villosa</u> L., Wild Yam.
  b. Leaves in whorls of 4 or more at one or more of the lower nodes; petioles often pubescent at base and apex. <u>D. quaternata</u> (Walt.) Gmel.

AMARYLLIDACEAE, Amaryllis Family

Herbs with bulbs, corms, or rhizomes; leaves linear, basal; flowers solitary or clustered, bisporangiate, epigynous; perianth regular or nearly so, of 6 segments; stamens 6; carpels 3, united; style 1; ovulary 3-loculed; fruit a capsule.

a. Flowers in an elongate spike. 1. <u>Agave</u>
a. Flowers solitary or few, not in a spike.
  b. Corolla with a corona. 2. <u>Narcissus</u>
  b. Corolla without a corona.
    c. Perianth yellow; plant pubescent. 3. <u>Hypoxis</u>
    c. Perianth white, sometimes tinged or spotted with red or green.
      d. Perianth segments equal. *<u>Leucojum</u>
      d. Inner perianth segments shorter. *<u>Galanthus</u>

1. <u>Agave</u> L., American Aloe

<u>A. virginica</u> L. Leaves lanceolate or oblanceolate, entire or denticulate, thick, in basal rosette; flowers fragrant, about 2-3 cm long, in a spike; perianth greenish-yellow; style and stamens exserted; anthers versatile; fruit a many-seeded capsule.

2. <u>Narcissus</u> L., Narcissus, Daffodil, Jonquil

Leaves linear, flat or terete; flower or small flower cluster subtended by a thin spathe; corona variable in shape, size, and color; fruit a capsule.

a. Corona 3-5 cm long; perianth usually yellow. *<u>N.</u> pseudo-narcissus L., Daffodil.
a. Corona 1 cm long or less; perianth usually white. *<u>N.</u> poeticus L., Poet's N.

3. Hypoxis L., Star-grass

<u>H. hirsuta</u> (L.) Coville Somewhat hairy; leaves linear; flowers few, in an umbel; perianth segments yellow, about 1 cm long; fruit indehiscent; seeds several, black, muricate.

IRIDACEAE, Iris Family

Mostly perennial herbs; leaves linear; flowers bisporangiate, epigynous, solitary or in spikes, racemes, or panicles, the solitary flower or the cluster subtended by spathelike bracts; perianth parts 6, petaloid; stamens 3, opposite the sepals and sometimes adnate to them; carpels 3, united; ovulary 3-loculed, placentae axile, or rarely 1-loculed, placentae parietal; style 1, sometimes 3-branched; stigmas 3, sometimes 3-lobed; fruit a capsule.

a. Style branches petal-like. 1. <u>Iris</u>
a. Style-branches slender.
  b. Plant without a stem above ground; flower solitary. *<u>Crocus</u>
  b. Plant with stem above ground; flowers in clusters.
    c. Perianth zygomorphic. *<u>Gladiolus</u>
    c. Perianth regular.
      d. Perianth 2-3 cm long, orange-yellow with red or purple spots. 2. <u>Belamcanda</u>
      d. Perianth about 1 cm long, blue or white. 3. <u>Sisyrinchium</u>

1. <u>Iris</u> L., Iris, Flag

Flowers large and showy, solitary or clustered; sepals spreading or reflexed, petals erect or arching; petaloid style-branches 2-lobed, stigma under a thin transverse flange below apex on dorsal side; one stamen under each style-branch; fruit a 3-loculed capsule.

a. Flowering stems usually not more than 15 cm tall, from superficial rhizomes; sepals not bearded; perianth blue or violet, rarely white.
  b. Sepals with 3-ridged, toothed, orange-white crest. <u>I. cristata</u> Ait., Crested Dwarf I.
  b. Sepals not crested, but with a puberulent yellow or orange band; rare or extinct in Ohio. <u>I. verna</u>
    L., Dwarf I.
a. Flowering stems taller than 15 cm, from usually deep rhizomes.
  b. Sepals bearded. *<u>I. germanica</u> L., Common I.
  b. Sepals not bearded.
    c. Perianth yellow, orange, or reddish-brown.
      d. Petals about as long as sepals; perianth reddish-brown to yellow. *<u>I. fulva</u> Ker
      d. Petals much shorter than sepals; perianth yellow. *<u>I. pseudacorus</u> L., Yellow I.
    c. Perianth blue, lilac, or rarely white, sometimes with spots of other color.
      d. Stem zigzag, shorter than leaves, sometimes decumbent; flowers short-stalked, from all
        but lowest axils; capsule 6-angled. <u>I. brevicaulis</u> Raf.
      d. Stem erect, with flowers on upper part; axillary flowers long-stalked; capsule 3-angled.
        e. Petals lanceolate, much shorter than sepals; sepals with or without minutely hairy green-
          ish yellow blotch at base; inner surface of capsule shining. <u>I. versicolor</u> L.
        e. Petals obovate, little shorter than sepals; sepals with bright yellow pubescent blotch at
          base; inner surface of capsule dull. <u>I. virginica</u> L. var. <u>shrevei</u> (Small) E. Anders.

2. <u>Belamcanda</u> Adans., Blackberry-lily

*<u>B. chinensis</u> (L.) DC. Perianth spreading; style 3-branched, each stigma 3-lobed; valves of capsule at maturity recurving and eventually falling away, the mass of fleshy black seeds attached to central column resembling a blackberry.

3. <u>Sisyrinchium</u> L., Blue-eyed-grass

Perennial herbs; stems flattened, slender, edges sometimes winged; leaves linear, grasslike; flowers ephemeral, in umbel-like clusters, each cluster enclosed by 2 spathelike bracts; hypanthium absent or very short; perianth blue, violet, or white, about 1 cm long, with yellow eye; filaments monadelphous; style 3-branched.

a. Umbels and enclosing spathelike bracts peduncled from axil of a leaflike bract.
  b. Peduncles flat, winged; plant deep-green, drying blackish. <u>S. angustifolium</u> Mill.

b. Peduncles filiform; plant pale green. S. atlanticum Bickn.
a. Umbels and enclosing spathelike bracts sessile in axil of a leaflike bract.
   b. Umbels 2 in axil of each leaflike bract; edges of outer spathelike bract free. S. albidum Raf.
   b. Umbels solitary in axil of each leaflike bract; edges of outer spathelike bract united above the
      base.
      c. Pedicels spreading or recurving; plants dark when dry. S. mucronatum Michx.
      c. Pedicels erect or ascending; plants pale when dry. S. montanum Greene

ORCHIDACEAE, Orchid Family

Perennial herbs, ours terrestrial, often with rhizomes or corms; roots sometimes thickened;
some species nongreen; leaves solitary or alternate on the stem (or rarely opposite), sometimes all
basal; flowers epigynous, bisporangiate, solitary or in clusters; perianth zygomorphic; sepals 3,
sometimes 2 of them united; petals 3, one of them, the lip, different from the others in size and shape,
sometimes spurred; carpels 3, united; ovulary 1-loculed with 3 parietal placentae; fertile stamens 1
or 2 (1 or more staminodes sometimes present), adnate to style and stigma and with them forming the
column; pollen in pollinia; fruit a many-seeded capsule.

a. Anthers 2; lip petal an inflated sac 2-7 cm long; flowers 1-3. 1. Cypripedium
a. Anther 1; lip petal not an inflated sac or, if somewhat inflated, smaller.
   b. Leaves white- or pale-reticulate, in basal rosette. 11. Goodyera
   b. Leaves not white- or pale-reticulate.
      c. Leaves in 1 whorl of usually 5 on the stem; sepals much longer than petals. 5. Isotria
      c. Leaf arrangement various, but leaves not in a single whorl on stem.
          d. Plants yellow, brown, or purple, nongreen; scales, but no foliage leaves, present.
             e. Perianth less than 1 cm long; lip sometimes with 1-3 short ridges. 13. Corallorhiza
             e. Perianth 1.5-2 cm long; lip with 5-7 longitudinal ridges. 18. Hexalectris
          d. Plants green, with 1 or more foliage leaves which are sometimes absent at anthesis.
             e. Flower with a spur.
                f. Leaves absent at anthesis; spur slender, 1.5-2 cm; perianth greenish-purple.
                   17. Tipularia
                f. Leaves present at anthesis.
                   g. Leaves 2, basal.
                      h. Perianth greenish-white or -yellow; leaves as wide as long or nearly so.
                         3. Habenaria
                      h. Lip white, remainder of perianth pink-purple; leaves longer than wide.
                         2. Orchis
                   g. Stem leafy. 3. Habenaria
             e. Flower without a spur.
                f. Lip above other perianth parts; flower apparently inverted. 7. Calopogon
                f. Lip below other perianth parts.
                   g. Flowers solitary or in clusters of 2-3, 1.5-5 cm long.
                      h. Flowers 2 or 3; cauline leaves less than 2 cm long, ovate, alternate, more
                         than 1 present at anthesis. 6. Triphora
                      h. Flowers solitary, rarely 2 or 3; leaves not as above.
                         i. One cauline leaf present at anthesis; floral bract(s) often leaflike; sepals and
                            lateral petals not united. 4. Pogonia
                         i. Only bracts present on stem at anthesis, the single linear leaf appearing
                            later; sepals and lateral petals united at base. 8. Arethusa
                   g. Flowers more numerous, in spikes or racemes.
                      h. Flowers in twisted spikes or spikelike racemes; perianth white or greenish-
                         white, except in one species less than 1 cm long; lip with 2 small basal cal-
                         losities. 10. Spiranthes
                      h. Flowers in racemes; set of other characters not as above.
                         i. Leaves several, alternate on the stem; perianth green and purple.
                            9. Epipactis
                         i. Leaves 2. (See third i.)
                            j. Leaves basal; lateral petals linear or filiform, about 1 cm long.
                               15. Liparis
                            j. Leaves at about middle of stem, opposite; lateral petals wider than lin-
                               ear, 2-3 mm long. 12. Listera

i. Leaf 1, sometimes developing after anthesis.
    j. Lip 3 mm long or less; leaf cauline, present at anthesis. 14. <u>Malaxis</u>
    j. Lip 1-1.5 cm long; leaf basal, appearing after anthesis. 16. Aplectrum

1. <u>Cypripedium</u> L., Lady's Slipper, Moccasin Flower

Roots fibrous, coarse; leaf-blades large, elliptic, ovate, or lanceolate; flowers solitary or few; sepals separate or the 2 lower united; lip an inflated sac 2-7 cm long; column bent downward, bearing an anther on each side and a bractlike staminode covering the tip.

a. Leaves 2, basal; lip pink, cleft on upper side. <u>C. acaule</u> Ait., Pink M.
a. Stem leafy; lip not cleft.
    b. Lateral petals white, not twisted; lip white or pink, streaked with rose or purple. <u>C. reginae</u>
        Walt., Showy L.
    b. Lateral petals yellow, green, brown, or purple, usually twisted.
        c. Lip yellow. <u>C. calceolus</u> L., Yellow L.
        c. Lip white, purple-striped within. <u>C. candidum</u> Muhl., Small White L.

2. <u>Orchis</u> L., Orchis

<u>O. spectabilis</u> L., Showy O. Plant 1-2 dm tall; leaves 2, basal, obovate or elliptic; flowers long-bracted, few, in a raceme; lip white, to 2 cm long, with conspicuous spur; sepals and lateral petals pink to lavender, connivent in a hood.

3. <u>Habenaria</u> Willd., Fringed Orchis, Orchis

Stem leafy or scapose; flowers in bracted spikes or racemes; lip spurred, its margin various; perianth of various colors; sepals and lateral petals separate.

a. Leaves 2, basal, broadly elliptic to circular; flowers greenish-white or -yellow.
    b. Flowers pediceled; scape bracted below inflorescence. <u>H. orbiculata</u> (Pursh) Torr.
    b. Flowers sessile; scape bractless below inflorescence. <u>H. hookeri</u> Torr.
a. Stem-leaves present.
    b. Lip entire, toothed, or crenulate, not fringed, not 3-parted.
        c. Lip entire; perianth greenish. <u>H. hyperborea</u> (L.) R. Br.
        c. Lip toothed or crenulate; perianth white, green, or yellow.
            d. Spur pouchlike, much shorter than lip. <u>H. viridis</u> (L.) R. Br. var. <u>bracteata</u> (Muhl.) Gray,
                Long-bracted Green O.
            d. Spur slender, equaling or longer than lip.
                e. Lip erose at tip, a small lobe on each side at base, a tubercle on median line at base.
                    <u>H. flava</u> (L.) R. Br.
                e. Lip 2-notched at tip, not lobed at base, without tubercle. <u>H. clavellata</u> (Michx.) Spreng.,
                    Green Woodland O.
    b. Lip fringed, not 3-parted, 1-1.5 cm long, spur much longer. (See third b.)
        c. Perianth white. <u>H. blephariglottis</u> (Willd.) Hook., White Fringed O.
        c. Perianth yellow or orange. <u>H. ciliaris</u> (L.) R. Br., Yellow Fringed O.
    b. Lip 3-parted, each part toothed or fringed.
        c. Perianth white, green, or yellow-green; lip deeply fringed.
            d. Lateral petals entire; spur 2 cm long or less. <u>H. lacera</u> (Michx.) Lodd.
            d. Lateral petals toothed or erose; spur 2-4 cm long; lateral petals and upper sepal forming a
                hood. <u>H. leucophaea</u> (Nutt.) Gray
        c. Perianth lavender to rose-purple, rarely white.
            d. Lip segments erose, terminal one deeply notched. <u>H. peramoena</u> Gray, Purple Fringeless O.
            d. Lip segments lacerate or with short fringe. <u>H. psycodes</u> (L.) Spreng., Purple Fringed O.
                e. Raceme usually less than 5 cm thick; lip 1.5 cm wide or less. var. <u>psycodes</u>
                e. Raceme 5 cm thick or more; lip 1.5-2.5 cm wide. var. <u>grandiflora</u> (Bigel.) Gray
                    (<u>H. fimbriata</u> (Ait.) R. Br.).

4. Pogonia Juss.

P. ophioglossoides (L.) Ker, Rose Pogonia. Stem with 1 ovate or elliptic leaf, 1-2 basal petioled leaves and a bracteal leaf under the flower sometimes present, also; flowers usually solitary (1-3); perianth about 1.5-2.5 cm long, rose-pink; lip crested with 3 rows of fleshy hairs.

5. Isotria Raf.

I. verticillata (Willd.) Raf., Whorled Pogonia. Stem with a single whorl of usually 5 obovate or elliptic leaves; flowers usually solitary; sepals to 6 cm long, green to purple-brown, lateral petals and lip much shorter, yellow-green; lip streaked with purple and crested.

6. Triphora Nutt.

T. trianthophora (Sw.) Rydb., Nodding Pogonia, Three Birds. Leaves alternate, sessile, ovate, 2 cm long or less; flowers few, lasting only a day; perianth pink or white, lip veined or tinged with purple or green, about 1.5 cm long.

7. Calopogon R. Br., Grass-pink

C. pulchellus (Salisb.) R. Br. Leaf 1, rarely 2, narrow, elongate, basal; flowers 2-several; perianth rose-purple or pink; lip uppermost, its upper surface bearing orange-tipped or red or purple hairs or papillae.

8. Arethusa L.

A. bulbosa L., Swamp-pink. Stem from a corm, scapose at first, a single leaf appearing from within the upper bract after anthesis; flowers 1, rarely 2; perianth rose-purple, 2.5-5 cm long; lip sometimes spotted, with fringed yellow and purple crests.

9. Epipactis Sw., Helleborine

*E. helleborine (L.) Crantz   Leaves ovate or lanceolate, smaller upward, several on the stem; flowers in a bracted raceme; perianth green and rose or purple, 1 cm long or more; basal half of lip saccate, apical half cordate-ovate.

10. Spiranthes Rich., Ladies' Tresses

Leaves usually narrow, basal or on lower part of stem; flowers in usually twisted spikes; perianth white or greenish-white, from about 0.5 to more than 1 cm long; lip with a callosity on each side at base.

a. Flowers in 1 rank, the rachis spirally twisted.
  b. Rachis, bracts, and ovularies densely pubescent; basal and usually some lower stem-leaves present at anthesis; flowers about 1 cm long. S. vernalis Engelm. & Gray
  b. Lower part of rachis and upper part of stem glabrous or nearly so; leaves often absent at anthesis.
    c. Lip with median band of green; roots clustered; inflorescence about 1 cm wide; perianth 4-6 mm long. S. gracilis (Bigel.) Beck
    c. Lip white; roots usually solitary; inflorescence about 0.5 cm wide; perianth 2-4 mm long. S. tuberosa Raf.
a. Flowers in more than 1 rank; spike crowded.
  b. Lip constricted at about the middle; sepals and lateral petals connivent in a hood; basal callosities minute; perianth 7-13 mm long. S. romanzoffiana Cham.
  b. Lip not or only slightly narrowed at middle.
    c. Spring-flowering; lip with wide central area of yellow; sepals and lateral petals connivent in a hood; perianth to 6 mm long. S. lucida (H.H. Eat.) Ames
    c. Autumn-flowering; sepals and lateral petals not connivent; lip not yellow.
      d. Perianth 6-12 mm long; callosities rounded. S. cernua (L.) Rich.
      d. Perianth about 5 mm long; callosities slender, incurved. S. ovalis Lindl.

11. Goodyera R. Br., Rattlesnake-plantain

Leaves ovate or ovate-lanceolate, with white or pale-green reticulate veins, in a basal rosette; flowers in bracted spikelike raceme, upper part of scape glandular-pubescent; perianth white, to about 6 mm long; lip saccate, with pointed tip.

a. Raceme densely flowered; tip less than half as long as sac. G. pubescens (Willd.) R. Br., Downy R.
a. Raceme loosely flowered, sometimes spiral; length of tip and of sac about equal; rare.
    G. tesselata Lodd.

12. Listera R. Br., Twayblade

L. cordata (L.) R. Br. Low slender herb with 2 opposite ovate leaves at about middle of stem; flowers small, purple to green, in a raceme, bracts minute; lip drooping, 2-toothed at base, deeply 2-lobed, much longer than sepals and lateral petals.

13. Corallorhiza Chat., Coral-root

Purple, yellow, or brown saprophytes with coralloid roots; scape bearing sheathing scales; flowers in racemes; perianth 8 mm long or less; at base of lateral sepals a short spur adnate to top of ovulary.

a. Lip with a tooth or lobe on each side at or below middle.
  b. Sepals and lateral petals yellow-green, lip white or somewhat spotted with red; spur of sepals very short or absent; lobes of lip short; capsule 8-12 mm long; rare. C. trifida Chat.
  b. Sepals, lateral petals, and lip white, spotted with red or purple; spur of sepals and lateral lobes of lip prominent; capsule 1.5-2.5 cm long. C. maculata Raf.
a. Lip entire, notched, or denticulate.
  b. Autumn-flowering; lip about 4 mm long, margined and spotted with purple, often wider than long; sepals and lateral petals connivent in a hood; capsule 6-8 mm long. C. odontorhiza (Willd.) Nutt.
  b. Spring-flowering; lip about 7 mm long, white dotted with purple; sepals and lateral petals spreading; capsule 9-11 mm long. C. wisteriana Conrad

14. Malaxis Sw., Adder's Mouth

M. unifolia Michx. Stem short, from a corm; leaf 1, sessile, about middle of stem, ovate; flowers in a bracted raceme; perianth white or green; lateral petals linear or filiform; lip 2-4 mm long, 3-toothed at apex.

15. Liparis Rich., Twayblade

Stem low; leaves 2, basal, shining, ovate or elliptic-lanceolate, to 15 cm long; flowers bracted, in a raceme; lateral petals linear or filiform.

a. Perianth yellow-green, about 5 mm long. L. loeselii (L.) Rich.
a. Sepals greenish-white, petals purple or mauve, about 1 cm long; lip with red-purple veins.
    L. lilifolia (L.) Rich.

16. Aplectrum (Nutt.) Torr., Putty-root

A. hyemale (Muhl.) Torr. Corms connected in a series; leaf 1, basal, petioled, elliptic, appearing late in season and sometimes shriveled before next year's flowering; scape with some bracts; flowers in a raceme, bracts small; perianth 1 cm long or a little more, green, yellow, and white, or tinged with purple.

17. Tipularia Nutt., Cranefly Orchid

T. discolor (Pursh) Nutt. Corms connected in a chain; leaf 1, basal, petioled, ovate, appearing in autumn after the flowers and living over winter; greenish-yellow or -purple sepals and lateral petals less than 1 cm long, lip shorter, slender spur to 2 cm.

18. Hexalectris Raf.

H. spicata (Walt.) Barnh., Crested Coral-root. Yellow, brown, and purple saprophyte with coral-like roots; stem with a few sheathing scales; flowers bracted, in a raceme; perianth about 2 cm long; lip 3-lobed, middle lobe with 5-7 longitudinal crests or ridges.

DICOTYLEDONS

SAURURACEAE, Lizard's Tail Family

Saururus L., Lizard's Tail

S. cernuus L. Perennial herb of marshes or shallow water; leaves alternate, simple, blades cordate-ovate; flowers hypogynous, bisporangiate, small, bracted, in peduncled nodding spikes; perianth none; stamens 6-8; filaments white; carpels 3-4, united only at base, indehiscent in fruit; styles short, curved outwardly.

SALICACEAE, Willow Family

Trees and shrubs with alternate simple stipuled leaves; diecious; both kinds of flowers in aments, each flower subtended by a small bract; perianth none; staminate flower of 1-many stamens; carpellate flower of 2-4 united carpels; style 1 or the 2-4 stigmas sessile; ovulary 1-loculed; placentae parietal; fruit a capsule; seeds many, comate.

a. Ament-bracts toothed, lobed, or fimbriate; buds with several exposed scales; pith 5-angled; flower
    with small cup at base. Populus
a. Ament-bracts entire or nearly so; buds with 1 scale; pith not 5-angled; flower or pedicel with 1 or
    2 small glands at base. Salix

Salix L., Willow

Trees and shrubs; leaves pinnately veined, blade-margin usually serrate, sometimes glands on petiole, base of blade, or tips of leaf-teeth; terminal bud self-pruned; twigs often self-pruned at basal brittle zones; staminate flower of 1-10 stamens; carpellate flower of 2 united carpels; style 1 or absent; stigmas 2, sometimes each 2-cleft.

1. KEY BASED ON CHARACTERS OF CARPELLATE PLANTS IN FLOWER AND IN FRUIT

In this key, "ament" refers to carpellate ament; "bract" refers to bract subtending carpellate flower.

a. Flowers appearing before leaves; ovulary and fruit pubescent; aments sessile or subsessile; bracts
    persistent until fruits are mature.
  b. Leaf-scars or many of them opposite or subopposite; fruit ovoid-conic, 3-4 mm long, sessile;
      bracts with rounded blackish tips. *S. purpurea
  b. Leaf-scars alternate.
    c. Scales brown; fruits white-tomentose, 6-8 mm long; branchlets white-tomentose; style red.
       S. candida
    c. Scales dark-brown to black or with dark tips; fruits and branchlets not white-tomentose.
      d. Fruit 3-5 mm long, blunt; style short; pedicel about 1/3 as long as ovulary, 0.5-1.5 mm
         long. S. sericea
      d. Fruit 6-9 mm long, gray-pubescent.
        e. Last year's stems pubescent or, at least, dull. S. humilis
        e. Last year's stems glabrous and lustrous.
          f. Style definite. S. discolor
          f. Stigma sessile or nearly so. *S. caprea
a. Flowers appearing with or after leaves; aments on short branches usually leafy at base.
  b. Flowers appearing after leaves; blades linear to linear-oblanceolate, glabrate or pubescent, with
    remote irregular fine sharp teeth; fruit glabrous or pubescent, 7-10 mm long; style absent;
    shrub. S. interior
  b. Without the above set of characters; flowers appearing with leaves, except in S. serissima.
    c. Ovulary and fruit glabrous.

d. Glabrous bog shrub; bracts yellowish-brown, persistent; fruit lance-ovoid, 6-8 mm long, pedicel 2-4 mm long, stigma sessile or nearly so; blades entire. S. pedicellaris
d. Without the above set of characters.
    e. Fruit conic, sessile; bracts pale, deciduous before fruits mature.
        f. Fruit 3-5 mm long; branches not decidedly drooping; glands sometimes present on petiole. *S. alba
        f. Fruit not more than 2.5 mm long; branches drooping. *S. babylonica
    e. Fruit pediceled.
        f. Flowers in whorls or approximately in whorls on ament-axis; bracts pale, deciduous before fruits mature.
            g. Capsule 7-12 mm long; flowers appearing in late summer, long after leaves; fruit blunt, pedicel stout. S. serissima
            g. Flowers appearing earlier; fruit 5.5 mm long at most, tapering to apex; vein islets very small.
                h. Young blades and branches pubescent; blades glaucous beneath; stigma almost sessile. S. caroliniana
                h. Young blades and branches glabrous or nearly so.
                    i. Blades linear-lanceolate, green beneath; stigma almost sessile; branches brittle. S. nigra
                    i. Blades lanceolate, glaucous beneath; style evident; branches not brittle. S. amygdaloides
        f. Flowers in spirals on ament-axis.
            g. Coarse glands present on summit of petiole (S. fragilis may sometimes belong here).
                h. Pedicel about 2 times length of gland. *S. pentandra
                h. Pedicel about 4 times length of gland. S. lucida
            g. Glands absent from petiole (except sometimes in S. fragilis).
                h. Bracts pale, deciduous before fruits mature; branches very brittle; fruit 4-5.5 mm long. *S. fragilis
                h. Bracts dark, brown to black, persistent until fruits mature; fruit 5-9 mm long. S. rigida and glaucophylloides
c. Ovulary and fruit pubescent.
    d. Pedicels 3-6 mm long; length of young blades not more than 3 times width, blades and young stems silky-pubescent; bracts yellow to brown. S. bebbiana
    d. Pedicels shorter; blades narrower.
        e. Style red, 1-1.5 mm long; young stems and under surface of blades white-tomentose; aments sessile or subsessile. S. candida
        e. Style short; young stems and under surface of blades not white-tomentose; aments on short leafy branches. S. petiolaris

2. KEY BASED ON CHARACTERS OF STAMINATE PLANTS IN FLOWER

In this key, "ament" refers to staminate ament; "bract" refers to bract that subtends each staminate flower.

a. Filaments, and sometimes also anthers, united, the 2 appearing as 1; leaf scars or some of them subopposite; flowers appearing before leaves. *S. purpurea
a. Filaments and anthers separate or nearly so; leaf scars or leaves alternate.
    b. Stamens 3-10; aments appearing with or after leaves, on short stems with some young leaves at base; bracts yellow or pale brown, soon deciduous.
        c. Flowers in whorls or approximately in whorls on axis of ament.
            d. Twigs brittle at base; blades linear-lanceolate.
                e. Young blades green on both sides, glabrous or glabrate. S. nigra
                e. Young blades glaucous and pilose beneath, hairs white. S. caroliniana
            d. Twigs not brittle at base; blades lanceolate or wider, becoming glaucous beneath. S. amygdaloides
        c. Flowers in spirals on ament-axis; glands present at summit of petiole.
            d. Aments 1-1.5 cm long, appearing in late summer; blades glaucous beneath; northern, rare, of swamps and bogs. S. serissima
            d. Aments longer, appearing in spring; blades green beneath. S. lucida and *pentandra
    b. Stamens 2 (rarely 3-4 in S. alba and S. fragilis).

c. Flowers appearing before leaves; ament-bracts persistent.
  d. Branches of previous year glabrous and lustrous; filaments pilose; staminate aments orna-
      mental. S. discolor and *S. caprea
  d. Branches of previous year pubescent or, if glabrous, then dull.
    e. Anthers yellow.
      f. Filaments hairy; bracts dark-tipped, long-villous. S. discolor
      f. Filaments glabrous or hairy; bracts brown to black. S. sericea
    e. Anthers red or purple; ament-bracts dark or dark-tipped.
      f. Branchlets, aments, and under surface of leaves white-woolly. S. candida
      f. Branchlets and young leaves hairy but not white-woolly. S. humilis
c. Flowers appearing with or after leaves.
  d. Bracts pale.
    e. Aments appearing after the linear-lanceolate, sharply and irregularly denticulate,
      leaves; leaves silky-pubescent to glabrous; filaments pubescent. S. interior
    e. Aments appearing with young leaves.
      f. Lower surface of young leaves and last year's stems hairy; bracts persistent.
        g. Young blades villous or tomentose, length three times the width; filaments pubes-
          cent. S. bebbiana
        g. Length of blades several times the width.
          h. Blades white-woolly beneath; filaments glabrous. S. candida
          h. Blades silky but not woolly beneath, white or tawny; filaments glabrous or hairy.
            S. petiolaris
      f. Leaves, stems, and filaments glabrous; bracts yellowish-brown, persistent; bog
        shrub. S. pedicellaris (See third f.)
      f. Last year's stems glabrous; filaments pubescent; bracts deciduous.
        g. Branches pendulous. *S. babylonica
        g. Branches not or little pendulous.
          h. Branches green or reddish, very brittle at base. *S. fragilis
          h. Branches green or yellow, not brittle at base. *S. alba
  d. Bracts dark or with dark tips, persistent. S. glaucophylloides and S. rigida

3. KEY BASED ON VEGETATIVE CHARACTERS PRESENT
WHEN LEAVES ARE MATURE

a. Leaves, or some of them, subopposite or opposite, glabrous, short-petioled, mostly linear-
  oblanceolate or spatulate. *S. purpurea L.
a. Leaves all alternate.
  b. Glands present on petiole usually near its summit; blades closely serrate.
    c. Blades green beneath or only paler than above, lustrous above, lanceolate to ovate-lanceolate.
      d. Blades long-attenuate at tip, glabrous or with pale or brown hairs; lower surface green and
        lustrous. S. lucida Muhl., Shining W.
      d. Blades acute to short-acuminate, lighter beneath, midvein yellow. *S. pentandra L.
    c. Blades more or less glaucous beneath.
      d. Blades elliptic to oblong-lanceolate, short-acuminate; shrubs, northern, of swamps and
        bogs; rare in Ohio. S. serissima (Bailey) Fern.
      d. Blades lanceolate, acuminate; trees.
        e. Branches pendulous; blades silky when young. *S. babylonica L., Weeping W.
        e. Branches little or not pendulous.
          f. Blades glabrous, margin with 5-6 serrations per cm; branches very brittle at base.
            *S. fragilis L., Crack W.
          f. Blade margins with 6-12 serrations per cm; branchlets green and blades silky below,
            or branchlets yellow or orange and blades glabrous or glabrate. *S. alba L.
  b. Petioles without glands.
    c. Blades linear-lanceolate to linear, almost sessile, silky-pubescent to glabrous, with small
      irregular remote sharp teeth; aments present in May and June, after the leaves; colonial
      shrubs. S. interior Rowles, Sandbar W.
    c. Plants not as above.
      d. Blade-margins rather evenly serrate, not revolute.
        e. Blades essentially glabrous on both sides.
          f. Blades green beneath or only slightly glaucous.

g. Blades green beneath, sometimes curved, narrow; vein islets very small, 4-5 per.mm. S. nigra Marsh., Black W.
g. Blades slightly glaucous beneath; vein islets larger. *S. fragilis L., Crack W.
f. Blades glaucous beneath.
  g. Branches pendulous. *S. babylonica L.
  g. Branches not pendulous.
    h. Petioles to 15 mm long, slightly twisted; blades lanceolate, long-acuminate, base not cordate; vein islets very small, 4-5 per mm. S. amygdaloides Anderss.
    h. Petioles usually shorter; vein islets larger.
      i. Trees; branches yellow or orange; blades tapering to base. *S. alba L.
      i. Shrubs; branches green or reddish; blades usually rounded to cordate at base, broadly lanceolate. S. glaucophylloides Fern.
e. Blades, at least those near ends of branchlets, hairy beneath and sometimes above.
  f. Vein islets tiny, 4-5 per mm. S. caroliniana Michx.
  f. Vein islets larger.
    g. Blades tapering to base; pubescence appressed.
      h. Blades usually not more than 8 cm long and 1 cm wide, thin; at least some of the hairs usually brown. S. petiolaris Sm.
      h. Blades often longer or wider than above; hairs not brown.
        i. Trees; hairs white. *S. alba L.
        i. Shrubs; hairs silvery. S. sericea Marsh., Silky W
    g. Blades usually rounded to subcordate at base; stipules on sprouts large, reniform or half-ovate. S. rigida Muhl.
d. Blade-margins entire, undulate, or remotely, irregularly, and often shallowly toothed, sometimes revolute.
  e. Blades glabrous.
    f. Blade-margin entire, often revolute; blade oblanceolate to oblong, apex obtuse to acute; stipules none; bog shrub. S. pedicellaris Pursh, Bog W.
    f. Blade margin somewhat crenate-serrate; blade elliptic to obovate, glaucous beneath. S. discolor Muhl., Pussy W.
  e. Blades pubescent.
    f. Blades linear-oblong to oblanceolate, revolute-margined, densely white-woolly beneath and sometimes above. S. candida Fluegge, Hoary W.
    f. Without the above set of characters; blades usually wider.
      g. Last year's twigs glabrous and lustrous; blades broadly ovate to elliptic, wider near middle. *S. caprea L., Goat W.
      g. Last year's twigs pubescent or, at least, not lustrous.
        h. Length of blades more than 3 times width, widest at or above middle, gray beneath and usually hairy. S. humilis Marsh., Pussy W.
        h. Length of blades less than 3 times width.
          i. Branchlets and under surface of blades with gray or whitish pubescence or tomentum; blades rugose beneath. S. bebbiana Sarg., Long-beaked W.
          i. Branchlets and under surface of blades with some brownish pubescence or tomentum; blades not rugose. S. discolor Muhl., Pussy W.

Populus L., Poplar, Aspen

Trees; twigs self-pruned; terminal bud present, often resinous; leaf-margin with gland-tipped teeth or small lobes, often 1 or more glands at base of blade; flower with a calyxlike cup or disk at base, in axil of an ament-bract; perianth none; staminate flower of 5-many stamens; carpellate flower of 2-4 united carpels.

1. KEY TO PLANTS IN FLOWER AND IN FRUIT

a. Stigmas or their lobes broad; stamens 12-many; cup or disk below carpellate flower symmetric or nearly so.
  b. Buds not or slightly viscid, scales pubescent toward base; stamens 12-20; cup of carpellate flower lobed, deciduous; style undivided at base. P. heterophylla
  b. Buds very viscid; cup of carpellate flower not lobed, not deciduous.
    c. Ament-bracts with many fringes; lobes of stigma with crenate margin.

  d. Stamens 20-30; buds fragrant. P. <u>balsamifera</u>
  d. Stamens usually 40-60; buds not fragrant. P. <u>deltoides</u>
 c. Ament-bracts with 9-11 segments; lobes of stigma with entire margin. *P. <u>nigra</u>
a. Stigmas or their lobes slender; stamens 6-12; cup of carpellate flower asymmetric; buds not or slightly viscid; ament-bracts long-bearded.
 b. Buds glabrous, lateral ones appressed; ovulary and capsule glabrous; ament-bracts 3-5-cleft. P. <u>tremuloides</u>
 b. Buds pubescent, the lateral not appressed.
  c. Buds finely pubescent; twigs sometimes pubescent; ament-bracts 5-7-cleft into long narrow segments. P. <u>grandidentata</u>
  c. Buds and twigs white-woolly; ament-bracts with many short triangular teeth. *P. <u>alba</u>

2. KEY TO PLANTS IN LEAF

a. Mature leaf-blades white-felted or white-tomentose beneath, with shallow lobes or a few coarse teeth.
 b. Blades white-felted beneath, shallowly lobed. *P. <u>alba</u> L., Silver P.
 b. Blades thinly white-tomentose beneath, coarsely toothed. *P. <u>canescens</u> (Ait.) Sm.
a. Mature leaf-blades neither white-tomentose nor white-felted beneath; margins serrate or dentate.
 b. Petioles flattened.
  c. Lowest lateral veins joining midvein at approximately a right angle; marginal teeth pointing forward.
   d. Blades truncate to subcordate at base, usually acuminate, usually longer than wide, glands present at base. P. <u>deltoides</u> Marsh., Cottonwood
   d. Blades truncate to cuneate at base, short-acuminate, usually wider than long, no glands at base. *P. <u>nigra</u> L., Black P. Lombardy Poplar, var. *<u>italica</u> Muench., is slenderly columnar in form.
  c. Lowest lateral veins joining midvein at other than a right angle; that is, ascending; blades rounded at base.
   d. Blade margins with coarse teeth, 5-15 on a side; expanding blades white-tomentose. P. <u>grandidentata</u> Michx., Large-tooth A.
   d. Blade margins with fine teeth, usually more than 15 on a side; blades glabrous or glabrate. P. <u>tremuloides</u> Michx., Quaking A.
 b. Petioles terete or only slightly flattened at summit.
  c. Tip of blade blunt, base pubescent. P. <u>heterophylla</u> L., Swamp Cottonwood
  c. Tip of blade acute or acuminate, base glabrous.
   d. Blades broadly ovate, pubescent; twigs pubescent. *X P. <u>gileadensis</u> Rouleau, Balm-of-Gilead
   d. Blades ovate, glabrous or nearly so, lustrous beneath; twigs glabrous. P. <u>balsamifera</u> L., Balsam P.

MYRICACEAE, Bayberry Family

<u>Myrica</u> L. (Incl. <u>Comptonia</u> L'Her.)

 Shrubs; leaves alternate, simple, gland-dotted, fragrant; monecious or diecious; the 2 kinds of flowers in separate aments, each subtended by an ament-bract and some other bracts; perianth none; staminate flower of few to several stamens; carpellate flower of 2 united carpels, ovulary 1-loculed, style short or absent, stigmas 2.

a. Leaf-blades entire to slightly serrate, obovate or oblanceolate; stipules none; carpellate aments becoming a cluster of a few small wax-covered globose fruits. <u>M</u>. <u>pensylvanica</u> Loisel., Bayberry
a. Leaf-blades pinnatifid, linear-oblong; stipules present; each carpellate flower and fruit exceeded by linear bracts, fruiting ament appearing burlike. <u>M</u>. <u>asplenifolia</u> L. (<u>C</u>. <u>peregrina</u> (L.) Coult.), Sweet-fern

JUGLANDACEAE, Walnut Family

 Trees; leaves alternate, pinnately compound; leaflets serrate, gland-dotted beneath; axillary buds superposed; monecious; flowers monosporangiate; corolla absent; staminate flowers in pendulous aments from buds below leaves of current season, each subtended by an ament-bract; carpellate flowers solitary or few in a cluster terminal on leafy stem, epigynous, each subtended by an involucre of bracts;

carpels 2, united, 1-loculed above, 2-4-loculed below; stigmas 2; fruit a nut, the hard shell (ovulary wall) enclosed in, and fused with, at least until maturity, the husk (accrescent bracts and hypanthium); seed one, 2-lobed.

a. Staminate aments solitary; leaflets usually 11 or more; pith diaphragmed. Juglans
a. Staminate aments in groups of 3; leaflets (except in one rare species) 9 or fewer; pith continuous, 5-pointed. Carya

Juglans L., Walnut

Leaves odd- or even-pinnate, elongate; perianth of staminate flower 3-6-lobed, of carpellate flower 4-lobed, minute; stamens several to many; husk fleshy; nut sculptured; seed edible.

a. Nut ellipsoid; a band of hairs at upper edge of leaf-scar; pith dark brown. J. cinerea L., Butternut, White W.
a. Nut globose; no band of hairs at upper edge of leaf-scar; pith light brown. J. nigra L., Black W.

Carya Nutt., Hickory

Leaves odd-pinnate; staminate calyx 2-3-lobed, carpellate 1-lobed; stamens 3-10; husk of nut woody; seed of most species edible.

a. Scales of terminal winter-bud valvate, yellow-hairy or yellow-scurfy; leaflets usually 7 or more, sometimes falcate.
  b. Staminate aments in sessile fascicle; stamens 3-6; leaflets usually 11-15; nut sweet; bud scales hairy. C. illinoensis (Wang.) K. Koch, Pecan.
  b. Staminate aments in peduncled fascicle; stamens 4; leaflets usually 5-9; nut bitter; bud scales scurfy. C. cordiformis (Wang.) K. Koch, Bitternut H.
a. Scales of terminal winter-bud imbricate; leaflets 3-9, not or little falcate.
  b. Terminal winter-bud large, 1-3 cm long; branchlets usually stout; husk thick, 4-10 mm.
    c. Bark of older trees shaggy; leaflets usually 5, rarely 7; a persistent tuft of hair on each side of each tooth of leaflets; branchlets, petioles, rachises, and lower leaf-surface hairy or glabrous. C. ovata (Mill.) K. Koch, Shagbark H.
    c. Leaflets usually more than 5, pubescent beneath and sometimes on margin but without the tufts of hair described above.
      d. Bark of older trees shaggy; leaf-rachis glabrous or the hairs mostly straight; leaflets 7-9; petiole persisting after abscission of leaflets; nut compressed. C. laciniosa (Michx. f.) Loud., Big Shellbark H.
      d. Bark not shaggy, dark; rachis, petiole, and lower surface of leaflets with curly fascicled hairs; petiole not persisting after abscission of leaflets. C. tomentosa Nutt., Mockernut
  b. Terminal winter-bud smaller, 0.5 to a little more than 1 cm long; bark not shaggy; branchlets glabrous; leaflets usually glabrous beneath but sometimes densely hairy; husk thin, 3 mm or less.
    c. Carpellate flowers and bracts hoary-pubescent; staminate ament-bracts hairy; leaflets usually 5; husk splitting along 1 or 2 sutures and only tardily. C. glabra (Mill.) Sweet, Pignut
    c. Carpellate flowers and bracts yellow-scurfy; staminate ament-bracts minutely pubescent; leaflets usually 7; husk splitting along all sutures. C. ovalis (Wang.) Sarg.

CORYLACEAE, Hazel Family

Shrubs and trees; leaves alternate, simple, stipuled; blades pinnately veined; monecious; flowers monosporangiate; perianth absent or calyx present; staminate flowers in drooping aments, 1 (apparently) to 3 in axil of each ament-bract, stamens (except in Alnus) bifid, each part bearing 1 anther-locule; carpellate flowers epigynous or perianth none, in aments or in small budlike clusters, 2-3 flowers in axil of each ament-bract; 1 or more small bracts at base of each flower; carpels 2, united, stigmas 2, ovulary 2-loculed below, 1-loculed above, ovules 4; seeds 1.

a. Leaves not 2-ranked; mature carpellate aments woody, conelike, persistent. 5. Alnus
a. Leaves 2-ranked except sometimes in Betula; carpellate aments not woody and conelike.
  b. Carpellate flowers in small scaly buds; fruit 1 cm or more in diameter, covered and exceeded by 2 accrescent bracts. 1. Corylus
  b. Carpellate flowers in cylindric or oblong aments; fruit smaller.

c. Three carpellate flowers in axil of each usually 3-lobed ament-bract; fruit winged, not enclosed in or subtended by an involucre. 4. Betula
c. Two carpellate flowers in axil of each ament-bract, the bracts soon deciduous; fruit not winged.
 d. Fruit subtended by and much exceeded by 3-lobed leaflike structure; staminate aments emerging in spring from scaly winter-buds; bark dark gray, smooth; trunk fluted. 3. Carpinus
 d. Fruit enclosed in a bladderlike sac; staminate aments present in winter; bark scaly; trunk not fluted. 2. Ostrya

1. Corylus L., Hazelnut

Ours shrubs; leaves 2-ranked, doubly serrate or incised; staminate aments emerging in autumn, flowering in spring, perianth absent, stamens 4; carpellate flower-cluster a scaly bud, 2 flowers under each scale, only the red stigmas protruding beyond scales; calyx minute; fruit enclosed by involucre of accrescent bracts.

a. Buds rounded at tip; twigs glandular; involucre prolonged beyond nut as two bladelike structures with laciniate margin. C. americana Walt., American H.
a. Buds pointed; twigs glandular only at nodes; involucre prolonged beyond nut as a tubular beak narrower and longer than nut. C. cornuta Marsh., Beaked H.

2. Ostrya Scop., Hop Hornbeam

O. virginiana (Mill.) K. Koch   Small tree; bark finely furrowed; twigs often glandular-pubescent; blades acuminate, serrate or doubly serrate; staminate flowers without perianth, 1-3 aments from buds on stems of previous year; carpellate aments solitary, terminal on branches of current year, each flower in a closed sac, the sacs in pairs in axil of an early-deciduous ament-bract; calyx minute; nut enclosed in accrescent sac, whole ament like inflorescence of Humulus, Hop.

3. Carpinus L., Hornbeam

C. caroliniana Walt., Blue-beech, American H.   Small tree; bark dark gray, smooth, trunk with fluted ridges; leaves 2-ranked, serrate or doubly serrate, acuminate; 1-3 staminate aments from buds on stems of previous year, flowers without perianth; carpellate aments terminal on branches of season, flowers in pairs in axil of bract, calyx very small; fruit a ribbed nutlet in axil of leaflike involucre.

4. Betula L., Birch

Trees and shrubs; staminate aments formed in autumn of previous season, flowers opening in spring; staminate flowers 3 in axil of each ament-bract, calyx very small; carpellate ament-bracts crowded, usually 3-lobed, each subtending three flowers, perianth none; fruit a nutlike samara with 2 lateral wings.

a. Leaf-blades rounded at tip, ovate to circular, white or gray beneath; bog shrubs. B. pumila L., Low B.
a. Leaf-blades acute or acuminate; trees.
 b. Blades rounded or subcordate at base, serrate almost to petiole; carpellate aments usually more than 1 cm wide; twigs with wintergreen flavor.
 c. Bark dark brown, not separating, resembling that of Prunus; carpellate ament-bracts glabrous, lobes short, divergent; leaves shining above. B. lenta L., Sweet or Cherry B.
 c. Bark yellow or silvery, separating in papery layers when old; carpellate ament-bracts pubescent, lobes ciliate; leaves dull above. B. lutea Michx. f., Yellow B.
 b. Blades cuneate or truncate at base, margins entire for some distance above petiole; carpellate aments, except in B. nigra, usually less than 1 cm wide.
 c. Aments in fruit thick-cylindric; wing of fruit not wider than body; blades with 8 or more pairs of lateral veins; lobes of bract oblong-linear; bark pink, tan, or red-brown, separating in papery layers. B. nigra L., River B.
 c. Aments in fruit slender-cylindric; wing of fruit wider than body; blades with fewer than 8 pairs of lateral veins; lateral lobes of bract wider than middle lobe, projecting laterally.

    d. Blades triangular with almost truncate bases, tapering from base or just above to long slender tips, glabrous or glutinous; branchlets glabrous; bark white with dark markings, not peeling. S. populifolia Marsh., Gray B.
    d. Blades ovate or rhombic, cuneate at base, not tailed at tip, pubescent beneath or on veins or in axils when young; branchlets pubescent; bark peeling.
        e. Winter buds resinous, shining; blades acute; bark white. B. alba L., White B.
        e. Winter buds not very resinous; blades acuminate; bark pink or yellow-white.
            B. papyrifera Marsh., Canoe B.

5. Alnus Ehrh., Alder

Shrubs and trees; pith 3-angled; winter-buds stalked; both kinds of aments formed the previous summer or autumn, flowers opening in spring; staminate aments pendulous, 3 flowers in axil of each ament-bract, calyx 4-parted, minute, stamens 4, filaments not divided; carpellate aments ellipsoid, each ament-bract subtending 2 flowers without calyx, the ament persistent, becoming woody.

a. Tree; leaf-blades broadly rounded at tip, somewhat emarginate, nearly as broad as long.
    *A. glutinosa (L.) Gaertn., European A.
a. Shrub; leaf-blades somewhat narrowed toward tip, longer than broad.
    b. Leaf-blades widest at or below the middle, bases rounded to subcordate, glaucous or paler beneath. A. rugosa (DuRoi) Spreng., Speckled A.
    b. Leaf-blades widest above middle, bases cuneate to somewhat rounded, not glaucous beneath.
        A. serrulata (Ait.) Willd., Common A.

FAGACEAE, Beech Family

Trees and shrubs with alternate simple pinnately-veined leaves with early-deciduous stipules; monecious; flowers monosporangiate, without petals; staminate flowers in aments or in capitate clusters; carpellate flowers epigynous, each flower or each small cluster of flowers surrounded by an involucre of bracts; carpels united, locules, styles, and stigmas as many; fruit a nut.

a. Staminate flowers in peduncled spherical clusters; pith not 5-angled; bark smooth. Fagus
a. Staminate flowers in slender aments; pith 5-angled.
    b. Carpellate flowers several in a globose long-prickly involucre that in fruit becomes a bur; staminate aments erect, flowers white. Castanea
    b. Each carpellate flower in a cup of many bracts that in fruit becomes the acorn cup; staminate aments pendulous, flowers greenish. Quercus

Fagus L., Beech

F. grandifolia Ehrh., American B. Trees; blades ovate, margins dentate, lateral veins straight and parallel, each ending in a tooth, as many teeth as veins; winter buds fusiform, long-pointed; flower-clusters in axils of leaves of the season; staminate flowers in small heads, perianth with several lobes, stamens several to many; carpellate flowers usually in 2's on a short peduncle, calyx 6-lobed, locules 3, 2 ovules in each locule, the 2 flowers and later the fruits enclosed in a 4-valved involucre; nut 3-angled.

Castanea Mill., Chestnut

Trees or shrubs; leaf margins toothed, lateral veins straight and parallel, each ending in a tooth, as many teeth as veins; calyx 6-lobed or -parted; flowers appearing after leaves, the staminate in elongate erect or spreading aments in the upper axils, stamens several to many; carpellate flowers usually in clusters of 3, each cluster in a 2- to 4-valved long-prickly involucre at base of some staminate aments or axillary, locules usually 6; 1-3 fruits within the involucre.

a. Leaf-blades green and glabrous beneath, margin coarsely serrate; once large trees, now present only as occasional sprouts. C. dentata (Marsh.) Borkh., Chestnut
a. Leaf-blades white-pubescent beneath, margin serrate or bristly; shrubs or small trees. C. pumila (L.) Mill., Chinquapin

Quercus L., Oak

Trees; leaves stipuled, usually lobed; staminate flowers in pendulous slender aments, ament-bracts soon deciduous or absent, calyx-lobes and stamens few to several; carpellate flowers epigynous, solitary, in clusters of usually 2, or in spikes, in axils of leaves of current season, each flower surrounded at base by an involucre of many small bracts; perianth usually 6-lobed, locules 3, each with 2 ovules; nut enclosed at base in cup or cupule formed from accrescent bracts. Our species may be separated into 2 groups: (1) White Oaks, acorns maturing the first season, leaf-blades lobed or toothed, lobes or teeth not bristle-pointed; and (2) Red Oaks and Black Oaks, acorns maturing the second season, leaf-blades either lobed, lobes bristle-pointed, or entire, midvein usually extending as a bristle.

a. Leaf-blades entire, oblong or lanceolate; acorn not more than 1.5 cm high and wide, cup shallow.
  b. Blades linear-oblong, glabrous beneath. *Q. phellos L., Willow O.
  b. Blades wider, pubescent beneath. Q. imbricaria Michx., Shingle O.
a. Blades obovate, broad near apex where width about equals length, entire or with a few small bristle-pointed lobes near tip, base narrow, rounded. Q. marilandica Muenchh., Blackjack O. (See third a.)
a. Blades toothed or crenate. (See fourth a.)
  b. Blades usually densely white-pubescent beneath but sometimes greenish and less pubescent; peduncles in fruit longer than petioles. Q. bicolor Willd., Swamp White O.
  b. Blades not densely white-pubescent beneath; fruits or fruit-clusters sessile or short-peduncled; lateral veins straight and parallel.
    c. Blades obovate, decidedly widest above middle; teeth rounded.
      d. Bark gray, flaky; blades softly pubescent beneath; petioles pubescent; sometimes planted in Ohio. *A. michauxii Nutt., Basket O.
      d. Bark darker, furrowed; blades sparsely pubescent beneath; petioles glabrous. Q. prinus L., Rock Chestnut O.
    c. Blades elliptic to obovate, often widest at middle or just above, teeth pointed and somewhat curved forward. Q. muehlenbergii Engelm., Yellow or Chestnut O.
a. Blades definitely lobed.
  b. Lobes of blades rounded, not bristle-tipped.
    c. Blades essentially glabrous beneath when mature.
      d. Blades usually whitened beneath, not auricled at base; acorn sessile. Q. alba L., White O.
      d. Blades blue-green beneath, auricled at base; peduncles usually several cm long. *Q. robur L., English O.
    c. Blades pubescent beneath when mature.
      d. Blades densely white-pubescent beneath or sometimes greenish and more sparsely pubescent; lobes shallow or irregular; peduncles usually 3 cm long or more. Q. bicolor Willd., Swamp White O.
      d. Lobes of blade deeper, at least some of the sinuses extending halfway to midvein; acorns sessile or nearly so.
        e. Two upper lateral lobes extending at right angle to midvein forming a cross with terminal lobe, their ends truncate or emarginate. Q. stellata Wang., Post O.
        e. Upper half of blade rounded, fan-shaped, usually shallowly-lobed, separated from narrower lower half by a deep sinus on each side; half or more of acorn covered by deep cup with fringed edge. Q. macrocarpa Michx., Bur O.
  b. Lobes of blades bristle-tipped.
    c. Blades with sinuses extending halfway, more or less, to midvein, often lustrous above, pubescent or glabrous beneath; winter buds 5-angled, yellow- or gray-pubescent, 6-10 mm long; cup of acorn rounded below, cuplike, covering half the nut; inner bark orange; petioles often yellow. Q. velutina Lam., Black O.
    c. Without the above combination of characters.
      d. Blades gray-pubescent beneath, usually some of the lobes elongate, narrow, falcate; acorn and cup at most 1.5 cm long and wide; cup deeply saucer-shaped. Q. falcata Michx., Spanish O.
      d. Blades glabrous beneath at maturity except for tufts of hairs sometimes present in vein-axils.
        e. Sinuses usually extending no more than halfway to midvein; lobes somewhat ascending; blades dull above; acorn cups either shallow, saucer-shaped, or deeper, cuplike. Q. borealis Michx. f. (Q. rubra L.), Red O.

e. Sinuses extending more than halfway to midvein; blades lustrous above.
  f. Cup of acorn shallow, saucer-shaped, truncate below, covering at most 1/3 of nut; axillary tufts of hair conspicuous.
   g. Winter buds light brown, tips puberulent; acorn 1.5 cm wide or less, about as long. Q. palustris Muenchh., Pin O.
   g. Winter buds gray, glabrous; acorn 2-3 cm wide, somewhat longer than wide. Q. shumardii Buckl.
  f. Cup of acorn cuplike or top-shaped, covering 1/3-1/2 of nut; winter buds somewhat pubescent or scales ciliate.
   g. Nut ovoid, usually with one or more rings around apex; cup brown, becoming lustrous; inner bark red. Q. coccinea Muenchh., Scarlet O.
   g. Nut narrowly ellipsoid to oblong, without apical rings; cup gray to brown, somewhat pubescent; inner bark yellow. Q. ellipsoidalis E. J. Hill

ULMACEAE, Elm Family

Trees and shrubs; leaves alternate, 2-ranked, simple, petioled, stipules early deciduous; flowers hypogynous or perigynous, bisporangiate or monosporangiate; petals none; stamens usually the same number as sepals and opposite them; carpels 2, united; styles short or absent; stigmas 2; ovulary 1-loculed, with 1 ovule; fruit a samara or a drupe.

a. Flowers in fascicles or racemes, appearing in spring before the leaves or in autumn; sepals united; leaves pinnately veined; fruit a samara. Ulmus
a. Flowers solitary or in clusters of a few in axils of leaves of the season; sepals separate; leaves with 3 main veins from top of petiole; fruit a drupe. Celtis

Ulmus L., Elm

Trees; leaves serrate or doubly serrate, often oblique at base; flowers bisporangiate, small, red-purple, yellow, or green, in fascicles or racemes, in most species appearing before the leaves; calyx 4- to 9-lobed; fruit a 1-seeded samara.

a. Lobes of calyx usually equaling or shorter than tube; flowers appearing in spring.
  b. Flowers and fruits long-pediceled, soon drooping; stigmas white or green; fruits ciliate.
   c. Flowers in umbellate clusters; fruits elliptic, about 1 cm long, the sides glabrous. U. americana L., American or White E.
   c. Flowers in racemes; fruits elliptic, 1.5 to 2 cm long, the sides pubescent. U. thomasi Sarg., Rock or Cork E.
  b. Flowers and fruits short-pediceled, in dense clusters, not drooping; fruits not ciliate.
   c. Fruits nearly circular, 1-2 cm wide, pubescent at middle of each side. U. rubra Muhl., Red or Slippery E.
   c. Fruits glabrous.
    d. Fruits nearly circular, slightly broader than long, 1-1.5 cm in diameter, seed near middle; small tree with slender branches. *U. pumila L., Siberian E.
    d. Fruits longer than broad, 2-2.5 cm long.
     e. Stigma red; fruits elliptic to slightly obovate, seed at middle. *U. glabra Huds., Wych E.
     e. Stigma white; fruit obovate, seed above middle. *U. procera Salisb., English E.
a. Lobes of calyx longer than tube; flowers appearing in autumn.
  b. Flowers in drooping racemes; fruit pubescent; leaf-blades doubly serrate. *U. serotina Sarg.
  b. Flowers in fascicles; fruit glabrous; leaf-blades simply serrate or nearly so. *U. parvifolia Jacq., Chinese E.

Celtis L., Hackberry

C. occidentalis L. Tree; leaf-blades serrate, 3 main veins from top of petiole; monecious; sepals 5, greenish, separate; staminate flowers in small clusters near base of branches of season, stamens 5, exsert; carpellate flowers solitary or 2 together in axils of upper young leaves, sometimes with anthers on short filaments; stigmas divaricate, recurved; fruit a drupe.

C. tenuifolia Nutt. (C. pumila Pursh). A shrub or small tree with leaf-blades entire or few-toothed; occurs rarely in southern Ohio.

MORACEAE, Mulberry Family

Small trees; sap milky; leaves alternate, simple, stipuled; flowers hypogynous, monosporangiate; calyx 4-parted; corolla none; stamens 4; carpels 2, united; ovulary usually 1-loculed, buried in or surrounded by the fleshy calyx, one carpel usually failing to develop; carpellate ament becoming a multiple fruit.

a. Leaf-blades entire; branches with axillary thorns; staminate and carpellate flower-clusters globular. 3. Maclura
a. Leaf-blades serrate, usually some of them lobed; branches without thorns; staminate flower-clusters cylindric.
   b. Carpellate flower-clusters globular. 2. Broussonetia
   b. Carpellate flower-clusters cylindric or ellipsoid. 1. Morus

1. Morus L., Mulberry

Diecious or imperfectly so; both kinds of flowers in cylindric or ellipsoid aments; leaves 2-ranked, ovate, cordate or truncate at base, all unlobed, or lobed and unlobed on same plant; fruit ellipsoid or cylindric, 2-3 cm long.

a. Leaf blades pubescent beneath, often all unlobed; fruit dark purple. M rubra L., Red M.
a. Leaf blades glabrous on both sides or pubescent on veins beneath, some usually lobed; fruit pale to dark. *M. alba L., White M.

2. Broussonetia L'Her., Paper Mulberry

*B. papyrifera (L.) Vent. Leaves and twigs pubescent; diecious; staminate flower-clusters cylindric, carpellate flower-clusters and multiple fruit globular, fruit 2-3 cm thick.

3. Maclura Nutt., Osage-orange

*M. pomifera (Raf.) Schneid. Diecious; leaf-blades entire, pinnately veined, acuminate, ovate; stems with axillary thorns; staminate flowers in loose somewhat globular clusters; carpellate flowers in dense globular clusters; multiple fruit 8-12 cm in diameter, orangelike in appearance.

CANNABINACEAE, Hemp Family

Erect or twining herbs; leaves petioled, palmately lobed, cleft, or compound, margins of lobes or leaflets serrate; diecious; flowers small, green; perianth regular; staminate flower of 5 separate sepals and 5 stamens; carpellate flower with cup-shaped calyx enclosing the 2 united carpels; styles and stigmas 2; ovulary 1-loculed, ovule 1; fruit an achene.

a. Leaves palmately compound or divided, the 3-7 leaflets or segments linear to lanceolate; plants erect. Cannabis
a. Leaves palmately lobed, the lobes broader; plants twining. Humulus

Cannabis L., Hemp

*C. sativa L. Tall annual; leaves stipuled, alternate above, opposite below; flowers in axillary paniculate clusters.

Humulus L., Hop

Perennial vines; leaves opposite, palmately lobed or some of them unlobed; staminate flowers in panicles, carpellate in short axillary spikes, 2 flowers together subtended by a foliaceous bract; fruit covered by the calyx, each two fruits covered by a bract.

a. Leaf blades usually 3-lobed, sinuses rounded, or upper leaves not lobed; bracts of inflorescence not ciliate in fruit. H. lupulus L., Common H.
a. Leaf blades usually 5- to 7-lobed, sinuses narrow; bracts of inflorescence ciliate in fruit.
   *H. japonicus Sieb. & Zucc., Japanese H.

URTICACEAE, Nettle Family

Ours herbs, sometimes with stinging hairs; leaves simple; stipules present or absent; flowers hypogynous, staminate and carpellate in separate or in the same inflorescences, sometimes mixed with bisporangiate ones, or staminate and carpellate flowers on separate plants; flowers in cymose clusters often aggregated in spikes, panicles, or glomerules; perianth (here called calyx) of 3-5 parts, separate or united; stamens usually 4; carpel 1; style 1 or absent; stigma often brushlike; ovulary 1-loculed, with 1 ovule; fruit an achene.

a. Leaves alternate.
    b. Blades ovate, large-toothed; flower-clusters elongate and branched, upper carpellate, lower staminate; with stinging hairs. 2. Laportea
    b. Blades lanceolate, entire to undulate; flower-clusters small, axillary, shorter than petioles; bracts longer than flowers. 5. Parietaria
a. Leaves opposite.
    b. Plants with stinging hairs; sepals 4 in carpellate flowers, the outer 2 smaller or sometimes absent. 1. Urtica
    b. Plants without stinging hairs; calyx of carpellate flowers 3-parted or tubular.
        c. Plants essentially glabrous; stem translucent; flowers in axillary panicles or glomerules; achene exserted beyond calyx. 3. Pilea
        c. Plants sometimes pubescent; stem not translucent; flowers in glomerules, the glomerules in spikes; achene enclosed by calyx. 4. Boehmeria

1. Urtica L., Nettle

With stinging hairs; leaves opposite, stipuled; monecious or sometimes diecious; flowers monosporangiate, small, green, in axillary clusters aggregated in panicles, spikes, or heads; staminate flower with 4 sepals, 4 stamens, and vestigial carpel; carpellate flower with 2-4 sepals and 1 carpel, stigma brushlike.

a. Flower-clusters 1-2 in each axil, mostly unbranched, usually not longer than petiole; annual.
    b. Blades oval or broadly elliptic, upper scarcely smaller than lower, margin incised or deeply serrate; flower-clusters slightly elongate. U. urens L.
    b. Lower blades broadly ovate, size gradually decreasing upward; flower-clusters globular.
        U. chamaedryoides Pursh
a. Flower-clusters longer than subtending petiole, branched; perennial.
    b. Blades lanceolate or lance-ovate, subcordate to rounded at base, sharply serrate; stinging hairs few. U. dioica L. var. procera (Muhl.) Wedd. (U. procera Muhl.)
    b. Blades broadly ovate, cordate at base, coarsely toothed; stinging hairs many. *U. dioica L.

2. Laportea Gaud., Wood-nettle

L. canadensis (L.) Wedd. With stinging hairs; flowers in axillary clusters; staminate flowers with 5 sepals and 5 stamens; carpellate flowers with 4 sepals, the 2 outer smaller or sometimes absent; style elongate, stigma slender; fruit asymmetric.

3. Pilea Lindl., Clearweed

P. pumila (L.) Gray Usually entirely glabrous, stem and leaves lustrous and translucent; leaves long-petioled, toothed; monecious or diecious; staminate flowers with 4-parted calyx and 4 stamens; carpellate flowers with usually 3-parted calyx and minute staminodes; stigma sessile, brushlike.

4. Boehmeria Jacq., False Nettle

B. cylindrica (L.) Sw. Blades long-petioled, ovate or ovate-lanceolate, acuminate, 3-ribbed; stipules present; monecious or diecious; carpellate and staminate flowers usually in different spikes; staminate flowers with 4 sepals, 4 stamens, and rudimentary carpel; carpellate flowers with tubular minutely-toothed calyx which is persistent around fruit; style exserted; stigma single.

5. Parietaria L., Pellitory

P. pensylvanica Muhl., Pennsylvania P. Low pubescent annual; flowers monosporangiate or bisporangiate; the three kinds mixed in close axillary clusters subtended by bracts; staminate flower with four-parted calyx and 4 stamens; carpellate flower with 4-lobed calyx; stigma nearly sessile, brush-like; calyx persistent, enclosing fruit.

SANTALACEAE, Sandalwood Family

Perennial herbs; usually partially parasitic on underground parts of other plants; leaves in Ohio species alternate, blades entire; flowers small, white, epigynous with epigynous hypanthium, in small clusters; sepals 5; petals none; stamens 5, at edge of disk; carpels 3-5, united; locule 1; fruit a drupe, 1-seeded, 4-10 mm wide.

a. Flowers in a terminal cluster, all bisporangiate. Comandra
a. Flowers in axillary clusters, some staminate. Geocaulon

Comandra Nutt.

C. umbellata (L.) Nutt. Flower-cluster corymbiform; fruit dry.

Geocaulon Fern.

C. lividum (Richards.) Fern. Clusters few-flowered, from upper axils, but stem elongates above them; only central flower of each cluster perfect; fruit red, juicy.

LORANTHACEAE, Mistletoe Family

Phoradendron Nutt.

P. flavescens (Pursh) Nutt., American Mistletoe. Small shrub growing upon and partially parasitic upon branches of trees; leaves green, thick, entire, short-petioled, opposite; diecious; flowers epigynous, small, in spikes; calyx regular, 3-lobed; corolla none; stamens 3; carpels united; ovulary 1-loculed; stigma 1; fruit white, berrylike.

ARISTOLOCHIACEAE, Birthwort Family

Low herbs or twining vines; leaves often few, simple, petioled, blades entire; flowers epigynous, bisporangiate, solitary; sepals 3, united, tubular below, 1-3-lobed above; petals none or vestigial; stamens 6 or 12; carpels 6 (in ours), united; style short and thick; stigma-lobes as many as carpels; ovulary 6-loculed; placentae central; ovules many; fruit a capsule.

a. Stamens 12; leaves 2; perianth regular, tube short and wide. Asarum
a. Stamens 6; leaves usually more than 2; perianth tube curved. Aristolochia

Asarum L., Wild Ginger

A. canadense L. Low herb with elongate rhizome bearing each year 2 large leaves 6-8 cm wide at anthesis, round-ovate, cordate at base, becoming larger; flowers solitary, red-brown, borne between petiole-bases near the ground; peduncle short, stout; sepals united below in a short tube, lobes variable; petals none or vestigial; seeds with a caruncle.

Intergrading varieties (or species) have been named, two of which are: var. canadense, with calyx-lobes longer than tube, spreading and curved upward, tapering to caudate or acuminate tips; and var. reflexum (Bickn.) Robins., with calyx-lobes about as long as tube, reflexed, abruptly narrowed to caudate tips.

Aristolochia L., Birthwort

A. serpentaria L., Virginia Snakeroot. Erect herb; leaves few, those near base of stem bract-like, blades ovate to lanceolate, bases cordate, truncate, or sagittate; flowers on short peduncles usually from near base of stem; calyx purple-brown, tube curved, limb flaring, 3-lobed; petals none; stamens adnate to style.

POLYGONACEAE, Buckwheat Family

Ours herbaceous, sometimes vines, or rarely shrublike; leaves alternate, simple; stipules sheathing the stem (ocreae), these sometimes obsolete; flowers hypogynous, usually bisporangiate, solitary, in clusters of a few, or in spikes, racemes, heads, or panicles; perianth regular or nearly so, deeply cleft into 3-6 parts (here called sepals) in 2 series; stamens 3-9; carpels 2-4, united; ovulary 1-loculed, ovule 1; styles 2-3, wholly or partially united, or absent; stigmas capitate or a tuft of branches; fruit an achene, 3-angled or lenticular.

a. Sepals 6; stigmas 3, brushlike; stamens 6-9.
    b. Achene 3-winged, not enclosed in fruit by sepals; calyx petaloid.   1. Rheum
    b. Achene 3-angled, enclosed in fruit by inner enlarged sepals; sepals usually greenish.   2. Rumex
a. Sepals usually 4-5; stigmas 2-3, usually minute or capitate; stamens 3-9.
    b. Perianth white; achene exserted from calyx; blades triangular-hastate or ovate and cordate.
        4. Fagopyrum
    b. Perianth white, green, rose, pink, or red; achene not exserted from calyx or, if so, then blades narrow; sometimes vines.   3. Polygonum

1. Rheum L., Rhubarb

*R. rhaponticum L. Basal leaves very large, petioles thick and fleshy, blades ovate-cordate; stem-leaves smaller; flowers in large terminal panicle; sepals 6, white or pink; stamens 9 or fewer; achene 3-winged. Cultivated.

2. Rumex L., Dock

Stems grooved; ocreae brittle; flowers sometimes monosporangiate, small, green, whorled, in racemes or panicles; the 3 inner sepals enlarged and closely covering fruit (these called valves), one or more of the 3 usually bearing a tubercle; stamens 6, styles 3; stigma tufted.

a. At least some of the leaves usually hastately lobed; diecious; sour to taste; plants usually less than 5 dm high. *R. acetosella L., Sheep-sorrel
a. Leaves not lobed; flowers usually bisporangiate.
    b. Valves of fruiting calyx toothed or fringed.
        c. Teeth of valves no longer than width of undivided portion; basal blades ovate to broadly oblong, usually red-veined; tubercle usually 1, ovoid. *R. obtusifolius L., Broad-leaf D.
        c. Teeth of valves bristle-form, longer than width of undivided portion; basal blades lanceolate; tubercles 3, lance-ovoid. *R. maritimus L., Golden D.
    b. Valves of fruiting calyx entire or nearly so.
        c. Blades usually pale-green, flat, entire, somewhat fleshy, narrowed to base.
            d. Pedicels strongly clavate, jointed at base, 2-5 times as long as fruiting calyx; tubercles 3, lance-ovoid. R. verticillatus L., Swamp D.
            d. Pedicels not strongly clavate, approximately equaling fruiting calyx.
                e. Tubercles 3; inflorescence usually leafy. R. mexicanus Meissn.
                e. Tubercles usually 1 or, if more, then often unequal; inflorescence not leafy.
                    R. altissimus Wood, Tall D.
        c. Blades dark-green, margin ruffled or crenulate or both.
            d. Tubercles 1 or none. *R. patientia L.
            d. Tubercles 3.
                e. Valves about twice as long as wide, achenes and tubercles at maturity fully or nearly as wide as valves. *R. conglomeratus Murr.
                e. Valves about as long as wide, much wider than achenes and tubercles at maturity.
                    f. Blades ruffled at edge; tubercle ellipsoid to subglobose, its base not above base of valve. *R. crispus L., Curled D.
                    f. Blades not ruffled at edge; tubercle narrow, pointed at tip, its base above base of valve. R. orbiculatus Gray, Great Water D.

3. Polygonum L. (Incl. Tovara Adans.)

Usually herbs, some twining; calyx 4-6-parted, the divisions often petaloid; stamens 3-9; carpels 2-3; styles 2-3, sometimes very short or absent, sometimes wholly or partially united; stigmas usually capitate.

a. Flowers axillary, solitary or in small clusters; ocreae hyaline, 2-lobed, becoming lacerate; petioles jointed; inner filaments wide. Knotweeds
    b. Leaf-blades with 2 lengthwise folds, linear, subulate-tipped; flowers 1 or 2 in axils of much reduced leaves forming spikelike inflorescence. P. tenue Michx.

92

b. Leaf-blades without lengthwise folds, not subulate-tipped; floral bracts little reduced, much exceeding flowers.
  c. Plants erect; sepals with yellow or green margins, outer ones longer than inner; pedicels usually longer than sheaths subtending them.
    d. Blades obovate or oblanceolate, about 2 times as long as wide. P. erectum L.
    d. Blades narrow, about 4 times as long as wide. P. ramosissimum Michx.
  c. Plants diffusely branched, prostrate to erect; sepals usually with pink or white margins; blades linear to oval; pedicels usually shorter than subtending sheaths. *P. aviculare L.
a. Flowers in terminal or axillary spikelike heads or racemes, or in panicles.
  b. Stems weak, reclining, 4-angled, with recurved prickles.
    c. Blades sagittate; achenes usually 3-angled. P. sagittatum L.
    c. Blades halberd-shaped; achenes lenticular. P. arifolium L.
  b. Stems erect or twining, without prickles.
    c. Styles 2, 2 mm long in flower, becoming longer in fruit, deflexed, hooked at tip, persistent; flowers remote, in long slender terminal racemes. P. virginianum L. (T. virginiana (L.) Raf.)
    c. Styles shorter, not deflexed, not elongate in fruit.
      d. Outer sepals not keeled or winged; erect herbs. Smartweeds
        e. Borders of ocreae spreading and flangelike.
          f. Tall erect annual; roots fibrous; spikes arching and drooping; blades ovate or oblong, long-petioled. *P. orientale L., Prince's-feather
          f. Perennial with long rhizomes; aquatic or terrestrial; spikes erect; blades lanceolate, short-petioled. P. amphibium L.
        e. Borders of ocreae not spreading and flangelike.
          f. Borders of ocreae not ciliate.
            g. Racemes many, terminal and lateral; perianth rose-color, pink, or white; annual, without extensive rhizomes.
              h. Racemes drooping; peduncles with sessile glands or glabrous; veins of outer sepals forked at apex, branches recurved. P. lapathifolium L.
              h. Racemes erect; peduncles with stalked glands or strigose or rarely glabrous; veins of outer sepals not as above. P. pensylvanicum L.
            g. Racemes terminal, rarely more than 1 or 2; perianth scarlet to pink; perennial with extensive rhizomes; aquatic or mud plants or of drier places; flowers dimorphic in relative length of styles and filaments.
              h. Racemes thick-cylindric or ovoid, 4 cm long or less. P. amphibium L.
              h. Racemes slender-cylindric, 4 cm long or more. P. coccineum Muhl.
          f. Borders of ocreae ciliate.
            g. Stems and peduncles glandular-hispid; leaves pubescent; spikes drooping. P. careyi Olney
            g. Stems and peduncles not glandular-hispid; leaves glabrous or nearly so.
              h. Calyx not gland-dotted (except sometimes on the upper half in P. hydropiperoides).
                i. Cilia of ocreae 5-10 mm long; cilia of ocreolae about as long as sheath; achene 3-angled; perianth rose-purple. *P. caespitosum Blume
                i. Cilia of ocreae 4 mm long or less; cilia of ocreolae shorter than sheath.
                  j. Racemes usually 2-4 cm long, compact; achene lenticular or 3-angled; leaves often with reddish spot; perianth rose-purple. *P. persicaria L.
                  j. Racemes elongate, slender, lax, often interrupted; achene 3-angled; perianth usually pink. P. hydropiperoides Michx.
              h. Calyx gland-dotted throughout.
                i. Racemes usually continuous; achene dull; usually some flowers in axil below topmost leaf; border of calyx red or green; stems often reddish. P. hydropiper L.
                i. Racemes often interrupted; achene shining; no flowers in axil below topmost leaf; border of calyx usually white; stems usually not reddish. P. punctatum Ell.
      d. Outer sepals keeled or winged.
        e. Twining vines; style short or none.
          f. Outer sepals more or less keeled, not winged, in fruit.
            g. Ocreae with ring of reflexed stiff hairs at base. P. cilinode Michx.
            g. Ocreae without stiff hairs at base; annual. *P. convolvulus L.

f. Outer sepals winged in fruit.
   g. Wings of sepals 1-3 mm wide at top, ruffled; achene 3.5 mm long or more.
      P. scandens L.
   g. Wings of sepals 1/4 to 1 mm wide at top; achene 3.5 mm long or less.
      P. cristatum Engelm. & Gray
e. Not vines; erect branched perennials, sometimes shrublike; styles divaricate.
   f. Blades truncate at base, abruptly cuspidate. *P. cuspidatum Sieb. & Zucc., Japanese Knotweed.
   f. Blades cordate at base, gradually tapering to tip. *P. sachalinense F. Schmidt

Fagopyrum Mill., Buckwheat

*F. sagittatum Gilib. Rather fleshy; stem striate; blades petioled; ocreae oblique; calyx 5-parted, petaloid; stamens 8; styles 3; stigmas capitate; achene acute, smooth and shining, exserted beyond calyx.

CHENOPODIACEAE, Goosefoot Family

Herbs; leaves alternate or opposite; flowers small, usually green, hypogynous, bisporangiate or monosporangiate; sepals and stamens 1-5; petals none; carpels 2-3 (5); styles or stigmas as many; fruit 1-seeded; seed lens-shaped, vertical (its long axis parallel to floral axis) or horizontal.

a. Leaves filiform or awl-shaped, spine-tipped. 7. Salsola
a. Leaves scalelike, opposite, small. 8. Salicornia (See third a)
a. Leaves not spine-tipped, not scalelike.
   b. Fruit enclosed by 2 triangular bracts; carpellate flowers without perianth. 4. Atriplex
   b. Fruit not enclosed by bracts; sepals 1 or more.
      c. Fruit longer than calyx and not enclosed by it; calyx of 1-3 sepals.
         d. Flowers in terminal spikes, solitary in axils of bracts; leaves linear, entire. 6. Coriospermum
         d. Flowers in dense small clusters in axils of leaflike bracts; leaves wider than linear, often with a large tooth on each side. 5. Monolepis
      c. Fruit either enclosed by calyx or not longer than calyx.
         d. Entire calyx or individual sepals horizontally winged in fruit.
            e. Entire calyx with continuous horizontal wing in fruit; blades coarsely sinuate-toothed. 1. Cycloloma
            e. Individual sepals horizontally winged in fruit; blades entire. 2. Kochia
         d. Neither entire calyx nor individual sepals horizontally winged in fruit. 3. Chenopodium

1. Cycloloma Moq., Winged Pigweed

C. atriplicifolium (Spreng.) Coult. Much-branched annual; leaves sinuate-toothed, early deciduous, petioled; flowers perfect or carpellate, in branched spikes; calyx lobes 5, incurved over ovulary, in fruit bearing a continuous horizontal wing; seed horizontal.

2. Kochia Roth

*K. scoparia (L.) Schrader, Summer Cypress. Annual; stem erect, much branched; leaves linear to lanceolate, entire, sessile, sometimes becoming red at maturity; flowers perfect or carpellate, in small clusters in axils of leaves, the clusters in axillary or terminal spikes; calyx 5-lobed, in fruit horizontally winged; stamens 5; styles 2-3; seed horizontal.

3. Chenopodium L., Goosefoot

Usually white-mealy or glandular; blades narrow to ovate, entire, toothed, or lobed; flowers perfect, the small clusters in spikes or panicles; calyx usually 5-parted (2-5); stamens usually 5 (1-5); styles 2-3; pericarp adherent to seed or free from it and easily removable.

a. Plants with glandular hairs or with sessile glands; not white-mealy; strong-scented.
   b. Calyx and other plant-parts with short gland-tipped hairs. *C. botrys L., Jerusalem-oak
   b. Calyx or blades or both gland-dotted; glandular hairs present or absent.
      c. Calyx gland-dotted; blades mostly 1-3 cm long, sinuate; rare. *C. pumilio R. Br.

c. Blades gland-dotted, longer than above, toothed to pinnatifid; pericarp gland-dotted, calyx usually not. *C. ambrosioides L., Mexican-tea

a. Plants without glandular hairs, without gland-dots, often white-mealy.
  b. Calyx-lobes sometimes fewer than 5; some or all seeds vertical.
    c. Flower-clusters capitate, 0.5-1 cm wide; calyx in fruit red and somewhat fleshy; blades triangular or hastate; seeds all vertical. C. capitatum (L.) Aschers, Strawberry Blite
    c. Flower-clusters and calyx not as above; blades oblong, sinuate-toothed. *C. glaucum L.
  b. Calyx-lobes usually 5; seeds all horizontal.
    c. Green throughout; flower-clusters in many axils; blades ovate, entire. *C. polyspermum L.
    c. More or less white-mealy or, if green, then blades large, with few large teeth or entire.
      d. Plants low, wide-branching; blades 1-3 cm long and wide, obviously white-mealy.
        e. Plant ill-scented; blades entire; pericarp adherent to seed. *C. vulvaria L.
        e. Not ill-scented; blades entire or toothed; seed free. *C. incanum (S. Wats.) Heller
      d. Plants erect, usually taller; blades longer than wide or larger than above.
        e. Blades large, ovate, base cordate or truncate, 1-5 large teeth on each side, rarely entire; sepals not keeled; seed adherent, 1.5-2.5 mm wide. C. hybridum L.
        e. Blades not as above; sepals often keeled, at least at tip; seed smaller.
          f. Blades lance-linear, entire, cuspidate, 1-3-veined from base. C. leptophyllum Nutt.
          f. Blades lanceolate to ovate, with midvein and evident branch-veins.
            g. Sepals strongly keeled, covering fruit; lines on seed, if present, radial; pericarp adherent; blades ovate to lanceolate, at least the lower usually toothed or lobed.
              *C. album L., Lamb's Quarters
            g. Sepals not or weakly keeled or keeled only at tip, not covering fruit.
              h. Blades lanceolate, acute, usually entire; lines on seed over lower part of embryo lengthwise; pericarp free. C. standleyanum Aellen (C. boscianum Moq.)
              h. Blades ovate, with many acuminate unequal teeth. *C. murale L.

4. Atriplex L., Orach

Leaves alternate or opposite or both on same plant, entire, toothed or lobed; monecious; staminate flowers of 3-5 united sepals and 3-5 stamens; carpellate flowers without perianth, enclosed by 2 bracts which enlarge in fruit; carpels 2; seeds vertical.

a. Leaves green or nearly so, lanceolate to triangular-hastate, opposite or alternate; calyx of staminate flowers 4-cleft; bracts entire or slightly toothed. A. patula L.
a. Leaves silver- or white-scurfy, alternate, ovate or narrower; bracts conspicuously toothed.
  b. Bracts widest at base, toothed along sides. *A. rosea L.
  b. Bracts tapering to base, toothed across the wide summit. *A. argentea Nutt.

5. Monolepis Schrad.

M. nuttalliana (Schult.) Greene   Low, branched; leaves alternate, lanceolate to ovate, usually with 1 large tooth on each side; flowers sessile in small dense clusters in axils of leaflike bracts, the clusters in terminal and axillary spikes; sepal 1; stamen 1; styles 2; seed vertical.

6. Coriospermum L., Bugseed

Leaves linear, entire, 1-nerved; flowers perfect, solitary in axils of bracts; sepals usually 1; stamens 1-2 (5); styles 2; seed vertical.

a. Lower bracts imbricate; fruit 3.5-5 mm long; spike dense. C. hyssopifolium L.
a. Lower bracts not imbricate; fruit 2-3 mm long; spike slender. C. nitidum L.

7. Salsola L.

S. kali L. var. *tenuifolia G. W. F. Meyer, Russian Thistle. Erect or ascending, much-branched; leaves stiff, narrow, spine-tipped; flowers perfect, solitary or in clusters of a few in upper axils; calyx deeply 5-lobed, horizontally keeled in fruit; seed horizontal.

8. Salicornia L., Glasswort, Samphire

S. europaea L.   Of salt marshes and saline soil; leaves scalelike, opposite; flowers in spikes, 3 in axil of each scale, central longest; fruit enclosed in unlobed calyx; seed vertical.

AMARANTHACEAE, Amaranth Family

Rather coarse herbs; leaves simple, opposite or alternate; flowers hypogynous, bisporangiate or monosporangiate, small, bracted, in spikes, panicles, or small axillary clusters; sepals 5 or fewer, separate or united, or none; petals none; stamens 5 or fewer, separate or united; styles and stigmas usually 2-3 or stigmas sessile; ovulary 1-loculed; fruit a dehiscent or indehiscent utricle.

a. Leaves alternate; flowers monosporangiate; plants not woolly. Amaranthus
a. Leaves opposite; flowers bisporangiate; plants woolly. Froelichia

Amaranthus L. (Including Acnida L.), Amaranth

Diecious or monecious; stems erect to prostrate, usually much branched; leaves alternate, blades usually entire; flowers green or reddish, monosporangiate, in axillary clusters or in simple or branched spikes; sepals 1-5 or none, separate, sometimes unequal; stamens 1-5; stigmas usually 2-3.

a. Species monecious.
    b. Flowers in small axillary clusters; stems usually much branched; stamens 2 or 3.
        c. Stems prostrate or decumbent; bracts acuminate, about equaling longest sepal and fruit; seed 1.5 mm wide; sepals 4-5. A. graecizans L., Mat A.
        c. Stems ascending to erect; bracts stiff, subulate, longer than sepals and fruit; seeds 1 mm wide; sepals of carpellate flowers usually 3. A. albus L.
    b. Flowers mostly in simple or panicled spikes; stamens 5.
        c. Leaf bases with a pair of spines; fruit indehiscent or irregularly dehiscent. *A. spinosus L., Spiny A.
        c. Leaf bases without spines; fruit dehiscent, top coming off as a lid.
            d. Terminal spike 1.5 cm thick, branches short, erect; calyx of carpellate flowers 3 mm long, tips rounded, truncate, or emarginate, mucronate, about half as long as the bracts. *A. retroflexus L., Rough Pigweed
            d. Terminal spike 1 cm thick or less, branches many, spreading to erect; calyx of carpellate flowers 2 mm long or less, acute, aristate, equaling or half as long as bracts. *A. hybridus L., Slender Pigweed
a. Species diecious; stamens 5; sepals of staminate flowers 5.
    b. Bracts 4-6 mm long, spine-tipped; sepals of carpellate flowers 5; terminal spike long and slender, lateral ones absent or few. *A. palmeri Wats.
    b. Bracts 2 mm long or less; sepals of carpellate flowers 1-2 or none.
        c. Bracts 1.5-2 mm long, they and sepals of staminate flowers with heavy excurrent midvein; utricle shorter than subtending bracts, circumscissile. A. tamariscinus Nutt. (Acnida tamariscina (Nutt.) Wood)
        c. Bracts 1-1.5 mm long, they and sepals of staminate flowers with slender midvein; utricle longer than bracts, splitting irregularly. A. tuberculatus (Moq.) Sauer (Acnida altissima Riddell)

Froelichia Moench

Leaves opposite; flowers perfect, in woolly spikes; calyx tubular, 5-lobed, in fruit ovoid; stamens 5, filaments united in a tube; fruit indehiscent.

a. Calyx in fruit bearing two lateral toothed, erose, or entire wings; often some tubercles or ridges on the face. F. floridana (Nutt.) Moq.
a. Calyx in fruit bearing two lateral rows of stiff processes somewhat united at base, and some spine-like processes or tubercles on the face. F. gracilis (Hook.) Moq.

NYCTAGINACEAE, Four-o'clock Family

Mirabilis L., Four-o'clock

Herbs; leaves opposite; blades entire; flowers hypogynous, perfect in ours, each flower or each cluster subtended by a calyxlike involucre; calyx regular, corollalike, funnelform, 5-lobed, base constricted above ovulary in such way that flower appears epignyous; stamens 3-5; carpel, locule, and ovule 1; style slender, stigma capitate; fruit indehiscent.

a. Calyx of various colors, 2-4 cm long; involucre less than 1 cm long, subtending individual flowers. *M. jalapa L., Common F.

a. Calyx about 1 cm long; involucre accrescent, veiny, subtending more than 1 flower.
   b. Main blades petioled, ovate to deltoid, base truncate to cordate (rarely cuneate); stem below
      inflorescence glabrous or nearly so. M. nyctaginea (Michx.) MacM.
   b. Main blades sessile or short-petioled, oblong or lanceolate, tapering to base; rare.
      c. Stem below inflorescence with long spreading hairs. M. hirsuta (Pursh) MacM.
      c. Stem below inflorescence with 2 bands of short curved hairs. M. albida (Walt.) Heimerl

PHYTOLACCACEAE, Pokeweed Family

Phytolacca L., Pokeweed

P. americana L. Large erect glabrous herb; leaves alternate, blades ovate or oblong, entire;
flowers usually bisporangiate, hypogynous, about 1 cm wide or less, in racemes; perianth regular, of
5 separate green, white, or pink sepals; petals none; stamens usually 10; carpels usually 10, united
in a ring; styles, stigmas, and locules 10; fruit a 10-seeded dark-purple berry.

AIZOACEAE, Carpetweed Family

Mollugo L., Carpetweed

*M. verticillata L. Annual prostrate branched herb; leaves whorled, 3-8 at a node, oblanceo-
late, 1-3 cm long; flowers hypogynous, perfect, pediceled, about 5 mm wide, a few together at nodes;
calyx regular, of 5 separate white or green sepals; petals none; stamens usually 3; styles short, 3;
capsule 3-loculed, many-seeded; placentae central.

PORTULACACEAE, Purslane Family

Herbs; leaves succulent; flowers hypogynous or partly epigynous, bisporangiate; perianth regular or nearly so; as referred to in this manual, sepals 2, petals 5 (6), usually separate (or these parts
interpreted as bracts 2, sepals 5 (6), petals none); stamens 4-many, opposite petals when of same
number, sometimes on corolla base; carpels united; styles separate above; ovulary 1-loculed; ovules
several to many; placentation central or basal; fruit a capsule.

a. Ovulary partly inferior; cauline leaves more than 2; capsule circumscissile. Portulaca
a. Ovulary superior; cauline leaves 2; capsule dehiscent lengthwise. Claytonia

Portulaca L., Purslane

Fleshy, sometimes prostrate; leaves alternate; ovulary partly inferior; sepals united at base;
stamens 6-many; style branches 3-several; seeds many, minute.

a. Leaves flat, obovate; flowers small, petals yellow; stamens 6-10. *P. oleracea L., Common P.
a. Leaves terete; flowers large, petals of various brilliant colors; stamens many. *P. grandiflora
   Hook., Rose-moss, Garden P.

Claytonia L., Spring Beauty

Perennial glabrous small herb; underground stem a corm, in ours; cauline leaves 2, usually opposite; basal leaf or leaves similar; flowers hypogynous, in bracted raceme; sepals ovate, separate,
persistent in fruit; petals usually 5, white or pink, with more deeply colored veins; stamens opposite
petals and attached to their bases; carpels 3, united; style 3-branched at tip.

a. Leaves linear to lanceolate, tapering to base, blade and petiole not differentiated. C. virginica L.
a. Leaves lanceolate, oblanceolate, or wider, definitely petioled. C. caroliniana Michx.

CARYOPHYLLACEAE, Pink Family

Herbs; leaves opposite, with or without stipules, blades entire; flowers hypogynous, usually bisporangiate, solitary or in cymose clusters; sepals usually 5 (4), separate or united; corolla regular,
of 5 separate petals, or lacking; stamens separate, usually as many as or twice as many as petals or
sepals, but sometimes not exactly; carpels 2-5, united; styles and stigmas as many, or rarely united;
ovulary 1-loculed, placenta free central or basal, or 3-5-loculed below, the partitions more or less
completely formed; fruit a utricle (1-seeded) or a capsule dehiscent by as many or twice as many valves
or teeth as styles.

a. Petals absent; fruit a utricle, 1-seeded; styles 2, separate or partly united; flowers inconspicuous.
  b. Leaves awl-shaped, without stipules. 1. Scleranthus
  b. Leaves small but flat, with conspicuous hyaline stipules. 2. Paronychia
a. Petals usually present; fruit a capsule, seeds several to many; styles 2-5, if 2, then petals conspicuous.
  b. Sepals separate or nearly so.
    c. Styles 5 or 4.
      d. Leaves linear-filiform, with or without stipules.
        e. Leaves several at a node; stipules present, minute. 4. Spergula
        e. Leaves opposite; stipules absent; plants very small. 5. Sagina
      d. Leaves ovate to linear but not linear-filiform, without stipules; capsule with twice as many teeth as styles; petals notched.
        e. Capsule ovoid, dehiscent by 5 valves, each 2-toothed; styles 5, alternate with sepals.
          7. Stellaria
        e. Capsule cylindric, dehiscent by 10 teeth, sometimes curved; styles opposite sepals.
          8. Cerastium
    c. Styles 3.
      d. Leaves linear-filiform, with stipules; petals entire. 3. Spergularia
      d. Leaves without stipules.
        e. Flowers in an umbel; stamens usually 3-5; petals erose at tip. 9. Holosteum
        e. Flowers in a cyme or solitary; stamens usually 8-10; petals entire, emarginate, or 2-lobed.
          f. Petals entire or barely emarginate. 6. Arenaria
          f. Petals 2-lobed or 2-parted. 7. Stellaria
  b. Sepals united forming a toothed or lobed tube.
    c. Styles 5.
      d. Calyx-lobes longer than calyx-tube; petals unappendaged, shorter than calyx. 10. Agrostemma
      d. Calyx teeth short; petals often appendaged, longer than calyx. 11. Lychnis
    c. Styles 2 or 3.
      d. Two or more closely appressed bracts immediately below calyx; styles 2; calyx with many fine nerves. 14. Dianthus
      d. Calyx without such bracts.
        e. Styles 2; capsule 4-valved; calyx with 5 prominent ribs or wings or with many obscure nerves. 13. Saponaria
        e. Styles 3; capsule 3- or 6-valved; calyx usually either inflated or with more than 5 prominent nerves. 12. Silene

1. Scleranthus L., Knawel

*S. annuus L. Small diffuse annual; leaves awl-shaped, without stipules; flowers small, green, clustered; sepals 5, united below, the tube becoming thick and enclosing the utricle; stamens usually 5-10, united with calyx; styles 2.

2. Paronychia Mill., Whitlow-wort

Ours annual; stems slender, forking; leaves with scarious stipules; flowers minute; sepals 5, separate or united at base, greenish; stamens 2-5, on base of calyx; style 2-branched; fruit a utricle.

a. Stem glabrous. P. canadensis (L.) Wood
a. Stem pubescent. P. fastigiata (Raf.) Fern.

3. Spergularia J. & C. Presl, Sand-spurrey

S. rubra (L.) J. & C. Presl Stem sometimes glandular-pubescent above; leaves linear-filiform, almost terete, with conspicuous acuminate stipules, often with fascicles of leaves in axils; flowers small, in terminal leafy cymes; sepals 5; petals 5, pink, entire, shorter than sepals; stamens 10 or sometimes fewer; styles 3; capsule 3-valved.

4. Spergula L., Spurrey

*S. arvensis L. Annual; leaves filiform, channeled at base, clustered at nodes; stipules minute, connate; flowers small, in terminal cymes; sepals 5; petals 5, white, about equaling sepals; stamens 10 or 5; styles 5; capsule 5-valved; seeds papillose, margined.

5. Sagina L., Pearlwort

Very small, tufted; leaves linear or subulate, without stipules; sepals 4-5; petals 4-5, entire or emarginate, or none; stamens as many as sepals or fewer or twice as many; styles and capsule-valves as many as sepals.

a. Petals and sepals usually 4; flowerless shoots or basal rosettes often present; not glandular.
   S. procumbens L.
a. Petals and sepals usually 5; flowerless shoots and basal rosettes absent; glandular or not. S. decumbens (Ell.) T. & G.

6. Arenaria L., Sandwort

Low herbs; leaves sessile or nearly so, without stipules; sepals 5; petals 5, white, entire or emarginate, rarely none; stamens 10; styles usually 3, opposite as many sepals; capsule-valves 3, entire or 2-toothed.

a. Leaves 2-4 cm long, oblong or oval, usually spreading, pellucid-punctate. A. lateriflora L.
a. Leaves either much shorter or linear or awl-shaped.
   b. Leaves ovate, about 5 mm long; bracts of cyme like foliage leaves. *A. serpyllifolia L., Thyme-leaf S.
   b. Leaves linear or filiform; bracts much smaller than foliage leaves.
      c. Leaves soft, slightly fleshy, midvein not evident at tip; pedicel and base of calyx often glandular. A. patula Michx.
      c. Leaves firm, midvein conspicuous; calyx and pedicel glabrous; axillary fascicles present.
         A. stricta Michx.
         d. Stems not rigid, 1-4 dm tall, leaves 1.5-3 cm long. var. stricta
         d. Stems rigid, 1-2 dm tall, leaves 5-10 (20) mm long. var. texana Robins.

7. Stellaria L. (Incl. Myosoton Moench), Chickweed

Low herbs; leaves without stipules; flowers solitary or in terminal cymes which sometimes appear lateral as result of stem elongation; sepals usually 5; petals 5, white, 2-cleft, sometimes absent; stamens 10, 8, or fewer; styles 3-5; capsule ovoid, opening by 1-2 times as many valves as styles.

a. Leaves ovate or elliptic; stem usually pubescent, often in lines.
   b. Styles 5; stem and sepals glandular-pubescent; petals longer than sepals. *S. aquatica (L.) Scop.
      (M. aquaticum (L.) Moench), Giant C.
   b. Styles 3; stem usually pubescent in lines.
      c. Blades usually 1-2 cm long, all, or all but the upper, petioled, ovate; annual. *S. media (L.)
         Cyrillo, Common C.
      c. Blades elliptic to oblanceolate, much longer than the above; perennial. S. pubera Michx.,
         Star C.
         d. Leaf-blades sessile. var. pubera
         d. Leaf-blades petioled. var. silvatica (Beguinot) Weath.
a. Leaves linear or lanceolate; stem glabrous.
   b. Inflorescence few-flowered, usually on a lateral branch; sepals with 3 faint nerves; seeds nearly smooth. S. longifolia Muhl.
   b. Inflorescence many-flowered, terminal; sepals with 3 strong nerves; seeds tubercled.
      *S. graminea L.

8. Cerastium L., Mouse-ear Chickweed

Pubescent, usually viscid; flowers in terminal cymes; sepals 5 or 4; petals 5 or 4, 2-lobed or -cleft, rarely entire or wanting; stamens 10, rarely fewer; styles 5 or 3-4; capsules usually cylindric, sometimes curved, opening by twice as many teeth as styles; seeds many.

a. Median and upper leaves obovate, ovate, or oblong, about 2 (3) times as long as wide; sepals about equaling petals; capsule not more than twice length of calyx.
  b. Inflorescence at maturity open, pedicels at maturity longer than calyx; bracts of cyme scarious at apex and margin; perennial. *C. vulgatum L.
  b. Inflorescence compact, pedicels at maturity shorter to scarcely longer than calyx; bracts of cyme herbaceous; annual. *C. viscosum L.
a. Median and upper leaves linear to lanceolate, more than 3 times as long as wide.
  b. Stems firm; petals 2-3 times as long as sepals; plants glandular, villous, or glabrous. C. arvense L.
  b. Stems weak; flowers nodding; petals equaling or much exceeding sepals; plant viscid-pubescent; capsule 2-3 times as long as sepals. C. nutans Raf., Nodding M.

9. Holosteum L., Jagged Chickweed

*H. umbellatum L. Glaucous annual; stem erect, glandular above; leaves oblong or oblanceolate, without stipules; flowers small, several in terminal long-peduncled umbelliform cluster; sepals 5; petals 5, white, tip erose; stamens 3-5 (10); styles 3; capsule dehiscent by 6 teeth.

10. Agrostemma L., Corn-cockle

*A. githago L. Tall, pubescent; leaves linear or lanceolate; long-peduncled flowers solitary at ends of branches; sepals 5, united, strongly 10-ribbed, the elongate slender lobes surpassing the petals; petals 5, purple-red, 2-3 cm long, slightly emarginate, oblanceolate; stamens 10; styles 5 (4), opposite petals; capsule ovoid or oblong, opening by as many teeth as styles.

11. Lychnis L., Campion

Pubescent, sometimes viscid; sepals 5, united below; petals 5, with narrow claws, often somewhat auricled at junction of claw and blade, each usually with 2 appendages; stamens 10; styles 5 (4); capsule 1- or 3-celled, opening by as many or twice as many teeth as styles.

a. White-woolly; petals purple-red. *L. coronaria (L.) Desr.
a. Not white-woolly.
  b. Usually diecious; calyx 10-nerved in staminate, 20-nerved in carpellate flowers; petals usually white, rarely pink, 2-cleft; flowers fragrant. *L. alba Mill., White C. Strongly resembles Silene noctiflora L.
  b. Flowers mostly bisporangiate; calyx 10-nerved; petals pink or rose-color, cleft into 4 narrow lobes. *L. flos-cuculi L., Cuckoo Flower

12. Silene L., Campion, Catchfly

Flowers bisporangiate or sometimes monosporangiate; sepals 5, united; petals 5, rarely absent, usually appendaged, claw narrow, usually auricles at junction of blade and claw, tips variously lobed or cleft, rarely entire; stamens usually 10; styles usually 3; capsule mostly stipitate, 1-loculed or more or less completely 3-loculed.

a. At least some leaves in whorls of 4; petals white, fringed. S. stellata (L.) Ait., Starry Campion
a. All leaves opposite; petals entire, emarginate, or 2-lobed.
  b. Calyx about 30-ribbed; petals reddish to white, appendages deeply 2-lobed. *S. conica L.
  b. Calyx usually 5-, 10-, or 20-ribbed.
    c. Petals white, pink, or rose-color, not red.
      d. Calyx glabrous.
        e. Inflorescence leafy-bracted; flowers few, corolla white, calyx inflated; blades thin, acuminate. S. nivea (Nutt.) Otth, Snowy Campion.
        e. Inflorescence small-bracted.
          f. Petals inconspicuous or wanting; stem usually with some dark glutinous areas; calyx ovoid. S. antirrhina L.
          f. Petals conspicuous; stem without glutinous areas.
            g. Calyx tube clavate; flowers in dense terminal clusters; corolla rose to white. *S. armeria L.
            g. Calyx tube and inflorescence not as above; calyx bladdery-inflated.

  h. Inflorescence or its branches erect, racemiform, elongate; rare. *S. cserei Baumg.

  h. Inflorescence spreading, not racemiform. *S. cucubalus Wibel, Bladder Campion

 d. Calyx pubescent, sometimes glandular.

  e. Short perennials with basal tufts of acute spatulate or oblanceolate leaves; petals pink, entire or retuse. S. caroliniana Walt.

   f. Upper stem and calyx glandular. var. pensylvanica (Michx.) Fern.

   f. Upper stem and calyx pubescent but not glandular. var. wherryi (Small) Fern.

  e. Tall plants without basal tufts at anthesis; petals deeply 2-lobed; capsule 3-loculed; calyx 10-ribbed.

   f. Ultimate branches of inflorescence racemes or spikes; flowers mostly bisporangiate; calyx cylindric, ribs not connected by cross-veins. *S. dichotoma Ehrh.

   f. Branches of inflorescence not racemes or spikes; flowers often monosporangiate; calyx in fruit ovoid, with cross-veins connecting the ribs. *S. noctiflora L. Strongly resembles Lychnis alba L. which, however, has 5 styles.

 c. Petals red.

  d. Petals entire or nearly so; stem erect, with 10-20 pairs of ovate to lanceolate, more or less clasping, leaves. S. regia Sims, Royal Catchfly.

  d. Petals 2-lobed; stem weak, leaves fewer.

   e. Cauline leaves 5-8 pairs, elliptic to round-ovate, upper sessile, lower on winged petioles; stipe of ovulary 6-8 mm long. S. rotundifolia Nutt.

   e. Cauline leaves 2-4 pairs, lower and basal ones oblanceolate, tapering to winged petioles; stipe of ovulary about 2 mm long. S. virginica L., Fire-pink

13. Saponaria L.

  Flowers in open or dense cymes; calyx 5-toothed; stamens 10; styles 2; capsule 1-loculed or incompletely 2-4-loculed, opening by 4 valves.

a. Calyx terete, not winged; petals appendaged, white or pink; flowers in dense clusters.
 *S. officinalis L., Bouncing Bet, Soapwort

a. Calyx 5-angled, winged in fruit; petals not appendaged, red or pink; inflorescence terminal, open.
 *S. vaccaria L., Cowherb

14. Dianthus L., Pink

  Flowers solitary or clustered; calyx cylindric, 5-toothed, with many fine nerves, subtended by bracts; petals long-clawed, dentate or crenate, not appendaged; stamens 10; styles 2; capsule 1-loculed, dehiscent by 4-5 teeth.

a. Leaves lanceolate to ovate-lanceolate; flowers in dense terminal clusters; bracts long-pointed, about equaling calyx. *D. barbatus L., Sweet William

a. Cauline leaves linear to linear-lanceolate.

 b. Flowers about 1 cm wide, in small dense clusters; bracts narrow, pubescent, about equaling calyx; petals rose-color, narrow. *D. armeria L., Deptford P.

 b. Flowers 1.5-2 cm wide, solitary or few; bracts broad, slender-pointed, scarcely pubescent, shorter than calyx; petals deep rose-color with darker area at base. *D. deltoides L., Maiden P.

CERATOPHYLLACEAE, Hornwort Family

Ceratophyllum L., Hornwort, Coontail

  Submersed aquatics; stems elongate, branched; leaves whorled, sessile, dichotomously dissected into filiform, often toothed, divisions; monecious; flowers minute, hypogynous, monosporangiate, solitary in axils; calyx (or calyxlike involucre) cut into 8-14 divisions; corolla none; stamens 10-20; carpel 1, style slender; ovule 1; fruit indehiscent.

a. Leaves forked mostly 1-2 times; fruit ovoid, with 1 terminal spine and 2 lateral spines near base.
 C. demersum L.

a. Leaves forked mostly 2-4 times; fruit slightly flattened with several lateral spines; apparently rare.
 C. echinatum Gray

NYMPHAEACEAE, Water-lily Family

Aquatic herbs; juice sometimes milky; leaf-blades submersed, floating, or emersed; flowers regular, perfect, emersed or floating; sepals, petals, and stamens sometimes intergrading, each 3-many, cyclic or spirally arranged; carpels 3-many, separate, sometimes embedded in the receptacle, or united; fruit an aggregate of follicles or of nutlets or a leathery berry.

a. Carpels united; leaf blades entire, with basal sinus.
  b. Petals and stamens inserted below the ovulary; flowers yellow. 1. Nuphar
  b. Petals and stamens inserted on sides of ovulary; flowers white or pink. 2. Nymphaea
a. Carpels separate; leaf-blades peltate or dissected.
  b. Carpels embedded in pits in an obconic flat-topped receptacle; flowers yellow; leaves peltate.
    3. Nelumbo
  b. Carpels above the receptacle and not embedded in it.
    c. Submersed leaves opposite or whorled, dissected; leaves with floating small blades usually alternate, peltate; perianth white or cream. 5. Cabomba
    c. Leaves alternate, peltate, blades entire, floating; perianth dull-purple. 4. Brasenia

1. Nuphar Sm., Yellow Water-lily, Spatter-dock

N. advena (Ait.) Ait. f. Leaves long-petioled; blades emersed, ovate or rounded, bases cordate-sagittate; flowers hypogynous, solitary in the axils, long-peduncled, emersed, 3–5 cm in diameter; outer perianth parts greenish, yellow, or purplish; petals many, yellow, smaller than the sepals, stamenlike; stamens many; carpels several to many, united; stigmas united in an almost sessile several- to many-rayed broad disk; fruit a leathery berry, locules and seeds several to many.

2. Nymphaea L., Water-lily

Leaf-blades floating, circular with a sinus, rarely sagittate, long-petioled; flowers large, on long flexible peduncles; sepals 4, green; petals and stamens many, inserted on sides of ovulary, outer ones white or pink and obviously petals, inner ones obviously stamens; a zone of intergrades between petals and stamens, the outer of which are more petal-like, the inner of which are more stamenlike; carpels many, united, ovulary subglobose, many-loculed; radiating stigmas many.

a. Flower fragrant; inner filaments narrower than the anthers; petals elliptic, tapering or acute at apex; petioles green. N. odorata Ait.
a. Flower not or only slightly fragrant; inner filaments wider than anthers; petals obovate, rounded at apex; petioles marked with purple. N. tuberosa Paine

3. Nelumbo Adans.

N. lutea (Willd.) Pers., American Water-lotus. Leaves long-petioled, peltate, the circular blades emersed; flowers 5–10 in. wide, solitary, long-peduncled; perianth parts many, the outer green, the inner yellow; stamens many; carpels many, the 1-ovuled ovularies and nutlike indehiscent fruits embedded in the obconic flat-topped receptacle.

4. Brasenia Schreb., Water-shield

B. schreberi Gmel. Stems elongate, branching; leaves peltate, blades elliptic, to 10 cm long, floating; flowers axillary, long-peduncled; sepals usually 3; petals dull-purple, usually 3, 1–1.5 cm long, separate; stamens several to many; 1-2-seeded carpels 4 or more, separate, clavate and indehiscent in fruit.

5. Cabomba Aubl.

C. caroliniana Gray, Fanwort. Submersed leaves opposite, palmately dissected; floating leaves few, usually alternate, peltate, blades linear-oblong; flowers long-peduncled from upper axils; sepals and petals each 3, separate, white or yellowish; stamens 6; carpels 3-6, separate, each usually with 3 ovules, indehiscent in fruit.

RANUNCULACEAE, Buttercup Family

Usually herbs; leaves usually alternate, often all basal, often compound or deeply divided; flowers hypogynous, bisporangiate or rarely monosporangiate; perianth regular or rarely zygomorphic; sepals present, separate, sometimes petaloid, sometimes early deciduous; petals absent or present and separate; staminodia sometimes present; stamens many or rarely few; carpels many to few, separate, fruit then an aggregate of achenes, follicles, or berries; or carpels rarely united, fruit then a capsule; or carpel 1, fruit then a berry or a follicle.

a. Perianth zygomorphic.
  b. Upper sepal spurred, upper petals with spurs extending into spur of sepal. 13. Delphinium
  b. Upper sepal hooded or saccate, 2 small long-clawed petals under the hood. 14. Aconitum
a. Perianth regular.
  b. Petals spurred. 12. Aquilegia
  b. Petals absent, or present and not spurred.
    c. Leaves simple, not lobed; flowers yellow.
      d. Petals none, sepals petal-like; leaf-blades reniform; fruit an aggregate of few to several follicles. 8. Caltha
      d. Petals and sepals present, petals with a nectar pit or scale on upper surface near base; leaf-blades narrow or cordate-ovate; fruit an aggregate of achenes. 1. Ranunculus
    c. At least some leaves lobed or compound; flowers yellow or not yellow.
      d. Leaves all basal; bracts of scape none, minute, or in a calyxlike whorl of 3 immediately beneath flower.
        e. Leaves ternately compound; bracts of scape none or minute; carpels stipitate, becoming follicles. 10. Coptis
        e. Leaves simple, 3-lobed; a calyxlike involucre immediately beneath flower; carpels sessile or nearly so. 4. Hepatica
      d. Cauline leaves present, alternate, opposite, or whorled.
        e. Climbing vines; leaves opposite; sepals usually 4, petals small or none. 6. Clematis
        e. Not vines.
          f. Leaves of stem opposite or whorled, subtending the peduncle or peduncles.
            g. Carpels pubescent; leaves palmately divided. 5. Anemone
            g. Carpels glabrous; basal leaves ternately decompound, stem-leaves simple to decompound. 3. Anemonella
          f. Leaves of stem alternate, or only 1 present.
            g. Small shrub with yellow wood; leaves clustered at top of stem, 1-2 times pinnately compound; sepals brown-purple. 18. Xanthorhiza
            g. Herbs.
              h. Petals and sepals present; petals yellow or rarely white, each with a nectariferous pit or scale on upper surface near base; staminodes absent; carpels becoming achenes. 1. Ranunculus
              h. Sepals present; petals neither yellow nor with a nectariferous pit or scale; or petals absent; sometimes small structures (here called staminodes, sometimes called petals) present between sepals and stamens, these sometimes nectariferous.
                i. Leaves decompound.
                  j. Flowers small, in racemes or panicles.
                    k. Flowers in short racemes; petals white; carpel single; fruit a berry. 16. Actaea
                    k. Flowers in elongate racemes or in large panicles; petals none.
                      l. Flowers often monosporangiate, in large panicles; carpels several, becoming achenes. 2. Thalictrum
                      l. Flowers bisporangiate, in simple or branched racemes; carpel usually 1, becoming a follicle; staminodes usually present. 15. Cimicifuga
                  j. Flowers 1.5-2 cm wide, solitary, axillary and terminal; carpels about 4-6, becoming follicles. 7. Isopyrum
                i. Leaves palmately parted.
                  j. Sepals 3, deciduous at anthesis; petals and staminodes none; fruit an aggregate of dark-red berries. 17. Hydrastis
                  j. Sepals 5 or more, large, not early deciduous; staminodes present; fruit an aggregate of follicles.

k. Sepals yellow; basal leaves 3-5-parted. 9. <u>Trollius</u>
k. Sepals green or tinged with red or purple; basal leaves 7-11-parted.
  11. <u>Helleborus</u>

1. <u>Ranunculus</u> L., Buttercup, Crowfoot

Leaves alternate, sometimes all basal, undivided or divided or compound; flowers solitary or in corymbose clusters; sepals 5 or fewer, green or yellow; petals 5 or rarely more or fewer, each bearing a small cup, pit, or scale near base on upper side; stamens many or rarely only 5; carpels many; fruit an aggregate of achenes.

a. Aquatics with submersed leaves dissected into filiform divisions; emersed leaves, if present, with broader segments.
  b. Petals yellow; emersed leaves, when present, about 3 times palmately lobed or parted; part of achene-margin corky-thickened. <u>R. flabellaris</u> Raf., Yellow Water C.
  b. Petals white; all leaves submersed, sessile or nearly so; achene wrinkled. <u>R. circinatus</u> Sibth. (R. longirostris Godr.), White Water C.
a. Plants of dry or wet places, not submersed; leaves not dissected into linear segments.
  b. All leaves unlobed.
    c. Blades all ovate to rotund, cordate; stamens many; petals yellow, 1-2 cm long; achenes pubescent; roots thickened. *<u>R. ficaria</u> L.
    c. Cauline blades lanceolate or narrower; achenes glabrous, sometimes roughened.
      d. Basal blades lanceolate; petals 5-10 mm long; achenes with horizontal beak 1 mm long or more; plants perennial, rooting from nodes. <u>R. ambigens</u> S. Wats.
      d. Basal blades ovate or oblong; petals about 2 mm long; achenes beakless or nearly so. <u>R. pusillus</u> Poir.
  b. Some or all of the blades lobed.
    c. Basal blades, or some of them, reniform, fan-shaped, or cordate, not lobed; flowers less than 1 cm wide.
      d. Beak of achene 0.7-1 mm long, hooked or curved; stem glabrous. <u>R. allegheniensis</u> Britt.
      d. Beak of achene minute or absent.
        e. Roots slender; stem glabrous or minutely pubescent above. <u>R. abortivus</u> L., Kidney-leaf C.
        e. Roots fusiform-thickened; stem villous below. <u>R. micranthus</u> Nutt.
    c. All blades lobed or compound.
      d. Flowers less than 1 cm broad; petals equaling or shorter than sepals.
        e. Stem glabrous, fleshy, hollow; beak of achene minute; aggregate of achenes cylindric. <u>R. sceleratus</u> L.
        e. Stem hairy; achene with obvious beak.
          f. Aggregate of achenes globose; beak of achene recurved. <u>R. recurvatus</u> Poir.
          f. Aggregate of achenes cylindric; beak of achene straight; leaves ternately compound, leaflets cleft. <u>R. pensylvanicus</u> L. f.
      d. Flowers more than 1 cm broad; petals longer than sepals.
        e. Style relatively short, stigma extending nearly to tip of ovulary, persistent on achene; beak of achene usually somewhat curved, 0.5-1.5 mm long.
          f. Base of stem enlarged and bulbous; sepals soon reflexed. *<u>R. bulbosus</u> L.
          f. Base of stem not bulbous; sepals spreading.
            g. Stem erect, tall; principal blades 5-7-sided in general outline, divided not quite to top of petiole, none of divisions stalked, ultimate ones linear. *<u>R. acris</u> L.
            g. Stem often elongate and trailing; at least some of the leaf-segments stalked (blades actually compound); blades often mottled. *<u>R. repens</u> L.
        e. Style relatively long, longer than the stigma, which is deciduous in fruit; beak of achene 1.4-3 mm long, straight or nearly so; some or all segments of blades stalked.
          f. Stem erect; some roots usually thickened.
            g. Leaves mostly basal, blades longer than wide, all divisions deeply divided, at least the terminal one stalked; petals more than twice as long as wide. <u>R. fascicularis</u> Muhl.
            g. Blades usually not longer than wide, divisions of basal ones sometimes not stalked; petals not more than twice as long as wide. <u>R. hispidus</u> Michx.
          f. Stem at first erect but eventually elongate and trailing; roots not thickened; lower leaves ternately compound, leaflets stalked; petals less than twice as long as wide. <u>R. septentrionalis</u> Poir., Swamp B.

2. Thalictrum L., Meadow-rue

Herbs, usually tall; leaves alternate, ternately decompound, leaflets often lobed; flowers small, monosporangiate or sometimes bisporangiate, in usually large panicles; sepals 4-5, white, green, or purplish, early deciduous; petals none; stamens many, filaments elongate; carpels several, separate, becoming achenes.

a. Flowers bisporangiate; filaments wide, white; anthers subglobose or oblong; achenes flat, curved, stipe about half as long as body. T. clavatum DC.
a. Plants diecious or some flowers bisporangiate; achenes not flat, sessile or stipe short.
  b. Stem-leaves petioled, scarcely expanded at anthesis; flowering in early spring; filaments slender, drooping. T. dioicum L., Early M.
  b. Stem-leaves sessile; flowering in late spring or in summer.
    c. Filaments widened upward, constricted below anthers; anthers to 1.5 mm long; stigma 1-2 mm long; achenes with short stipe, the lower deflexed, aggregate globose; carpellate plants usually with some bisporangiate flowers. T. polygamum Muhl.
    c. Filaments slender or only slightly enlarged upward, drooping; anthers more than 2 mm long; stigma 2 mm long or more.
      d. Blades with sessile or stalked glands beneath; anthers blunt or apiculate. T. revolutum DC.
      d. Blades pubescent or glabrous beneath but not glandular; anthers apiculate; lower achenes somewhat deflexed. T. dasycarpum Fisch. & Lall.

3. Anemonella Spach, Rue-anemone

A. thalictroides (L.) Spach   Small slender glabrous herb with fusiform-thickened tubers; basal leaves ternately decompound; stem-leaves 2-3, opposite or whorled, usually ternately compound, subtending a single flower or a few-flowered umbel; sepals 5-10, petaloid, white or pink, 1-1.5 cm long; petals none; stamens many; carpels few to several, becoming ribbed achenes.

4. Hepatica Mill., Liverleaf

Leaves basal, 3-lobed, evergreen, new ones of the season appearing after the flowers; flowers several, each solitary on a hairy scape, an involucre of 3 bracts so near base of flower as to appear calyxlike; sepals petaloid, several, blue, purple, pink, or white, about 1 cm long; petals none; stamens many; carpels many; fruit an aggregate of pubescent achenes.

a. Lobes of leaves rounded at tip or very blunt. H. americana (DC.) Ker
a. Lobes of leaves acute at tip. H. acutiloba DC.

5. Anemone L., Anemone

Leaves palmately compound or divided, the basal ones long-petioled; stem-leaves 2 or more, opposite or whorled in an involucre subtending the one or more peduncles which may bear secondary involucres and peduncles; sepals in ours greenish, white, or pink, several, petaloid; petals none; stamens many; carpels many, separate; achenes pubescent.

a. Involucral leaves sessile. A. canadensis L.
a. Involucral leaves petioled.
  b. Plant 1-2 dm tall, 1-flowered; aggregate fruit not elongate. A. quinquefolia L., Wood A.
  b. Plant taller, usually more than 1-flowered; fruit elongate; secondary involucres often present.
    c. Fruit cylindric, length 2-3 times width; involucral leaves irregularly toothed or cleft usually above middle only; styles 0.5 mm long. A. cylindrica Gray
    c. Fruit ellipsoid, length 1.5-2 times width; involucral leaves serrate or doubly serrate to below middle; styles 1 mm long or more. A. virginiana L., Thimbleweed

6. Clematis L., Clematis

Ours herbaceous or slightly wood vines with twining petioles or leaf rachises; leaves opposite, compound; flowers monosporangiate or bisporangiate, solitary or in panicles; sepals usually 4; petals none; staminodes sometimes present; stamens many; carpels many, separate, becoming achenes with long plumose persistent styles.

a. Flowers in panicles; sepals white, 6-12 mm long; almost diecious. C. virginiana L., Virgin's-bower
a. Flowers solitary on long peduncles, sepals longer.
   b. Calyx thick, urn-shaped, 15-25 mm long, blue- or purple-tinged. C. viorna L.
   b. Calyx thin and veiny, spreading, 3-5 cm long, pink-purple; staminodes present. C. verticillaris DC.

7. Isopyrum L., False Rue-anemone

I. biternatum (Raf.) T. & G. Slender glabrous herb; leaves alternate, ternately compound or decompound, leaflets 3-lobed; flowers solitary, axillary and terminal; sepals 5, about 1 cm long, white, petaloid; petals none; stamens 10-many; carpels usually 3-6; follicles spreading, 2-3-seeded.

8. Caltha L., Marsh-marigold

C. palustris L. Succulent perennial herb of wet places; leaves alternate, petioled, blades crenate to entire, round-cordate to reniform; flowers conspicuous; sepals yellow, 5-9, 1-2.5 cm long; stamens many; carpels few to several; follicles several-seeded.

9. Trollius L., Globe Flower

T. laxus Salisb. Herb; leaves palmately-parted, basal ones long-petioled, cauline ones alternate, short-petioled; flowers solitary, terminal; sepals 5 or more, greenish-yellow, 1-2 cm long; staminodes several to many, small; stamens many; carpels several to many, becoming follicles.

10. Coptis Salisb., Goldthread

C. groenlandica (Oeder) Fern. Low glabrous perennial with slender yellow rhizomes; leaves basal, evergreen, ternately compound, leaflets obovate, toothed, obscurely 3-lobed; scape 1-flowered; sepals 5 or more, white, petal-like; staminodes 5-7, small, hollow at summit; stamens many; carpels few to several, stipitate; follicles tipped by persistent styles.

11. Helleborus L., Hellebore

*H. viridis L., Green H. Basal leaves large, palmately parted into oblong or oblanceolate toothed segments; cauline leaves smaller; sepals 5, greenish, about 3 cm long; petals none; stamens many; several clavate hollow tubular staminodes present; carpels few to several; follicles transversely veined.

12. Aquilegia L., Columbine

Herbs; lower leaves ternately decompound, leaflets lobed; flowers showy, solitary at ends of branches; perianth regular; sepals 5, same color as petals; petals 5, spurred, spurs projecting backward between sepals; stamens many; carpels usually 5, styles slender; fruit an aggregate of follicles.

a. Perianth red and yellow. A. canadensis L., Wild C.
a. Perianth blue, purple, pink, or white. *A. vulgaris L., Garden C.

13. Delphinium L., Larkspur

Leaves palmately divided; flowers in racemes or panicles; perianth zygomorphic; sepals 5, petal-like, upper one spurred; petals 2 or 4, all or the 2 upper prolonged into spur of sepal; stamens many; carpels 1-5, becoming follicles.

a. Carpel and follicle 1; petals 2, united; flowers blue, violet, pink, or white; leaves dissected into linear segments; annual. *D. ajacis L.
a. Carpels 3; petals 4, separate; flowers blue, purple, or white; perennial.
   b. Follicles divergent at maturity; roots thickened; raceme few-flowered, usually not branched; middle portion (between two deepest sinuses) of middle and lower cauline blades cleft into narrow segments with parallel sides. D. tricorne Michx., Dwarf L.

b. Follicles erect at maturity; roots not thickened; flowers many; middle portion (described on the opposite page) of middle and lower cauline blades with triangular broad-based lobes or teeth. D. exaltatum Ait., Tall L.

14. Aconitum L., Monkshood

A. noveboracense Gray   Herb; leaves alternate, palmately cleft; flowers 1.5-2 cm wide, usually in a raceme; perianth blue, zygomorphic; sepals 5, petal-like, upper largest and hooded or spurred on back; upper 2 petals small, concealed under hooded sepal, other 3 minute or absent; stamens many; carpels 3-5, separate, becoming follicles.

15. Cimicifuga L.

C. racemosa (L.) Nutt., Black Cohosh.   Tall herb; leaves large, ternately and pinnately decompound, leaflets toothed, incised, or lobed; flowers small, in elongate terminal simple or branched racemes; perianth regular; sepals 4-5, white, petaloid, early deciduous; petals none; staminodes 1-several, clawed, 2-lobed; stamens many; carpels usually 1 (2-3), style short and thick, stigma truncate; follicle transversely-veined.

16. Actaea L., Baneberry

Herbs; leaves large, ternately decompound, leaflets ovate, sharply toothed or incised; flowers small, white, in short dense terminal raceme, axis and pedicels of which elongate in fruit; perianth regular; sepals 3-5, early deciduous; petals few to several; stamens many, filaments long and slender; carpel single; stigma sessile, broad, 2-lobed; fruit a several-seeded berry.

a. Fruiting pedicels about as thick as axis of fruiting raceme; fruit usually white with red stigma.
   A. pachypoda Ell., White B.
a. Fruiting pedicels slender; fruit red.   A. rubra (Ait.) Willd., Red B.

17. Hydrastis Ellis, Golden Seal

H. canadensis L.   Herb with yellow rhizome; leaves 3, 1 basal, 2 near summit of stem, alternate, rounded, cordate at base, palmately 5-7-lobed, lobes serrate or incised; flowers solitary, terminal; calyx regular, of 3 greenish-white early-deciduous sepals; petals none; stamens several to many; carpels several to many, 2-ovuled; fruit an aggregate of dark-red 1-2-seeded berries.

18. Xanthorhiza Marsh., Yellow-root

*X. simplicissima Marsh.   Low shrub; stem unbranched; leaves pinnately compound, crowded on short terminal stem-segment; flowers small, purple-brown, in racemes or panicles; sepals 5, about 3 mm long; petals none; staminodes 5; carpels several, becoming follicles.

BERBERIDACEAE, Barberry Family

Herbs and shrubs; leaves alternate, sometimes basal; flowers hypogynous, bisporangiate; perianth (or each of its whorls) regular; sepals 4-6 or more, often early deciduous; petals 6-9, separate, sometimes glandlike; except in Podophyllum, stamens as many as petals and opposite them and anthers opening by uplifted lids; ovulary 1-loculed, placenta 1, parietal, or ovules basal; fruit a berry or capsule.

a. Shrubs; flowers yellow.
   b. Leaves pinnately compound. *Mahonia
   b. Leaves simple; branches with spines. 4. Berberis
a. Herbs.
   b. Leaves all basal, leaflets 2; flowers on leafless scapes. 2. Jeffersonia
   b. Flowering stem with 2 palmately divided leaves. 1. Podophyllum
   b. Leaves 2, ternately compound or decompound. 3. Caulophyllum

1. Podophyllum L., May-apple

P. peltatum L. Flowering stem with 2 leaves deeply palmately parted; flowerless plants with one peltate, radially parted, leaf; flowers solitary, waxy white, 3-5 cm wide; sepals 6, early deciduous; petals 6-9; stamens more than petals, often twice as many; stigma sessile, large and thick; fruit a fleshy berry 4-5 cm long.

2. Jeffersonia Bart., Twinleaf

J. diphylla (L.) Pers. Glabrous; leaves all basal, immature at anthesis, long-petioled, with 2 half-ovate leaflets; flowers 2-3 cm wide; perianth early deciduous; sepals 4; petals 8, white; stigma 2-lobed; capsule obovoid, half-circumscissile, the top a lid; seeds many, in several rows on placenta, ariled.

3. Caulophyllum Michx., Blue Cohosh

C. thalictroides (L.) Michx. Glabrous, with thick rhizome; stem simple; leaves 2, sessile, ternate, the lower large, decompound, a smaller one subtending the short raceme; flower 1 cm wide or less, yellow, green, or purple; sepals 6, bracts 3 or 4; petals 6, glandlike, smaller than sepals; stamens 6; ovulary soon bursting, the 2 drupelike blue seeds maturing uncovered.

4. Berberis L., Barberry

Shrubs; foliage leaves on very short stems in axils of simple or 3-branched spines; sepals 6, petaloid, subtended by some bracts; petals 6, yellow, each with 2 glands at base; stamens 6; fruit a berry, 1-few-seeded.

a. Leaf-blades entire; flowers in small umbels or solitary; spines unbranched or lateral branches
   short. *B. thunbergii DC., Japanese B.
a. Leaf-blades toothed; flowers in racemes; spines often branched.
   b. Racemes few-flowered, short; petals notched. B. canadensis Mill., American B.
   b. Racemes many-flowered, elongate; petals entire. *B. vulgaris L., Common B.

MENISPERMACEAE, Moonseed Family

Menispermum L., Moonseed

M. canadense L. Slender twining woody vine; leaves alternate, simple, blades broadly ovate to orbicular, cordate or truncate at base, entire to palmately-lobed, peltate, petiole attached near base of blade; diecious; flowers minute, greenish, in axillary panicles; sepals 4-8; petals 6-8; stamens 12-24; carpels 2-4, separate; fruit an aggregate of drupes.

MAGNOLIACEAE, Magnolia Family

Deciduous or evergreen trees and shrubs; leaves alternate, simple, stipules covering vegetative buds and leaving complete rings as scars; flowers large, hypogynous, bisporangiate, solitary; perianth parts similar, separate, in circles of 3, each circle regular; stamens many, spirally arranged; carpels many, on an elongate receptacle, separate or somewhat united; fruit a conelike aggregate.

a. Leaf-blades broadly emarginate or truncate at apex, usually shallowly lobed. Liriodendron
a. Leaf-blades rounded to acute at apex, not truncate or emarginate, entire. Magnolia

Magnolia L., Magnolia

Deciduous or evergreen trees and shrubs; leaf-blades entire; perianth of 9-18 similar segments; fruit an aggregate of follicles, each with 1 or 2 red berrylike seeds which remain on the threadlike attachments for some time after dehiscence of follicles:

a. Flowers appearing after or with the leaves.
   b. Leaves evergreen, coriaceous; flowers white, to 2 dm wide; trees. *M. grandiflora L., Bull Bay
   b. Leaves deciduous.
      c. Shrub or small tree; blades glaucous beneath, elliptic to oblong, to about 1 dm long; buds slender, pointed; flowers white, to 7 cm wide. *M. virginiana L., Sweet Bay

c. Trees, native; leaves larger.
    d. Blades cordate or auricled at base, white and finely pubescent beneath, to 1 m long; buds and twigs downy; flowers to 3 dm wide, petals white, purplish at base. <u>M</u>. <u>macrophylla</u> Michx., Bigleaf **M.**
    d. Blades rounded or tapering at base, paler but not white beneath.
        e. Blades 3-6 dm long; buds and twigs glabrous; flowers white, to 2.5 dm wide, with unpleasant odor. <u>M</u>. <u>tripetala</u> L. Umbrella **M.**
        e. Blades usually less than 2.5 dm long; twigs glabrous or nearly so; buds silky-pubescent; flowers greenish-yellow, less than 10 cm wide. <u>M</u>. <u>acuminata</u> L., Cucumber **M.**
a. Flowers appearing before the leaves; buds densely pubescent; blades obovate.
  b. Sepals and petals alike, 12-18, white. *<u>M</u>. <u>stellata</u> (Sieb. & Zucc.) Maxim.
  b. Sepals 3, smaller than the petals; petals purplish, rarely white. *<u>M. X soulangeana</u> Soul.

<u>Liriodendron</u> L., Tuliptree, Tulip-poplar

<u>L</u>. <u>tulipifera</u> L. Tree; sepals 3; petals 6, greenish-yellow with orange blotch, upright, 4-5 cm long; mature carpels samaralike, 1-2-seeded.

CALYCANTHACEAE, Calycanthus Family

<u>Calycanthus</u> L., Sweet Shrub, Carolina Allspice

Shrubs; leaves opposite, simple; blades ovate, entire; flowers perigynous, bisporangiate, regular, solitary; perianth parts many, separate, similar, lanceolate, maroon, attached at or below rim of the concave hypanthium (receptacle); stamens several to many; carpels several to many, separate, within the concavity; fruit an aggregate of achenes enclosed by the hypanthium.

a. Leaves softly pubescent beneath; flowers very fragrant. <u>C</u>. <u>floridus</u> L.
a. Leaves glabrous or almost so beneath, sometimes glaucous; flowers somewhat fragrant. <u>C</u>. <u>fertilis</u> Walt.

ANNONACEAE, Custard-apple Family

<u>Asimina</u> Adans., Pawpaw

<u>A</u>. <u>triloba</u> (L.) Dunal Small tree with alternate 2-ranked simple leaves; blades entire, obovate to oblanceolate; flowers hypogynous, bisporangiate, short-peduncled, 3-4 cm wide, from buds in axils of last year's leaves, appearing before or with young leaves of the season; perianth regular; sepals 3; petals 6, in 2 whorls, separate, maroon; stamens many, spirally arranged, filaments short, anthers 4-loculed; carpels few to several, separate, each of the 1-few that mature becoming a fleshy several-seeded berry 5-12 cm long.

LAURACEAE, Laurel Family

Trees and shrubs; spicy-aromatic; leaves alternate, simple, petioled; flowers small, yellow or greenish-yellow, usually monosporangiate, the carpellate with vestigial stamens, the staminate with vestigial carpel; perianth (here called calyx) regular, of 6 segments; staminate flowers with 9 stamens, the 3 inner ones with glands at base; anthers opening by uplifted lids; carpel 1; fruit a 1-seeded ellipsoid drupe.

a. Blades unlobed, 2-lobed, and 3-lobed on same tree; flowers pediceled, in terminal clusters; stems up to several years old, green; anthers 4-loculed; diecious. <u>Sassafras</u>
a. Blades not lobed; flowers sessile or nearly so, in lateral clusters; stems older than 1 year, not green; anthers 2-loculed; diecious or imperfectly so. <u>Lindera</u>

<u>Sassafras</u> Nees

<u>S</u>. <u>albidum</u> (Nutt.) Nees Small trees; twigs green; dioecious; flowers from buds at apex of branches of previous year, appearing before or with young leaves; calyx persistent in fruit; anthers opening by 4 uplifted lids; carpellate flowers with 6 rudimentary stamens; fruit blue.

Lindera Thunb.

L. benzoin (L.) Blume, Spicebush. Shrubs; leaf-blades obovate or elliptic, acute at base; mostly diecious; flowers appearing before the leaves; calyx deciduous in fruit; anthers opening by 2 valves; carpellate flowers with several staminodes in 2 forms; fruit red.

PAPAVERACEAE, Poppy Family

Ours herbs, sometimes with milky juice; leaves alternate or basal, rarely opposite; flowers hypogynous, bisporangiate; sepals 2 or 3, minute or early deciduous; corolla regular (or each whorl regular); petals 4 to many or rarely none, separate; stamens several to many; carpels 2 to many, united, stigmas as many; ovulary 1-loculed or occasionally 2-loculed as result of a false septum; placentae parietal; ovules several to many; fruit a capsule.

a. Flowers scapose, solitary; petals white; leaf one, basal. 1. Sanguinaria
a. Stem leafy; petals usually red, yellow, or cream-color, or none.
   b. Petals none; flowers in elongate terminal panicles. 4. Macleaya
   b. Petals present; flowers solitary or in umbels.
      c. Stigmas and placentae 4 to many; capsule opening by pores or slits near the summit; petals of
         various colors. 5. Papaver
      c. Stigmas and placentae 2-4; capsule opening by longitudinal slits; petals yellow.
         d. Ovulary hairy; stigmas and placentae 2-4; fruit ellipsoid. 2. Stylophorum
         d. Ovulary glabrous; stigmas and placentae 2; fruit linear. 3. Chelidonium

1. Sanguinaria L., Bloodroot

S. canadensis L. Low herb with stout rhizome; juice red; leaf 1, blade somewhat palmately lobed, circular; flower scapose, solitary, 2-5 cm wide; sepals 2, separate, early deciduous; petals white, early deciduous, 8-12 or fewer or more; stamens many; style short; carpels, stigmas, and placentae 2; capsule fusiform.

2. Stylophorum Nutt.

S. diphyllum (Michx.) Nutt. Wood-poppy. Rhizome stout; leaves pinnatifid, several basal, 2 cauline, opposite; flowers about 5 cm wide; sepals 2, pubescent; petals 4, deep yellow; stamens many; carpels 2-4; capsule ellipsoid, pubescent.

3. Chelidonium L., Celandine

*C. majus L. Leaves glaucous beneath, pinnatifid, alternate; flowers pediceled, several in peduncled umbels; sepals 2; petals 4, yellow, about 1 cm long; stamens many; carpels 2; capsule linear-cylindric.

4. Macleaya R. Br., Plume Poppy

*M. cordata (Willd.) R. Br. Leaves palmately lobed, glaucous beneath, petioled; flowers in elongate terminal panicles; sepals 2, cream-color; petals none; stamens many; carpels 2, united; stigma sessile; capsule oblanceolate, 2-loculed.

5. Papaver L., Poppy

Sepals 2; petals usually 4; stamens many; carpels 4 to many; stigmas in a sessile crown with rays as many as carpels; capsule many-seeded, dehiscent near the top with slits or pores.

a. Cauline leaves cordate, clasping, incised and toothed, glabrous. *P. somniferum L., Common or
   Opium P.
a. Cauline leaves pinnatifid, not clasping, segments lanceolate, incised and toothed.
   b. Capsule obovoid, about twice as long as wide; rays of stigma 5-9. *P. dubium L.
   b. Capsule subglobose; rays of stigma 8-14. *P. rhoeas L., Corn P.

FUMARIACEAE (PAPAVERACEAE, in part), Fumitory Family

Herbs; leaves alternate or all basal, usually decompound or dissected; flowers hypogynous, bisporangiate, bracted, in racemes; sepals 2, sometimes early deciduous, minute; petals 4, one or both of the 2 outer ones spurred, the 2 inner narrow below, wider and somewhat connivent above; stamens 6, in 2 sets of 3 each, in each set filaments united, anther of middle one 2-loculed, of the outer ones 1-loculed; carpels 2, style 1, ovulary 1-loculed; placentae 2, parietal.

a. Corolla isobilateral; both outer petals spurred or saccate at base.
  b. Slender climbing vines; stem leafy; leaves alternate. 1. Adlumia
  b. Not climbing; leaves all basal; racemes scapose. 2. Dicentra
a. Corolla zygomorphic; only 1 outer petal spurred.
  b. Sepals not toothed; ovulary and capsule elongate; seeds several. 3. Corydalis
  b. Sepals toothed; ovulary and indehiscent fruit subglobose; seed 1. 4. Fumaria

1. Adlumia Raf., Climbing Fumitory

A. fungosa (Ait.) Greene Glabrous vine with alternate decompound leaves; flowers in drooping axillary panicles; petals pink, united in a tube; stamens united in a tube adherent to corolla; capsule slender, covered by persistent corolla.

2. Dicentra Bernh., Bleeding-heart

Delicate, somewhat fleshy, herbs; leaves dissected, all basal in our species; corolla about 2 cm long, white or tinged with pink; capsule linear.

a. Spurs of corolla short, rounded; flowers fragrant; underground stems yellow pea-shaped tubers.
  D. canadensis (Goldie) Walp, Squirrel-corn
a. Spurs of corolla long, divergent, tapering to base; underground stems white or pink granulate bulbs.
  D. cucullaria (L.) Bernh., Dutchman's-breeches

3. Corydalis Medic., Corydalis

Stems leafy; leaves decompound; flowers zygomorphic; capsule many-seeded; seeds with crests or arils.

a. Flowers pink with yellow tips; stem erect. C. sempervirens (L.) Pers., Pink C.
a. Flowers yellow; stems diffuse.
  b. Outer petals keeled but not crested on back; corolla 12-15 mm long, golden yellow. C. aurea Willd., Golden C.
  b. Outer petals with toothed crest on back; corolla 6-9 mm long, pale yellow. C. flavula (Raf.) DC., Pale C.

4. Fumaria L., Fumitory

*F. officinalis L. Stem diffusely branched; leaves decompound; flowers zygomorphic; sepals somewhat toothed; petals pink- or red-purple with darker tips; fruit 1-seeded, globose, 2-3 mm wide, indehiscent.

CAPPARIDACEAE, Caper Family

Herbs; leaves alternate, palmately compound, petioled; flowers hypogynous, bisporangiate, zygomorphic, in racemes; receptacle bearing a gland; sepals 4; petals 4, clawed; stamens 6 to many, rarely fewer than 6, filaments elongate; carpels 2, united; locule 1, style 1; placentae 2, parietal; fruit an elongate capsule; seeds many, reniform.

a. Stamens 8 to many; fruit hardly stipitate in the calyx. Polanisia
a. Stamens usually 6, rarely fewer; fruit long-stipitate in the calyx. Cleome

Polanisia Raf., Clammy-weed

P. graveolens Raf. Viscid-pubescent; leaflets 3, oblong, entire; petals emarginate at tip, white or nearly so; stamens several, filaments purple; at least the lower flowers of raceme subtended by leaflike bracts.

111

Cleome L., Spider Flower

Leaflets 3-7, lanceolate or oblong; bracts subtending flowers smaller than the leaves; flowers long-stalked; petals white or pink; filaments very long.

a. Leaflets 3, entire or nearly so; plant glabrous or pubescent but not viscid. C. serrulata Pursh
a. Leaflets 5-7, often serrulate; plant viscid-pubescent; spines usually present at base of petioles.
   *C. spinosa L.

CRUCIFERAE, Mustard Family

Herbs; leaves alternate or rarely opposite or whorled, simple or compound, without stipules; flowers hypogynous, bisporangiate, in racemes or corymbs; perianth regular or rarely zygomorphic; sepals 4, separate; petals 4, separate, rarely none; stamens 6, 2 shorter than the others, or rarely stamens 4 or 2; carpels united, style 1, stigma usually 2-lobed; ovulary 2-loculed; placentae parietal; fruit linear and much longer than wide (a silique) or short (a silicle); seeds 1 or 2 to many, when several, attached in 2 rows in each locule but lying in either 1 or 2 rows.

Approximate size of flowers is indicated in keys and descriptions as follows: small, petals 4 mm long or less; medium, petals 5-8 mm long; large, petals 9 mm long or more. The term beak, as used here, refers to the terminal projection beyond tip of valves of fruit.

KEY TO PLANTS WITH BOTH FLOWERS AND FRUITS

a. Fleshy plants of lake shores; fruit with a cross-septum, the two portions indehiscent, the upper separating, usually containing a seed, lower often seedless; petals purple. 14. Cakile
a. Plants not as above; fruit (with rare exceptions) with lengthwise septum separating 2 locules side by side.
  b. Fruit (exclusive of beak) less than 3 times as long as wide.
    c. Fruit flattened parallel to septum which is as wide as fruit.
      d. Fruit 1.5 cm wide or more; corolla pink to purple; flowers large; lower leaves opposite or nearly so. 13. Lunaria
      d. Fruit less than 1.5 cm wide; leaves alternate or all basal.
        e. Sepals persistent in fruit; fruit circular; plant stellate-pubescent. 4. Alyssum
        e. Sepals deciduous in fruit.
          f. One seed in each locule; petals white, entire. 3. Lobularia
          f. More than 1 seed in each locule; petals bifid, notched, or entire.
            g. Stem less than 3 dm tall, sometimes scapose; seeds wingless; rosette present at anthesis. 1. Draba
            g. Stem taller, leafy; seeds narrowly winged; plant canescent; no well-developed rosette present at anthesis. 2. Berteroa
    c. Fruit not, or slightly, flattened, or flattened at right angle to narrow septum.
      d. Fruit rough and wrinkled or reticulate and pitted, indehiscent.
        e. Petals white, small; fruit notched at apex, wider than long; leaves 1-2-pinnatifid; stem decumbent or spreading. 8. Coronopus
        e. Petals yellow; stem and leaves not as above.
          f. Fruit reticulate, pitted, globose; leaves lanceolate, sagittate-clasping at base; pubescent with forked hairs. 11. Neslia
          f. Fruit wrinkled, ovoid with pointed beak; coarse plant; basal leaves large, lyrate-pinnatifid. 12. Bunias
      d. Fruit not rough and wrinkled, not reticulate and pitted.
        e. Fruit flattened.
          f. Fruit triangular at maturity (oblong to globose when young), emarginate at apex, narrowed to base; plant with forked hairs. 9. Capsella
          f. Fruit circular, ovoid, or obovoid.
            g. Corolla somewhat zygomorphic, the 2 petals on outer side of flower larger. *Iberis
            g. Corolla regular or petals absent.
              h. More than 1 seed in each locule of fruit. 5. Thlaspi
              h. One seed in each locule of fruit.
                i. Fruit broadly ovoid or heart-shaped, tapering to short point and beaked at apex; cauline leaves clasping. 7. Cardaria
                i. Fruit almost circular, notched at apex, style, if present, within the notch; leaves clasping or not. 6. Lepidium

    e. Fruit not, or only slightly, flattened.
        f. Plant with stellate or forked hairs; petals yellow or cream; stem-leaves sagittate-clasping. 10. <u>Camelina</u>
        f. Plants glabrous or hairs simple.
            g. Petals purple, pink, or white.
                h. Petals white, small; basal leaves 1-2 dm long, or plant aquatic with dissected leaves; fruit ellipsoid. 28. <u>Armoracia</u>
                h. Petals pink-purple to white, large; fruit oblong, ending in conical beak, 0.5-1 cm thick. 15. <u>Raphanus</u>
            g. Petals yellow.
                h. Fruit with flattened beak about as long as pubescent body; flowers large. 16. <u>Brassica</u>
                h. Fruit ovoid, ellipsoid, or short-cylindric, beak small; flowers small or medium. 26. <u>Rorippa</u>
b. Fruit (exclusive of beak) 3 to many times as long as wide.
    c. Cauline leaves clasping by cordate, auricled, or sagittate bases.
        d. Petiole-base clasping; blades acuminate, sharply dentate, sometimes with 1-2 small basal lobes on each side; petals purple. 30. <u>Iodanthus</u>
        d. Blades without petioles.
            e. Petals yellow.
                f. Blades entire, rounded at apex; plant glabrous; petals pale yellow; fruit 4-angled, to 1 dm long. 19. <u>Conringia</u>
                f. At least the lower blades incised or pinnatifid.
                    g. Essentially glabrous; beak of fruit not more than 3 mm long; leaves of basal rosette lyrate-pinnatifid, terminal lobe large and rounded. 29. <u>Barbarea</u>
                    g. Glabrous or pubescent; beak of fruit more than 3 mm long, usually 8 mm or more; rosette leaves not as above or absent at anthesis. 16. <u>Brassica</u>
            e. Petals white, greenish, or cream yellow; plant glabrous or pubescent, usually some hairs, when present, forked; fruits 1.5-10 cm long, appressed or spreading; seeds often winged. 36. <u>Arabis</u>
    c. Cauline leaves absent or not clasping; if sessile, then bases not cordate, auricled, or sagittate.
        d. Petals purple, pink, white, or creamy-white.
            e. Leaves, at least the lower, compound, deeply cleft, or lobed.
                f. Leaves palmately compound or palmately cleft. 33. <u>Dentaria</u>
                f. Leaves, or some of them, pinnately compound or pinnately cleft or lobed.
                    g. Petals 1-2 cm long; fruit 5-10 mm thick, with conical pointed beak, indehiscent, or fruit more slender and cross-septate. 15. <u>Raphanus</u>
                    g. Petals and fruit not as above.
                        h. Blades, at least the lower, 2-3 times pinnate. 23. <u>Descurainia</u>
                        h. Blades once pinnate or only lobed or cleft.
                            i. Aquatic; leaves pinnately compound, leaflets oval to circular; fruiting pedicels spreading; 2 rows of seeds in each locule. 27. <u>Nasturtium</u>
                            i. Not aquatic; if leaves actually compound, then pedicels ascending; 1 row of seeds in each locule.
                                j. Cauline blades not pinnatifid or pinnate; fruit pendent, curved, flattened, 7-10 cm long, petals creamy-white, <u>or</u>, fruit 2-5 cm long, spreading or ascending, petals white, pink, or purple. 36. <u>Arabis</u>
                                j. Cauline blades pinnatifid or pinnate, or leaves all basal.
                                    k. Scapes several from a basal rosette, each bearing 1 to few flowers; fruit 4-5 mm wide, 2-3 cm long. 32. <u>Leavenworthia</u>
                                    k. At least some cauline leaves present; fruit narrower.
                                        l. Seed narrowly winged; fruit 1.5-2 mm wide; petals 3 mm long. 35. <u>Sibara</u>
                                        l. Seed wingless; either fruit more slender or petals much longer. 34. <u>Cardamine</u>
            e. Leaves simple, not lobed or cleft (or, rarely, with 1-2 small basal lobes on each side).
                f. Foliage with odor of onion; stem-blades triangular-ovate. 20. <u>Alliaria</u>
                f. Foliage not with odor of onion; blades not triangular-ovate.
                    g. Basal rosettes present at anthesis.

113

h. Fruits long, recurved, and flattened, or fruits 2-5 cm long and petals 5-8 mm long. 36. Arabis
h. Fruits not more than 2 cm long; petals smaller.
  i. Fruit not more than 5 times as long as wide, flattened parallel to septum; seeds in 2 rows in each locule. 1. Draba
  i. Fruit much more than 5 times as long as wide; seeds in 1 row in each locule. 22. Arabidopsis
g. Basal rosettes not present at anthesis but some long-petioled basal leaves may be present.
  h. Fruit with beak 1 cm long or more, at maturity constricted and cross-septate at intervals; petals large, blue-purple. 31. Chorispora
  h. Fruit with short beak or none, not cross-septate.
    i. Fruit at maturity 2-5 mm long and about 1 mm wide; plant small. 1. Draba
    i. Fruit larger than above.
      j. Cauline blades entire, undulate, or remotely toothed; fruit elastically dehiscent; glabrous or hairs simple. 34. Cardamine
      j. Cauline leaves sharply, sometimes doubly, toothed; fruit not elastically dehiscent.
        k. Glabrous or hairs simple; blades sometimes with small lobes at base. 30. Iodanthus
        k. At least some of the hairs forked; blades not lobed at base. 24. Hesperis
d. Petals yellow or rarely orange.
  e. Pubescence at least partly of forked or stellate hairs.
    f. Blades pinnately compound or decompound. 23. Descurainia
    f. Blades simple, sometimes pinnatifid. 25. Erysimum
  e. Pubescence of only simple hairs, or plants glabrous.
    f. Raceme bracted with leaflike pinnatifid bracts. 17. Erucastrum
    f. Raceme bractless or bracted only at very base.
      g. Beak of fruit 5 mm long or more, sometimes half as long as body or more; flowers large.
        h. Petals 1.5-2 cm long; fruit with transverse septa. 15. Raphanus
        h. Petals about 1 cm long; fruit without transverse septa, each valve strongly 3-nerved, or strongly 1-nerved with anastomosing side nerves. 16. Brassica
      g. Beak of fruit less than 5 mm long.
        h. Fruits closely appressed to axis at maturity.
          i. Petals about 3 mm long; fruit 1.5 cm long, tapering from base to tip. 21. Sisymbrium
          i. Petals about 1 cm long; body of fruit not tapering, each valve with 1 prominent midvein and anastomosing side veins. 16. Brassica
        h. Fruits not closely appressed to axis at maturity.
          i. Fruit linear to ellipsoid, 1.5 cm long at most, sometimes curved, valves nerveless or obscurely nerved; flowers small; seeds in 2 irregular rows or in 1 row in each locule. 26. Rorippa
          i. Fruit linear; flowers medium to large.
            j. Fruit 1-10 cm long; valves 3-nerved; seeds in 1 row. 21. Sisymbrium
            j. Fruit 2-3.5 cm long; valves 1-nerved; seeds in 2 rows. 18. Diplotaxis

1. Draba L., Whitlow-grass

Low plants; hairs stellate, forked, or simple or of more than one type; blades entire or dentate; flowers usually small; petals white to yellow, bifid, emarginate, or entire, sometimes lacking; fruit oval or oblong, flattened parallel to the broad septum; seeds in 2 rows in each locule, wingless.

a. Petals lobed to the middle; leaves all basal. *D. verna L., Early W.
a. Petals entire or only notched; at least 2 stem-leaves present.
  b. Fruit about 1 mm wide, 2-4 mm long; basal leaves few or absent. D. brachycarpa Nutt.
  b. Fruit wider and usually longer; basal rosettes present.
    c. Blades entire; inflorescence-axis and pedicels glabrous. D. reptans (Lam.) Fern.
    c. Blades dentate; inflorescence-axis and pedicels pubescent; petals sometimes absent. D. cuneifolia Nutt.

114

2. Berteroa DC., Hoary Alyssum

*B. incana (L.) DC. Plant gray-green, canescent, the pubescence stellate; blades entire, lanceolate, acute; flowers small; petals white, bifid; fruit ellipsoid, 5-8 mm long, about half as wide, somewhat flattened parallel to septum; seeds several in each locule.

3. Lobularia Desv., Sweet Alyssum

*L. maritima (L.) Desv. Stem weak, somewhat ascending; blades entire, linear to oblanceolate, acute; hairs of stem and leaves forked; flowers small, fragrant; petals white; fruit oval or circular, flattened parallel to septum; 1 seed in each locule.

4. Alyssum L.

*A. alyssoides L. Stellate-pubescent annual; stems usually branched at base; leaves oblanceolate or linear, obtuse, entire; flowers very small; sepals persistent; petals yellow or whitish; fruit circular, flattened parallel to septum; 2 seeds in each locule.

5. Thlaspi L., Penny-cress

Cauline leaves sessile or clasping, with sagittate bases; flowers small; petals white in ours; fruit circular, obovate or obcordate, winged, flattened at right angle to septum; seeds 2-several in each locule.

a. Plants glabrous.
  b. Stem to 5 dm tall; stem-blades oblong or lanceolate, auricles acute; fruit 10-14 mm long, winged all around. *T. arvense L., Field P.
  b. Stem to 3 dm tall; stem-blades ovate, auricles blunt; fruit 4-6 mm long, wing wider at summit. *T. perfoliatum L., Perfoliate P.
a. Stems hairy below, grooved; stem-leaves with acute auricles; fruit narrowly winged; with odor of garlic; rare. *T. alliaceum L.

6. Lepidium L., Pepper-grass

Glabrous or pubescent with simple hairs; leaves entire to pinnatifid; flowers small; petals white, yellow, or green, sometimes none; stamens 6, 4, or 2; style short or none; fruit circular or slightly elongate, flattened at right angle to septum, wingless or somewhat winged; 1 seed in each locule.

a. Stem-leaves cordate or sagittate-clasping; stamens usually 6.
  b. Upper leaves broadly ovate, deeply cordate-clasping; basal and lower stem-leaves dissected into linear divisions. *L. perfoliatum L.
  b. Stem-leaves lanceolate or oblong, many, crowded, clasping by sagittate bases; basal leaves entire to pinnatifid. *L. campestre (L.) R. Br., Field P.
a. Stem-leaves not cordate- or sagittate-clasping; stamens 4 or 2.
  b. Petals present, about as long as sepals. L. virginicum L., Virginia P.
  b. Petals absent or rudimentary and shorter than sepals.
    c. Plants fetid; basal leaves bipinnatifid; petals none. *L. ruderale L.
    c. Plants nearly scentless; basal leaves toothed or pinnatifid; petals none or rudimentary.
      *L. densiflorum Schrad.

7. Cardaria Desv., Hoary-cress

*C. draba (L.) Desv. Leaves oblong, usually dentate, the cauline auriculed, clasping; flowers small, racemes several; petals white; fruit ovoid, somewhat cordate or reniform, tapering at top to persistent style about 1 mm long; seed 1 in each locule.

8. Coronopus Zinn., Wart-cress

*C. didymus (L.) Sm. Diffuse; leaves once to twice pinnatifid; flowers small; petals white; fruit rugose, notched at base and apex, indehiscent, wider than high, style and stigma shorter than notch, 1 seed in each locule.

9. Capsella Medic., Shepherd's Purse

*C. bursa-pastoris (L.) Medic. Hairs forked; rosette leaves oblong, toothed or pinnatifid; cauline leaves lanceolate or linear, sessile with auricled bases; flowers small; petals white; fruits triangular, emarginate or obcordate, flattened at right angle to septum, several seeds in each locule.

10. Camelina Crantz, False-flax

Cauline leaves linear to lanceolate, sagittate-clasping, basal ones spatulate, entire or repand; flowers small; petals yellow; fruit obovoid; style persistent, conspicuous in fruit; seeds many, 2 rows in each locule.

a. Stem and leaves rough-pubescent with shorter stellate hairs and longer simple ones. *C. microcarpa Andrz.
a. Stem glabrous or pubescent with short stellate and simple hairs. *C. sativa (L.) Crantz

11. Neslia Desv., Ball-mustard

*N. paniculata (L.) Desv. Pubescent with branched hairs; leaves oblong, scabrous, nearly entire, sagittate-clasping; inflorescence branched; flowers small; petals yellow; fruit globose, slightly flattened, indehiscent, reticulate.

12. Bunias L.

*B. orientalis L. Tall; sparsely covered with low rounded protuberances; leaves dentate to pinnatifid, the lower large; petals yellow, medium-sized; fruit asymmetric, rugose, ovoid, 7-10 mm long.

13. Lunaria L., Honesty

*L. annua L. Lower leaves sometimes opposite; blades ovate, cordate, dentate; petals large, pink-purple; fruit broadly elliptic, flat, rounded at ends, to 5 cm long, about two-thirds as wide; silvery septum persistent after valves have fallen.

14. Cakile Mill., Sea-rocket

C. edentula (Bigel.) Hook. Fleshy glabrous annual of lake shores; branched stems spreading or ascending; leaves obovate or oblanceolate, entire, sinuate-toothed, or lobed; flowers of medium size; petals purple; fruits jointed, the upper segment longer and separating from the lower at maturity, both indehiscent, 1-loculed, the upper 1-seeded, the lower often seedless.

15. Raphanus L., Radish

Erect; at least the lower leaves lyrate-pinnatifid; flowers large; petals purple, white, or yellow, obovate, clawed; fruit indehiscent, long-beaked, actually with transverse septum near base, separating it into 2 parts, the lower abortive or small and usually seedless, the upper spongy or constricted between the seeds.

a. Petals yellow, in some forms violet or white; fruit 4-6 mm thick, when ripe constricted between the seeds. *R. raphanistrum L., Wild R.
a. Petals pink or pale purple; fruit 5-10 mm thick, little if at all constricted between the few seeds. *R. sativus L., Garden R.

16. Brassica L., Mustard

Coarse; at least the lower leaves incised or pinnatifid, the upper usually entire or dentate; flowers large; petals in ours yellow; fruits terete or angled, often constricted between the seeds, tipped with a conspicuous 2-edged or conical, sometimes 1-seeded, beak; seeds globose, wingless, 1 row in each locule.

a. Upper leaves clasping.
  b. Leaves glaucous, glabrous or nearly so; annual; root slender. *B. campestris L., Field M.
  b. Leaves green, lower ones pubescent; biennial; root thickened. *B. rapa L., Turnip.

116

a. Upper leaves not clasping.
   b. Beak of fruit flat or 4-angled, 2-edged, 3-nerved on each half, sometimes 1-seeded, almost half
      to as long as body of fruit; each valve 3-nerved.
      c. Fruit hispid-pubescent; beak flat, as long as body of fruit. *B. hirta Moench, White M.
      c. Fruit glabrous or sparsely pubescent; beak somewhat 4-angled, much shorter than body of
         fruit. *B. kaber (DC.) L. C. Wheeler
   b. Beak of fruit slender, not 2-edged, 1-nerved on each half or nerveless, without a seed, less than
      half as long as body; each valve with 1 strong nerve.
      c. Plant green, pubescent; pedicels and fruits appressed. *B. nigra (L.) Koch, Black M.
      c. Plant glaucous, glabrous; pedicels spreading, fruits ascending. *B. juncea (L.) Coss., In-
         dian M.

17. Erucastrum Presl

*E. gallicum (Willd.) O. E. Schulz   Stem pubescent, hairs simple; leaves pinnatifid or bipin-
natifid; racemes leafy; flowers of medium size; petals yellow; pedicels and fruits ascending; fruit
somewhat curved, as much as 3 cm long, each valve 1-nerved; seeds many, 1 row in each locule.

18. Diplotaxis DC.

Glabrous or nearly so; leaf-blades toothed to deeply pinnatifid; flowers medium-sized to large;
petals yellow; fruit linear, elongate, somewhat flattened; seeds in 2 rows in each locule.

a. Leaves extending well up the stem; fruit with a stipe 1-2 mm long above base of perianth.
   *D. tenuifolia (L.) DC.
a. Leaves mostly near base of stem; fruit sessile above base of perianth. *D. muralis (L.) DC.

19. Conringia Adans., Hare's-ear-mustard

*C. orientalis (L.) Andrz. Glabrous, somewhat succulent, annual with erect stem; leaves en-
tire, the upper elliptic or oblong and cordate-clasping, the lower tapering to base; flowers medium to
large; petals pale yellow; fruit slender, 4-angled, 4-12 cm long, ascending; seeds many, 1 row in
each locule.

20. Alliaria B. Ehrh., Garlic-mustard

*A. officinalis Andrz. Tall; glabrous or slightly pubescent with simple hairs; with odor of gar-
lic; leaves petioled; cauline blades deltoid-ovate, dentate, bases cordate or rounded, lower blades
reniform; flowers of medium size; petals white; fruits up to 6 cm long, linear, spreading, 4-angled,
on thick pedicels, each valve with 1 strong nerve; seeds short-cylindric, 1 row in each locule.

21. Sisymbrium L.

Tall annuals with pinnatifid leaves; flowers small or medium-sized; petals yellow; fruits terete,
rather short and tapering to a point, or linear-cylindric; seeds oblong, 1 row in each locule.

a. Petals small; fruits 1 1/2 cm long, closely appressed, almost sessile, tapering from base to tip.
   *S. officinale (L.) Scop., Hedge-mustard
a. Petals of medium size; fruits linear, spreading or ascending.
   b. Segments of upper leaves linear or filiform; fruits 5-10 cm long, 5-10 times as long as the thick
      pedicels. *S. altissimum L., Tumble-mustard
   b. Segments of upper leaves broader; fruits 1-3.5 cm long, about 1-3 times as long as the slender
      pedicels. *S. loeselii L.

22. Arabidopsis Heynh., Mouse-ear Cress

*A. thaliana (L.) Heynh. Stem erect, slender, rather low, often much branched; hairs forked;
leaves mostly in basal rosette, entire or toothed, oblong or oblanceolate, sessile, up to 5 cm long;
petals white, small; fruits slender, about 1.5 cm long, pedicels spreading or ascending; seeds many,
1 row in each locule.

117

23. <u>Descurainia</u> Webb & Berthelot, Tansy-mustard

Stems erect; leaves 1-3 times pinnate; pubescence of branched or of both branched and simple hairs; flowers small; petals yellow, sometimes pale; fruits linear or clavate, each valve with prominent midnerve.

a. Seeds in 1 row in each locule; fruit linear, about 1 mm wide; axis of raceme usually finely pubescent but not glandular.   *<u>D</u>. <u>sophia</u> (L.) Webb
a. Seeds in 2 rows; fruits clavate, 1-2 mm wide; axis of raceme often glandular. <u>D</u>. <u>pinnata</u> (Walt.) Britt. var. <u>brachycarpa</u> (Richards.) Fern.

24. <u>Hesperis</u> L.

*<u>H</u>. <u>matronalis</u> L., Dame's Rocket. Tall; pubescence of simple and of branched hairs; leaves entire or toothed, lanceolate, acuminate; flowers large, fragrant; petals purple, pink, or white; fruits spreading, 5-10 cm long, linear; seeds many, oblong, somewhat angled in cross section, 1 row in each locule.

25. <u>Erysimum</u> L.

Pubescent with 2- or 3-parted hairs; leaves lanceolate, entire or toothed; flowers medium to large; petals yellow or orange, long-clawed; fruit linear, 4-sided, each valve with prominent nerve; stigma lobed; 1 row of seeds in each locule.

a. Sepals about 1 cm long; petals 1.5 cm long or more, orange-yellow or bright yellow.
   <u>E</u>. <u>arkansanum</u> Nutt., Western Wallflower
a. Sepals and petals smaller; petals yellow.
   b. Fruit 1-3 cm long, not more than about twice as long as the slender pedicels; pedicels spreading, fruit erect or ascending; petals bright yellow, 3-6 mm long. *<u>E</u>. <u>cheiranthoides</u> L.
   b. Fruit much more than twice as long as pedicels; petals pale yellow, 6-10 mm long.
      c. Divergently branched; axis of inflorescence zigzag; fruits divergent, to 1 dm long, somewhat constricted between seeds at maturity; pedicels as thick as fruits. *<u>E</u>. <u>repandum</u> L.
      c. Unbranched or branches mostly ascending; fruits ascending, to 4 cm long, not constricted between seeds; pedicels more slender than fruits. <u>E</u>. <u>inconspicuum</u> (Wats.) MacM.

26. <u>Rorippa</u> Scop., Yellow-cress

Aquatic to terrestrial; stems branching; leaves dentate to pinnatifid; glabrous or pubescent with simple hairs; flowers small or of medium size; petals yellow; fruit globose, ovoid, or linear-cylindric; seeds many, 2 irregular rows in each locule or rarely 1 row.

a. Pedicels 1-3 mm long; mature fruits usually 7-10 mm long, mostly 4 or more times as long as pedicels; upper leaves entire to crenate, lower sometimes pinnatifid; stamens 4. <u>R</u>. <u>sessiliflora</u> (Nutt.) Hitchc.
a. Pedicels longer than above; mature fruits varying from twice as long as pedicels to shorter than pedicels.
   b. Fruits linear-cylindric, mostly about 1.5 cm long; leaves deeply pinnatifid; stems spreading or ascending; petals longer than sepals; with rhizomes. *<u>R</u>. <u>sylvestris</u> (L.) Bess.
   b. Fruits mostly 3-6 mm long; leaves coarsely toothed or pinnatifid; petals shorter than sepals.
      <u>R</u>. <u>islandica</u> (Oeder) Borbas
      c. Stem and leaves glabrous or nearly so; fruits 2-3 times as long as thick. var. <u>fernaldiana</u> Butt. & Abbe
      c. Stem and leaves frequently pubescent; fruits less than twice as long as thick. var. <u>hispida</u> (Desv.) Butt. & Abbe

27. <u>Nasturtium</u> R. Br., Water Cress

*<u>N</u>. <u>officinale</u> R. Br. Glabrous herbs growing in water or on mud; leaves pinnately compound, the 3-11 circular to oval leaflets entire; flowers small; petals white; fruits broadly linear, on spreading pedicels; seeds many, 2 rows in each locule.

28. Armoracia Gaertn., Mey. & Scherb.

Glabrous; aquatic with dissected submersed leaves and oblong emersed ones, or terrestrial with oblong toothed to pinnatifid leaves; flowers of medium size; petals white; seeds in 2 rows in each locule.

a. Aquatic; submersed leaves pinnately dissected into filiform segments; emersed leaves, if present, lanceolate to oblong, to 8 cm long; persistent style half as long as body of fruit or longer. A. aquatica (Eat.) Wieg., Lake Cress.
a. Terrestrial; basal leaves with blades as much as 3 dm long, toothed to pinnatifid; fruits seldom formed; style short. *A. lapathifolia Gilib., Horseradish.

29. Barbarea R. Br., Winter Cress

Glabrous or nearly so; stems striate; basal leaves lyrate-pinnatifid, cauline leaves entire to pinnatifid, bases clasping; flowers large or of medium size; petals yellow; fruits linear, terete or somewhat 4-angled; seeds in 1 row in each locule.

a. Basal leaves with 4-8 pairs of lateral leaflets; cauline leaves pinnatifid; fruit 4.5-7 cm long. *B. verna (Mill.) Aschers., Early W.
a. Basal leaves with 1-4 pairs of lateral leaflets; cauline leaves usually not pinnatifid; fruit 1.5-3 cm long. *B. vulgaris R. Br., Yellow Rocket.
   b. Fruits erect or appressed. var. *vulgaris
   b. Fruits spreading or ascending. var. *arcuata (Opiz.) Fries

30. Iodanthus T. & G., Purple Rocket

I. pinnatifidus (Michx.) Steud. Erect, glabrous; cauline leaves ovate to lanceolate, dentate, doubly dentate, or pinnately lobed, tapering to winged petiole, base usually auricled; flowers large or of medium size; petals purple, long-clawed; fruits linear, spreading; seeds wingless, 1 row in each locule.

31. Chorispora R. Br.

*C. tenella (Pall.) DC. Branched annual, somewhat glandular; blades lanceolate or oblanceolate, undulate-toothed or the lower incised; petals blue or purple, large; fruit 3-4 cm long, cross-septate and constricted, beak 1 cm long or more.

32. Leavenworthia Torr.

L. uniflora (Michx.) Britt. Low, glabrous, scapose; leaves basal, lyrate-pinnatifid or pinnately compound, the segments or leaflets ovate to reniform; flowers of medium size, solitary on leafless scapes; petals purple; fruits flat, broadly linear or oblong, 4-5 mm wide; seeds flat, margined, 1 row in each locule.

33. Dentaria L., Toothwort

Stems erect; leaves palmately divided or palmately compound, the 2 or 3 cauline ones in ours opposite, whorled, or alternate; flowers large, few, in a short raceme or corymb; petals white, pink, or purple; fruits linear, flattened, dehiscent from base; seeds wingless, 1 row in each locule.

a. Segments or leaflets of cauline leaves linear or narrowly lanceolate; rhizome of fusiform separable tubers.
   b. Leaflets of basal leaf broader than of cauline; cauline leaves 2, rarely 3, opposite or nearly so. D. heterophylla Nutt.
   b. Basal leaves like the cauline ones, or not present with the flowering stem.
      c. Cauline leaves usually 3, each with 3 toothed, incised, or variously cleft leaflets; rachis of inflorescence more or less pubescent. D. laciniata Muhl., Cut-leaved T.
      c. Cauline leaves 2 or 3, each with 2 or 3 long-stalked leaflets cut into 3-7 linear segments; rachis of inflorescence glabrous. D. multifida Muhl.
a. Segments or leaflets of the 2 cauline leaves ovate to lanceolate; rhizome continuous. D. diphylla Michx., Two-leaved T.

119

34. Cardamine L., Bitter Cress

Glabrous or pubescent with simple hairs; stems erect or sometimes decumbent; leaves entire to pinnately compound; flowers small to large; siliques linear, elastically dehiscent from the base; seeds in 1 row in each locule.

a. Blades entire or dentate, or rarely with 1 or 2 small lobes at base.
  b. Stem decumbent or erect, without tuberous base; blades of basal and stem-leaves circular or somewhat angled, sometimes with 1 or 2 small lobes at base. C. rotundifolia Michx., Round-leaf B.
  b. Stem erect from tuberous base; basal leaves petioled, blades circular, cauline leaves shorter-petioled and narrower upward, upper sessile.
    c. Petals purple or purple-tinged; stem pubescent, especially above, purple; flowering in early April. C. douglassii (Torr.) Britt., Purple B.
    c. Petals white; stem glabrous; flowering later. C. bulbosa (Schreb.) BSP., Bulbous B.
a. Leaf-blades pinnately divided or pinnately compound.
  b. Petals 8 mm long or more, pink or white, 3 times as long as calyx. C. pratensis L., Meadow B.
  b. Petals smaller, white.
    c. Petioles ciliate; basal leaves many at anthesis, stem-leaves relatively few; stem glabrous; stamens usually 4. *C. hirsuta L.
    c. Petioles not ciliate; basal leaves present or absent at anthesis, stem-leaves several.
      d. Stem, when emersed, pubescent at base; rachis of cauline leaves winged, terminal leaflet usually much broader than lateral ones. C. pensylvanica Muhl.
      d. Stem glabrous; rachis of cauline leaves little or not winged, terminal leaflet little broader than lateral ones. C. parviflora L. var. arenicola (Britt.) O. E. Schulz

35. Sibara Greene, Virginia Rock Cress

S. virginica (L.) Rollins   Leaves deeply pinnatifid, the basal ones with many segments; flowers small; petals white or pinkish; pedicels somewhat spreading; fruit linear, flattened, nerveless or faintly nerved, usually ascending; seeds narrowly winged, nearly circular, 1 row in each locule.

36. Arabis L., Rock Cress

Leaves entire, or dentate to pinnatifid; pubescence, when present, of simple, of forked, or of stellate hairs; flowers small or of medium size; petals white, pink, pale purple, or pale yellow; fruits linear, usually flattened; seeds flattened, winged or wingless, in 1 or 2 rows in each locule.

a. Cauline leaf-blades narrowed to base, not clasping; seeds in 1 row in each locule.
  b. Stem usually less than 3 dm high, much branched; basal blades lyrate-pinnatifid, cauline entire or few-toothed; fruits spreading-ascending, 2-5 cm long; seeds wingless. A. lyrata L.
  b. Stem mostly 3-10 dm high, unbranched or nearly so; basal blades dentate or lobed, cauline toothed, tapering to both ends; fruits flattened, becoming pendulous, 5-8 cm long, seeds broadly winged. A. canadensis L.
a. Cauline leaf-blades clasping with auricled or sagittate bases.
  b. Fruits erect or appressed at maturity.
    c. Stem and leaves more or less pubescent or upper leaves glabrous; fruits flattened, 3-5 cm long, about 1 mm wide, seeds in 1 row, winged. A. hirsuta (L.) Scop.
      d. Hairs of stem spreading, mostly simple. var. pycnocarpa (Hopkins) Rollins
      d. Hairs of stem appressed, mostly forked. var. adpressipilis (Hopkins) Rollins
    c. Upper and middle cauline leaves and stem glabrous, usually base of stem and lowest leaves pubescent; fruit 4-9 cm long.
      d. Fruit flattened, 1.5-2.5 mm wide, seeds in 2 rows. A. drummondii Gray
      d. Fruit not flattened, about 1 mm wide, seeds in 1 or 2 rows, beak thick, 2-lobed. A. glabra (L.) Bernh.
  b. Fruit ascending, spreading, or curved downward at maturity.
    c. Stem and blades pubescent; fruits 1.5-4 cm long, seeds in 1 row.
      d. Pedicels of fruits 1 cm long or more; hairs of leaf-blades mostly simple. A. patens Sulliv.
      d. Pedicels of fruits 5 mm long or less; hairs of upper surface of blades mostly simple, of lower surface mostly stellate. A. perstellata E. L. Br. var. shortii Fern.
    c. Middle and upper stem and blades glabrous; lower stem and blades sometimes pubescent; fruit 3-10 cm long; seeds winged.

d. Fruits eventually curved downward, seeds in 1 row. A. laevigata (Muhl.) Poir.
d. Fruits straight, ascending or spreading, seeds in 2 rows when young, in 1 row at maturity; basal leaves stellate-pubescent. A. divaricarpa A. Nels.

RESEDACEAE, Mignonette Family
Reseda L., Mignonette

Herbs; leaves alternate, stipules in form of glands; flowers hypogynous, bisporangiate, zygomorphic, in dense terminal racemes or spikes; sepals about 6; petals 4-7, lobed; stamens several to many; asymmetric hypogynous disk present; carpels several, united; stigmas as many, sessile; ovulary 1-loculed, opening at top before maturity; placentae parietal; fruit a capsule.

a. Petals 4, yellow; leaves entire, lanceolate. *R. luteola L., Dyer's Rocket
a. Petals usually 6, greenish-white; leaves pinnately divided. *R. alba L.

SARRACENIACEAE, Pitcher-plant Family
Sarracenia L., Pitcher-plant

S. purpurea L. Bog herb; leaves basal, pitcher-shaped, spreading or somewhat ascending; flowers solitary on leafless scapes, 5-7 cm wide, purple-red, hypogynous, regular, bisporangiate; sepals and petals 5; stamens many; carpels 5, united; style 1, apex with a 5-rayed expansion, a stigma under each ray; ovulary 5-loculed; fruit a many-seeded capsule; placentae central.

DROSERACEAE, Sundew Family
Drosera L., Sundew

Small bog herbs; leaves basal or, in submersed plants, cauline, bearing reddish glandular hairs from which a viscid secretion exudes; flowers hypogynous, bisporangiate, 4-7 mm wide; perianth regular; sepals, petals, and stamens 4-8, petals separate; carpels united; styles separate, 3-5, each forked; ovulary 1-loculed with 3-5 parietal placentae.

a. Leaf-blades orbicular or wider than long, abruptly narrowed to hairy petioles. D. rotundifolia L.
a. Leaf-blades spatulate, much longer than wide, tapering to non-hairy petioles. D. intermedia Hayne

PODOSTEMACEAE, Riverweed Family
Podostemum Michx., Riverweed

P. ceratophyllum Michx. Aquatic herb with aspect of a bryophyte, growing in running water attached to stones; leaves alternate, usually repeatedly forked, lobes narrow; flowers minute, axillary, on a stalk from a sessile involucre, without perianth, consisting of ovulary, 2 stamens with filaments united and 2 staminodes; ovulary 2-loculed, stigmas 2; placentae central; capsule ellipsoid or obovoid; seeds many.

CRASSULACEAE, Orpine Family

Herbs; flowers hypogynous or slightly perigynous, bisporangiate; perianth regular; sepals 4-5; petals 4-5, separate or slightly united at base, or none; carpels 4-5, separate or united at base; stamens 8-10; fruit an aggregate of follicles or a beaked capsule.

a. Carpels united to about the middle; petals usually none; leaves not fleshy. Penthorum
a. Carpels separate; petals 4-5; leaves fleshy. Sedum

Penthorum L., Ditch Stonecrop

P. sedoides L. Erect; leaves alternate, sessile, lanceolate, serrate; flowers greenish, short-pediceled, in forked 1-sided cymes; sepals 5; petals usually none; stamens 10; carpels 5, united below; capsule opening by the circumscissile dehiscence of the 5 beaks which are the separate upper portions of the carpels.

Sedum L., Stonecrop

Fleshy, glabrous; stems erect or decumbent; flowers in cymes; petals present; fruit an aggregate of follicles.

a. Leaves terete, alternate; stems creeping, mat-forming.
  b. Petals yellow; inflorescence branches few; leaves closely imbricated, 6 mm long or less.
    *S. acre L.
  b. Petals white or pink; inflorescence repeatedly branched; leaves not closely imbricated, to about 12 mm long. *S. album L.
a. Leaves flat, fleshy.
  b. Leaves, at least the lower, opposite or in whorls of 3; inflorescence not compact, of 2-5 divergent branches, flowers sessile or subsessile, leafy-bracted.
    c. Petals white, anthers purple; upper leaves of flowering stems alternate or opposite; leaves crowded at tips of nonflowering stems. S. ternatum Michx.
    c. Petals yellow; leaves opposite or in whorls of 3. *S. sarmentosum Bunge
  b. Leaves alternate or opposite; inflorescence compact, many-flowered; flowers often not sessile, not leafy-bracted.
    c. Leaves oblong or ovate, toothed toward apex; petals pink or rose-purple. *S. telephium L.
    c. Leaves obovate or spatulate, entire or remotely dentate; petals white or pale pink.
      S. telephoides Michx.

SAXIFRAGACEAE, Saxifrage Family

Herbs and shrubs; leaves alternate or opposite; flowers usually bisporangiate, ranging from scarcely perigynous through perigynous, partially epigynous, wholly epigynous, to epigynous with epigynous hypanthium; perianth regular or slightly zygomorphic; sepals and petals 4-5 or petals none; stamens as many as petals, twice as many, or more; carpels 2-4, united to top of styles, or only the styles separate, or only bases of carpels united, then stigmas, styles, and upper part of ovularies separate; locules as many as carpels or locule 1 with parietal placentae as many as carpels; fruit a capsule or of separate follicles.

a. Herbs; flowers perigynous or partially epigynous.
  b. Petals none; small decumbent plants of wet soil, the small leaves partly opposite; flowers nearly sessile, small; sepals greenish; anthers yellow to red. 6. Chrysosplenium
  b. Petals present; stems not decumbent.
    c. Flowers solitary, peduncle bearing 1 leaf; leaves entire. 7. Parnassia
    c. Flowers in clusters; leaves not entire.
      d. Stamens 5 or as many as the petals.
        e. Hypanthium almost free from carpels; petals white; capsule 2-loculed; stamens shorter than sepals. 1. Sullivantia
        e. Hypanthium adnate to approximately lower half of carpels; petals often green or purple; capsule 1-loculed; stamens often as long as or longer than sepals. 4. Heuchera
      d. Stamens 10 or twice as many as the petals.
        e. Inflorescence a panicle; basal leaf-blades ovate or oblong, tapering to base, pinnately veined. 2. Saxifraga
        e. Inflorescence a raceme; basal leaf-blades ovate or circular, cordate at base, palmately veined.
          f. Petals entire; carpels unequal. 3. Tiarella
          f. Petals pinnatifid; carpels equal. 5. Mitella
a. Shrubs; flowers epigynous.
  b. Flowers with tubular epigynous hypanthium; leaves alternate; ovulary 1-loculed; stamens as many as petals. 10. Ribes
  b. Flowers without epigynous hypanthium; leaves opposite; ovulary 2-4-loculed; stamens twice as many as petals or more.
    c. Stamens twice as many as petals; ovulary 2-loculed; flowers, at least central ones of cluster, small. 9. Hydrangea
    c. Stamens 20 or more; ovulary 4-loculed; flowers 2 cm wide or more. 8. Philadelphus

122

1. Sullivantia T. & G.

S. sullivantii (T. & G.) Britt. Slender herbs; leaves mostly basal, long-petioled, palmately-veined, blades reniform or circular, crenate to slightly lobed; flowers small, in a panicle, peduncle with 1 or 2 leaves near base; perianth regular; sepals 5; petals 5, white, about 3 mm long; stamens 5; hypanthium adnate to base of carpels; carpels 2, united below, upper portions separate and appearing as beaks on the 2-loculed capsule; seeds many, winged.

2. Saxifraga L., Saxifrage

Leaves in basal rosettes; flowers on leafless scapes; perianth regular or almost regular; sepals 5; petals 5; stamens 10; in our species hypanthium adnate to bases of carpels; carpels 2, somewhat united at base, summits divergent and tapering; capsule 2-loculed at base, each carpel dehiscing as a follicle; ovules many, on axile placentae.

a. Basal leaves crenate or dentate, the larger not more than 5 cm long; sepals about as long as hypanthium, spreading or ascending; petals white, 4-6 mm long. S. virginiensis Michx., Early S.
a. Basal leaves denticulate or repand, the larger 10-20 cm long; sepals somewhat longer than hypanthium, reflexed; petals yellow, green, or purple, 2-3 mm long. S. pensylvanica L.

3. Tiarella L., False Mitrewort, Foamflower

T. cordifolia L. Slender herb; leaves basal, long-petioled, blades round-ovate, shallowly lobed, cordate at base, palmately veined; flowers slightly perigynous, raceme usually scapose; hypanthium short; sepals 5; corolla regular, of 5 white petals 3-5 mm long; stamens 10; carpels 2, united below, unequal; capsule 1-loculed; placentae parietal.

4. Heuchera L., Alum-root

Erect or ascending; leaves mostly basal, long-petioled, blades round-ovate, toothed or lobed, palmately veined, bases cordate; flowers small, sometimes irregular, white, green, or purple, in panicles; peduncle naked or with a few alternate leaves; sepals, petals, and stamens 5; hypanthium adnate to bases of carpels; carpels 2, united below, free upper portions forming 2 beaks on capsule; ovulary 1-loculed below; placentae 2, parietal.

a. Hypanthium minutely glandular-pubescent on outer surface. H. americana L.
a. Hypanthium villous on outer surface.
  b. Lobes or teeth of blades acute; seeds echinate. H. villosa Michx.
  b. Lobes or teeth of blades rounded, short-mucronate; seeds smooth, slightly ridged. H. parviflora Bartl. var. rugelii (Shuttlew.) R. B. & L.

5. Mitella L., Mitrewort, Bishop's-cap

Low herbs; leaves mostly basal, blades round-ovate to reniform, lobed or crenate, cordate at base; flowers in racemes, sometimes spikelike; hypanthium adnate to base of carpels; sepals 5; petals 5, white, pinnatifid, small; stamens 10, in ours; carpels 2, united; ovulary 1-loculed; placentae parietal, 2.

a. Peduncle bearing 2 opposite leaves; petals white; terminal lobe of blades longest. M. diphylla L.
a. Peduncle leafless or bearing 1 leaf; petals greenish-yellow; leaves obscurely lobed. M. nuda L.

6. Chrysosplenium L., Golden Saxifrage

C. americanum Schwein. Decumbent herb; leaves small, short-petioled, upper alternate, lower opposite, round-ovate, entire or obscurely toothed; flowers small, solitary, nearly sessile, green, yellow, or red; sepals 4-5; petals none; stamens usually 8, in notches of disk; styles 2; ovulary flat, 1-loculed; placentae 2, parietal.

7. Parnassia L., Grass-of-Parnassus

P. glauca Raf. Glabrous herb; cauline leaf 1, sessile, basal leaves long-petioled; blades entire, coriaceous, round-ovate; flowers solitary; perianth regular, strongly veined; sepals 5; petals 5, white;

staminodes 5, opposite petals, each apparently 3 sterile stamens united at base; stamens 5; carpels united, ovulary superior, 1-loculed; placentae 4, parietal; stigmas 4.

8. Philadelphus L., Mock-orange

*P. coronarius L. Shrub; leaves opposite, short-petioled; blades ovate to oblong, somewhat dentate; flowers fragrant, epigynous, in short racemes; perianth regular; sepals 4-5; petals 4-5, white, about 2 cm long; stamens many; ovulary usually 4-loculed; styles separate above; fruit a capsule.

9. Hydrangea L.

H. arborescens L. Shrub; leaves opposite, long-petioled, blades ovate to oblong, cordate to tapering at base, serrate; flowers small, in rounded or flattened compound cymes, epigynous, sometimes the marginal ones consisting of only a flat enlarged calyx; sepals 4 or 5; petals 4 or 5, white; stamens 8-10; carpels 2, united below, separate above; ovulary 2-loculed below; styles 2; fruit a capsule.

10. Ribes L., Currant, Gooseberry

Shrubs, sometimes with prickles; leaves alternate, palmately veined and lobed; flowers epigynous, usually bisporangiate; perianth regular; sepals 5; petals 5, smaller than sepals; stamens 5; epigynous hypanthium present; carpels usually 2; ovulary 1-loculed with 2 (rarely 3) parietal placentae; styles separate or united; fruit a berry.

a. Pedicels not jointed; flowers solitary or in clusters of 2-5. Gooseberries
  b. Ovulary and fruit prickly; hypanthium longer than sepals; stamens about equaling petals.
   R. cynosbati L.
  b. Ovulary and fruit not prickly, either glabrous or pubescent; hypanthium shorter than or equaling sepals.
   c. Stamens and style long-exserted beyond sepals. R. missouriense Nutt.
   c. Stamens and style little if any exserted beyond sepals.
    d. Hypanthium pubescent; ovulary pubescent or glandular. *R. grossularia L., Garden G.
    d. Hypanthium and ovulary glabrous. R. hirtellum Michx.
a. Pedicels jointed beneath the ovulary; flowers in racemes of usually more than 5. Currants
  b. Ovulary and fruit covered with glandular hairs.
   c. Stems prickly; fruit purple-black. R. lacustre (Pers.) Poir., Swamp G.
   c. Stems not prickly; fruit dark-red. R. glandulosum Grauer, Skunk C.
  b. Ovulary and fruit without glandular hairs (glandular dots may be present).
   c. Hypanthium several times as long as ovulary, longer than calyx; racemes leafy-bracted; flowers fragrant. *R. odoratum Wendl., Buffalo C.
   c. Hypanthium little if any longer than ovulary; racemes not leafy-bracted.
    d. Blades resin-dotted beneath.
     e. Flowers 8-10 mm long; bracts longer than pedicels. R. americanum Mill., Wild Black C.
     e. Flowers 5-6 mm long; bracts shorter than pedicels. *R. nigrum L., Black C.
    d. Blades not resin-dotted; fruits bright red. *R. sativum Syme, Currant

HAMAMELIDACEAE, Witch Hazel Family

Trees and shrubs with alternate simple leaves; flowers monosporangiate or bisporangiate; perianth regular, of 4 sepals and 4 petals, or none; stamens 4-many; carpels 2, united, styles 2, ovulary 2-loculed, partly inferior.

a. Leaves pinnately veined, margins repand; petals 4, yellow, linear; flowers in small axillary clusters. Hamamelis
a. Leaves palmately veined, palmately lobed; petals none; flowers in dense globular clusters. Liquidambar

Hamamelis L., Witch Hazel

H. virginiana L. Shrub; leaves 2-ranked, short-petioled, blades oblique and rounded or cordate at base; buds stalked; flowers bisporangiate or some of them monosporangiate, in axillary clusters, appearing in autumn; sepals, petals, stamens, and staminodes 4; petals yellow, 1.5-2 cm long, ribbon-like; hypanthium adnate to base of carpels; fruit a 2-seeded capsule.

Liquidambar L., Sweet or Red Gum

L. styraciflua L. Tree; leaves palmately veined, lobes 5, serrate, narrowly triangular; pith 5-angled; monecious; flowers monosporangiate, without perianth, but with minute scales, the staminate of many stamens, in heads, the heads in racemes; carpellate flowers cohering in axillary peduncled spherical heads, consisting of 2 styles and ovulary; ovules many in each locule, 1 or 2 maturing; fruit multiple.

PLATANACEAE, Plane Tree Family

Platanus L., Plane Tree   Sycamore

Large trees with peeling bark; leaves alternate, simple, palmately veined and lobed; stipules foliaceous, encircling twig; flowers monosporangiate, the 2 kinds in separate spherical heads on same tree; staminate flower with small calyx, 3-7 stamens, and minute petals, filaments short or none, a fleshy cap at top of anthers; carpellate flower hypogynous, calyx small, petals present or absent, staminodes present, carpels several, separate; fruit an aggregate of achenes.

a. Exposed inner bark white; fruiting heads usually solitary. P. occidentalis L., Sycamore
a. Exposed inner bark greenish or yellowish; fruiting heads 1-3 per peduncle; often planted.
    *P. acerifolia Willd., London Plane Tree

ROSACEAE, Rose Family

Trees, shrubs, and herbs, sometimes climbing; leaves alternate (rarely opposite), simple or compound, stipules or stipule-scars usually present; rarely diecious; flowers perigynous or epigynous (rarely merely hypogynous), bisporangiate or rarely monosporangiate; perianth usually regular; sepals usually 5 (3-8), sometimes subtended by a circle of bracts; petals usually 5, separate, rarely none; stamens 5-many, separate; carpels 1 or few to many, separate or united; when solitary, the carpel is free from the hypogynous hypanthium and fruit is a drupe or rarely an achene; when 2 to many, separate, carpels are free from the hypogynous hypanthium and fruit is an aggregate of achenes, follicles, or drupelets; when 2-5 carpels are united, the flower is epigynous (hypanthium is adnate to the ovulary), styles or style-branches, stigmas, and locules are as many as carpels (except when, rarely, locules become twice as many by growth of a false septum in each carpel), placentae are axile, and fruit is a pome.

a. Flowers perigynous (or rarely almost hypogynous); hypanthium sometimes completely enclosing
    carpels but not adnate to them; carpels few to many, rarely only 1, separate or slightly united at
    base.
  b. Trees and shrubs.
    c. Carpel 1; fruit a drupe; blades simple, serrate; often glands present on petiole; flowers
        sometimes appearing before leaves. 22. Prunus
    c. Carpels more than 1.
      d. Carpels more than 10; if rarely fewer, then flowers 2 cm wide or more.
        e. Leaves pinnately compound with 5-7 entire leaflets; stipules dry, sheathing; petals yel-
            low. 14. Potentilla
        e. Leaves compound with serrate leaflets, or simple and palmately lobed; plants usually
            prickly or bristly.
          f. Hypanthium globose or urn-shaped, narrowed at mouth, becoming fleshy around bony
              achenes in fruit; flowers large and showy, of various colors. 21. Rosa
          f. Hypanthium shallow, not narrowed at mouth, not becoming fleshy; fruit an aggregate
              of druplelets; petals white to rose-color. 17. Rubus
      d. Carpels 10 or fewer; flowers small, white or pink.
        e. Leaves pinnately compound; flowers in large panicles. 4. Sorbaria
        e. Leaves simple, sometimes not yet present at anthesis.

    f. Leaves shallowly lobed, present at anthesis; carpels sometimes somewhat united at base, at maturity 7-10 mm long; seeds 1.5-2 mm long, obovoid. 1. <u>Physocarpus</u>
    f. Leaves entire or toothed, sometimes appearing after the flowers; carpels separate; follicles and seeds much smaller than above. 2. <u>Spiraea</u>
  b. Herbs.
   c. Low scapose herbs with simple or 3-foliolate basal leaves.
    d. Carpels more than 10; flowers subtended by sepaloid bracts; basal blades 3-foliolate.
     e. Petals white; bracts about equaling sepals. 11. <u>Fragaria</u>
     e. Petals yellow; bracts 3-lobed or -toothed, broader than sepals. 12. <u>Duchesnea</u>
    d. Carpels 10 or fewer; bracts present or absent.
     e. Petals white; leaves simple, blades round-ovate, crenate, cordate at base; flowers usually solitary. 18. <u>Dalibarda</u>
     e. Petals yellow; leaves 3-foliolate, leaflets irregularly lobed or toothed; flowers in clusters of a few. 13. <u>Waldsteinia</u>
   c. Not scapose.
    d. Flowers scarcely perigynous; leaves pinnately compound, terminal leaflet large and palmately lobed; petals white to pink; inflorescence a panicle; carpels 5-15. 15. <u>Filipendula</u>
    d. Without the above set of characters.
     e. Carpels more than 10, becoming achenes.
      f. Styles persistent in fruit, jointed near tip, becoming hooked. 16. <u>Geum</u>
      f. Styles deciduous, not jointed, not becoming hooked. 14. <u>Potentilla</u>
     e. Carpels 10 or fewer.
      f. Leaves, except sometimes lowest, 3-foliolate or 3-parted; petals 1-2 cm long, white or pink; carpels 5. 5. <u>Gillenia</u>
      f. Leaves pinnately compound or decompound; petals much less than 1 cm long.
       g. Leaves decompound; diecious; petals white; flowers in spikelike racemes aggregated in panicles; carpels about 3. 3. <u>Aruncus</u>
       g. Leaves compound; carpels 1 or 2, ovularies concealed within hypanthium.
        h. Petals yellow; flowers in narrow racemes; hypanthium, in fruit, with band of hooked prickles. 19. <u>Agrimonia</u>
        h. Petals absent, sepals 4, petaloid; flowers in heads or dense spikes; hypanthium without prickles. 20. <u>Sanguisorba</u>
a. Flowers epigynous; carpels united.
 b. Ovules many in each locule of ovulary; blades tapering at base; stipules conspicuous. 8. <u>Chaenomeles</u>
 b. Ovules usually 1-2 in each locule of ovulary.
  c. Usually with typical lateral thorns; ripe carpels bony.
   d. Deciduous; 1 or 2 ovules in each locule, if 2, then 1 sessile and 1 stalked and sterile; blades various. 9. <u>Crataegus</u>
   d. Evergreen; 2 like ovules in each locule; blades oblanceolate. 10. <u>Cotoneaster</u>
  c. Without typical lateral thorns, but some branches may end in thorns; ripe carpels papery or leathery.
   d. Flowers in racemes; pome small, 1 cm in diameter or less; leaves simple; no glands on upper side of midvein. 7. <u>Amelanchier</u>
   d. Flowers in simple or compound corymbs; pomes large, sometimes with grit cells, or if 1 cm in diameter or less, then leaves either with a row of toothlike glands on upper side of midvein or pinnately compound. 6. <u>Pyrus</u>

1. <u>Physocarpus</u> Maxim., Ninebark

 <u>P. opulifolius</u> (L.) Maxim. Shrub; older bark peeling in thin layers; leaves simple, blades shallowly palmately 3-lobed; flowers 1 cm wide or less, in rounded corymbs; sepals 5; petals 5, white; stamens many; carpels 2-5, separate or united at base, becoming few-seeded follicles.

2. <u>Spiraea</u> L.

 Shrubs; leaves simple; flowers small, in terminal or lateral umbels, corymbs, or panicles; sepals 5; petals 5, white to rose-color; stamens 10-many; carpels usually 5, separate, 2-several ovules in each; fruit an aggregate of follicles.

126

a. Flowers in umbels, corymbs, or racemes.
  b. Leaves elliptic or oblong, denticulate; flowers often double, in 3-6-flowered sessile umbels.
     *S. prunifolia Sieb. & Zucc.
  b. Leaves rhombic-ovate or obovate, toothed toward tip; flowers in corymbs or racemes.
     *S. vanhouttei (Briot) Zab.
a. Flowers in terminal panicles.
  b. Blades white- or brownish-tomentose beneath, coarsely serrate; panicles brownish-tomentose; petals usually rose-pink. S. tomentosa L., Hardhack, Steeplebush.
  b. Blades green, not tomentose, beneath; petals white to pink.
    c. Hypanthium, inflorescence-axis, and pedicels puberulent; leaves finely toothed. S. alba DuRoi, Meadow-sweet
    c. Hypanthium, inflorescence-axis, and pedicels glabrous; leaves coarsely toothed. S. latifolia (Ait.) Borkh.

3. Aruncus Adans., Goat's-beard

A. dioicus (Walt.) Fern. Tall herb; leaves alternate, 2-3 times compound, without stipules; leaflets ovate to lanceolate, acuminate, doubly serrate; diecious; flowers small, each kind with rudiments of the other sporophylls; sepals 5, petals 5, white, about 1 mm long; stamens 15 or more; carpels usually 3, separate, becoming follicles.

4. Sorbaria A. Br., False Spiraea

*S. sorbifolia (L.) A. Br. Shrub with pinnately-compound leaves, leaflets lanceolate-oblong, serrate, acuminate, sessile; flowers small, in terminal panicles; sepals 5; petals 5, white; stamens many, filaments elongate; carpels 5, united at base, each with several ovules, dehiscent along 2 sutures.

5. Gillenia Moench, Indian-physic

Herbs; leaves trifoliolate, almost sessile, stipuled, leaflets lanceolate, acuminate, serrate or incised or of lower leaves sometimes pinnately divided; flowers in open terminal panicles; hypanthium cylindrical to campanulate; sepals 5, petals 5, 1-2 cm long, white or pink; stamens 10-20; carpels 5, separate, slightly united at first, becoming 2-4-seeded follicles.

a. Stipules linear, awl-shaped, entire or slightly incised; leaflets serrate. G. trifoliata (L.) Moench, Bowman's root
a. Stipules ovate, incised; leaflets serrate or incised, of the lower leaves often pinnately divided.
  G. stipulata (Muhl.) Trel., American Ipecac

6. Pyrus L.

Trees and shrubs, sometimes with branches ending in thorns; leaves alternate, simple or pinnately compound; hypanthium united with carpels to or almost to summit of ovularies; sepals and petals 5; stamens 15 or more; carpels 2-5, united; styles as many; fruit a pome.

a. Leaves pinnately compound; styles separate; flowers about 1 cm wide, in dense compound clusters; fruit about 1 cm wide. (Sorbus) Mountain-ash
  b. Twigs, pedicels, and lower surface of leaf-blades pubescent; winter buds not glutinous.
     *P. aucuparia (L.) Gaertn., European M.
  b. Twigs, pedicels, and lower surface of leaf-blades glabrous or glabrate; winter buds glutinous.
     P. decora (Sarg.) Hyland
a. Leaves simple; flowers in umbels, corymbs, or racemes.
  b. A row of slender toothlike glands on upper side of midvein of leaf; fruit purple or black, 1 cm or less wide; without thorns. (Aronia) Chokeberry
    c. Inflorescence and lower leaf-surface glabrous, or midvein pubescent; fruit black.
     P. melanocarpa (Michx.) Willd., Black C.
    c. Inflorescence and lower leaf-surface pubescent; fruit dark purple. P. floribunda Lindl.
  b. Glands of midvein of leaf sometimes absent; flowers 2 cm wide or more; branches sometimes ending in thorns; fruit large.
    c. Mouth of hypanthium closed around styles; styles separate to base; fruit with grit cells; petals white; blades finely serrate, glabrous at maturity. *P. communis L., Pear

c. Mouth of hypanthium open; styles united at base; fruit without grit cells; petals pink or white; blades sometimes lobed. (Malus) Apple
   d. Young branches, petioles, pedicels, lower leaf-surface, and calyx pubescent; anthers yellow; blades usually crenate-serrate. *P. malus L., Common A.
   d. Young branches, petioles, pedicels, and lower leaf-surface glabrous or glabrate; calyx glabrous or pubescent.
      e. Blades ovate, serrate or lobed, mostly acute; general. P. coronaria L. Wild Crab.
      e. Blades lanceolate, serrate or dentate, blunt; rare. P. angustifolia Ait.

7. Amelanchier Medic., Juneberry, Shadbush

Shrubs and small trees; leaves alternate, simple, blades pinnately veined, serrate; flowers in short racemes at end of leafy branches of current season, appearing with young leaves; sepals 5; petals 5, white or pink; stamens usually 20; carpels 5, united; styles 5, united below; locules 5, 2 ovules in each; later, by false septation, locules 10, 1 ovule in each; fruit a small pome.

a. Summit of ovulary pubescent; blades rounded at base, lower 1/3-1/2 of margin sometimes entire.
   b. Blades oblong; petals narrow; raceme loose; lateral veins relatively straight and ending in teeth. A. sanguinea (Pursh) DC.
   b. Blades ovate-oblong; petals about half as wide as long; raceme dense; lateral veins upcurving and anastomosing. A. spicata (Lam.) K. Koch
a. Summit of ovulary glabrous; blades ovate to obovate, acute to acuminate, finely serrate nearly to the rounded or cordate base.
   b. Blades at anthesis small and white-pubescent beneath, glabrous or nearly so at maturity; pedicels sometimes pubescent. A. arborea (Michx. f.) Fern.
   b. Blades at anthesis half grown, glabrous or nearly so, red, purple, or bronze; pedicels glabrous. A. laevis Wieg.

8. Chaenomeles Lindl.

*C. lagenaria (Loisel.) Koidz., Japanese Quince. Shrub; branches often ending in thorns; blades lustrous above, serrate; flowers epigynous, 3-5 cm wide; petals scarlet, pink, or white; stamens many; pome 3-7 cm wide.

9. Crataegus L., Hawthorn

Shrubs and small trees; leaves simple, blades serrate or lobed; axillary thorns usually present; flowers epigynous, in corymbs; petals and sepals 5; stamens 5-many; carpels bony, 2-5, united, or rarely 1; fruit a pome.

A very difficult genus taxonomically; no key to species is included.

10. Cotoneaster Ehrh.

*C. pyracantha (L.) Spach, Fire-thorn. Shrub, with thorns; blades crenate-serrate; flowers small, in compound corymbs; petals white; stamens 20; pome red or orange.

11. Fragaria L., Strawberry

Herbs with stolons; leaves basal; leaflets 3, obovate, cuneate at base, serrate; flowers in cymose clusters; sepals 5, alternating with bracts, sepals and bracts persistent in fruit; petals 5, white; stamens many; carpels many, separate; style attached to side of ovulary; fruit consisting of fleshy receptacle with dry achenes scattered over surface.

a. Sepals appressed around young fruit; achenes in pits on receptacle; leaflets obviously stalked. F. virginiana Duchesne
a. Sepals spreading or reflexed in fruit; achenes not in pits; leaflets nearly sessile. F. vesca L.
   b. Peduncles and petioles with abundant spreading hairs. var. vesca
   b. Peduncles and petioles with sparse ascending hairs. var. americana Porter

12. Duchesnea Smith, Indian Strawberry

*D. indica (Andr.) Focke   Leaves basal, alternate on stolons; leaflets 3, crenate; flowers solitary, axillary, on horizontal stems; sepals 5, alternating with foliaceous 3-toothed bracts; petals 5, yellow; stamens many; carpels many; fruit resembling that of strawberry but insipid in taste and not juicy.

13. Waldsteinia Willd., Barren Strawberry

W. fragarioides (Michx.) Tratt.   Low herb; leaves basal, leaflets 3, cuneate at base, toothed, laciniate, or shallowly lobed; flowers few to several in a corymb; sepals 5; bracts small and sometimes deciduous; petals 5, yellow, 1 cm long or less; stamens many; carpels 2-5, separate, becoming achenes.

14. Potentilla L., Cinquefoil, Five-finger

Herbs or shrubs; stems upright or trailing; leaves pinnately or palmately compound; sepals 5 or 4, alternating with as many bracts, sepals and bracts persistent in fruit; petals 5 or 4; stamens 5-many; carpels many, separate, each 1-ovuled, on a dry receptacle; style attached to side or tip of ovulary, sometimes nearly basal; fruit an aggregate of achenes.

a. Shrub; leaves pinnately compound; leaflets 5-7, narrow, entire; stipules dry, sheathing; petals yellow; achenes pubescent. P. fruticosa L., Shrubby C.
a. Herbs; leaves pinnately or palmately compound; leaflets toothed; achenes glabrous.
  b. Principal leaves pinnately compound, with 5 to several leaflets.
    c. Petals red-purple, shorter than sepals; style lateral. P. palustris (L.) Scop., Marsh C.
    c. Petals yellow or white.
      d. Flowers solitary; leaves silvery-pubescent below; small leaflets between larger leaflets of basal leaves; style lateral; petals yellow; achene with dorsal groove. P. anserina L.
      d. Flowers clustered; leaves not silvery-pubescent below; basal leaves without interspersed small leaflets.
        e. Petals white or cream-white; style nearly basal, thickened near middle; plant glandular, hairs brownish. P. arguta Pursh
        e. Petals yellow; style terminal; plant pubescent, not glandular; achene with large corky protuberance on ventral side. P. paradoxa Nutt.
  b. Leaves palmately compound.
    c. Flowers in cymes; stems erect; petals yellow.
      d. Leaves 3-foliolate. P. norvegica L.
      d. Leaves, or the principal ones, 5-foliolate.
        e. Leaflets silvery-pubescent beneath, laciniate-toothed, teeth with revolute margins; stipules small. *P. argentea L., Silvery C.
        e. Leaflets pubescent beneath but not silvery, crenate-dentate; stipules large.
          f. Inflorescence nearly leafless; petals about 1 cm long, pale yellow; stamens 25-30; achene with network of ridges. *P. recta L., Upright C.
          f. Inflorescence leafy; petals about 0.5 cm long, deep yellow; stamens fewer; achene with longitudinal ridges. *P. intermedia L.
    c. Flowers solitary, or rarely 2, in axils; stems trailing.
      d. Flowers 18-25 mm wide; perianth 4-5-merous; leaflets glabrous or only slightly pubescent beneath. *P. reptans L.
      d. Flowers 15 mm wide or less; petals and sepals 5; leaflets pubescent beneath.
        e. Leaves not fully grown at anthesis; leaflets obovate, upper 1/2 rounded, fan-shaped, lower 1/2 entire, cuneate; dense pubescence of lower side of bracts obscuring veins; leaves mostly basal, stem soon prostrate; hairs white. P. canadensis L.
        e. Leaves fully grown at anthesis; leaflets, except of lowest leaves, elliptic or oblanceolate, upper 3/4 of margin toothed, lower 1/4 entire; veins on lower side of distal half of bracts evident; several lower internodes of stem erect; hairs not noticeably white. P. simplex Michx., Common C.

15. Filipendula Adans.

Tall herbs; leaves pinnately compound, in ours with terminal leaflet large, palmately lobed and veined; stipules large; hypanthium small; flowers scarcely perigynous, 1 cm wide or less, in large panicles; sepals and petals 4-7; stamens many; carpels 5-15, separate, 2-ovuled, indehiscent and usually 1-seeded in fruit.

a. Petals pink; terminal leaflet 7-9-lobed; carpels not twisted in fruit. F. rubra (Hill) Robins.,
   Queen-of-the-prairie
a. Petals white; terminal leaflet 3-5-lobed; carpels twisted in fruit. *F. ulmaria (L.) Maxim.,
   Queen-of-the-meadow

16. Geum L., Avens

Herbs; basal and lower cauline leaves various, simple or often pinnatifid or pinnately compound, middle and upper cauline leaves usually 3-foliolate; sepals 5; petals 5; stamens many; carpels many, separate; styles persistent in fruit, in ours jointed and becoming hooked; fruit an aggregate of achenes.

a. Calyx without bracts; fruit on a stipe above hypanthium; petals 1-2 mm long, yellow. G. vernum
   (Raf.) T. & G., Spring A.
a. Calyx with bracts.
   b. Calyx usually purple, erect or spreading; petals 7-10 mm long, usually suffused with purple;
      fruit stalked above hypanthium; style plumose above joint, hirsute on lower portion. G. rivale
      L., Purple A.
   b. Calyx green, spreading; petals white or yellow; fruit sessile in the hypanthium.
      c. Petals yellow or orange-yellow, about as long as calyx; receptacle long-hairy; fruit globose-
         ovoid; stem stout, hirsute. G. aleppicum Jacq. var. strictum (Ait.) Fern.
      c. Petals white or, if not, then much shorter than the calyx.
         d. Receptacle glabrous or minutely pubescent; pedicels stout, copiously hirsute; petals white,
            shorter than the calyx. G. laciniatum Murr.
         d. Receptacle densely hairy; pedicels slender, minutely pubescent; fruit obovoid; achenes,
            except sometimes uppermost, reflexed.
            e. Petals white, about as long as sepals; teeth and tips of leaves or leaflets acute; stem
               glabrous or sparingly pubescent. G. canadense Jacq.
            e. Petals pale yellow, much shorter than sepals; teeth and tips of leaves or leaflets blunt;
               stem usually hirsute below. G. virginianum L.

17. Rubus L., Blackberry, Raspberry, Dewberry

Usually shrubs; stems erect or trailing, often prickly; stems of first year usually unbranched and not flowering (primocanes); from these, the second year, short flowering branches (floricanes) grow; leaves lobed or pinnately or palmately compound, sometimes different on the two kinds of stems; flowers bisporangiate or monosporangiate, in terminal or axillary racemes, corymbs, or panicles, or rarely solitary; sepals usually 5; petals 5, white, rarely rose-color or pink; stamens many; carpels many, separate, each with 2 ovules, on a convex or elongate dry or fleshy receptacle; fruit an aggregate of 1-seeded drupelets readily separable from a dry receptacle or remaining attached to a fleshy receptacle.

a. Leaves simple, palmately 3-5-lobed; petals rose-color; flower 3-5 cm wide. R. odoratus L.,
   Rose-flowered R.
a. Leaves, or most of them, compound.
   b. Blades white-downy or white-felted beneath; stems bristly or prickly or both.
      c. Drupelets separating from dry receptacle; blades white-downy beneath.
         d. Calyx equaling or slightly longer than petals; stems bristly or prickly or both, or nearly
            smooth; fruit red (rarely yellow). R. idaeus L. var. strigosus (Michx.) Maxim., Red R.
         d. Calyx much longer than petals.
            e. Stems very glaucous, not bristly but with stout recurved prickles; pedicels prickly;
               fruit purple-black. R. occidentalis L., Black R.
            e. Stems, pedicels, and calyx with abundant long gland-tipped reddish bristles.
               *R. phoenicolasius Maxim., Wineberry
      c. Drupelets not separating from receptacle; blades white-felted beneath; stems with strong flat
         curved prickles. *R. procerus P. J. Muell., Himalaya Berry
   b. Blades not white-downy beneath; stems prickly or occasionally unarmed, not glaucous; drupelets
      not separating from receptacle.
      c. Bladelets laciniate-cleft; calyx lobes with foliaceous appendages. *R. laciniatus Willd.,
         Cutleaf B.

c. Bladelets not laciniate-cleft.
  d. Stems trailing or low-arched, often rooting at tips, with erect or ascending flowering branches; flowers solitary or clusters few-flowered.
    e. Stems herbaceous, slender, usually without prickles; stipules oblanceolate; fruit red-purple. R. pubescens Raf., Dwarf R.
    e. Stems somewhat woody, prickly or bristly; stipules linear to setaceous.
      f. Stems with bristles; leaves coriaceous, evergreen or nearly so, central leaflet not acute or acuminate; petals 5-9 mm long. R. hispidus L.
      f. Stems with a few thick-based slightly curved prickles; leaves not coriaceous, central leaflet acute; petals 1-1.5 cm long. (See third f.) R. flagellaris Willd., Common D.
      f. Stems with both bristles and prickles; leaves of primocanes 5-foliolate, central leaflet with bristly and prickly petiolule; petals 7-10 mm wide. R. trivialis Michx., Southern D.
  d. Stems erect or ascending; inflorescence with several to many flowers. Blackberries
    e. Inflorescence conspicuously stipitate-glandular, not leafy. R. allegheniensis Port.
    e. Inflorescence glandless or nearly so.
      f. About half the pedicels subtended by leaflike bracts. R. frondosus Bigel.
      f. None of the pedicels or only the lowest few subtended by leaflike bracts.
        R. pensilvanicus Poir.

18. Dalibarda Kalm

D. repens L. Low pubescent herb with prostrate stem; leaves basal, blades simple, round-ovate, cordate, 3-5 cm long; flowers of 2 kinds, usually sterile petaliferous ones solitary or 2 together on upright scape, and fertile apetalous cleistogamous ones on recurved scapes; sepals 5 or 6, bract-less; petals 5, white, 4-8 mm long; stamens many; carpels 5-10; fruit an aggregate of nearly dry drupes.

19. Agrimonia L., Agrimony

Erect herbs; leaves pinnately compound, small leaflets interspersed with larger ones; stipules large, toothed or laciniate; flowers small, in narrow, often interrupted, spikelike racemes; hypanthium hemispheric or turbinate, at maturity with band of hooked bristles just below base of sepals; sepals 5; petals 5, yellow; stamens 5-15; carpels 2, separate, enclosed by but free from persistent hypanthium, which becomes hardened in fruit around the 2 achenes.

a. Leaves with 11-15 larger leaflets and many smaller ones; leaflets lanceolate or elliptic, gland-dotted below, those of upper and middle leaves acuminate; stem densely long-hairy. A. parviflora Ait.
a. Leaves with 5-9 larger leaflets, these ovate, obovate, or elliptic.
  b. Axis of inflorescence glandular; lower surface of leaflets gland-dotted and glabrous or almost glabrous between veins, sometimes pubescent on veins.
    c. Stem with scattered long hairs; bristles of fruit in several rows on projecting flange, lower bristles becoming wide-spreading, lowest somewhat reflexed; roots not thickened.
      A. gryposepala Wallr.
    c. Stem nearly glabrous except for glands; bristles of fruit rather scanty, in narrow band, not on a flange, erect or ascending, not overtopping beak of calyx; roots thickened. A. rostellata Wallr.
  b. Axis of inflorescence pubescent but not or little glandular; blades pubescent beneath; bristles of fruit ascending.
    c. Leaflets acuminate, gland-dotted beneath; roots not thickened; mature hypanthium 4-6 mm long. A. striata Michx.
    c. Leaflets acute or obtuse, not or obscurely gland-dotted beneath; roots thickened; mature hypanthium less than 4 mm long. A. pubescens Wallr.

20. Sanguisorba L., Burnet

Herbs; leaves pinnately compound; flowers small, monosporangiate or bisporangiate, in dense spikes or heads; hypanthium narrowed at mouth, somewhat 4-angled; sepals 4, petaloid; petals none; stamens 4 or many; carpels 2, separate, or 1, enclosed by hypanthium but free from it; hypanthium persistent in fruit, becoming hardened around the achene or achenes.

a. Stamens many; carpels 2; flowers in dense heads; leaflets ovate, 2 cm long or less. *S. minor Scop.

a. Stamens 4; carpel 1; flowers in cylindric spikes; leaflets oblong, 3-10 cm long. S. canadensis L., American B.

21. Rosa L., Rose

Shrubs, erect, trailing, or climbing, usually prickly; leaves pinnately compound; stipules adnate to petiole; flowers large, showy, solitary or in corymbs; hypanthium urn-shaped, narrowed at mouth, becoming fleshy in fruit; sepals 5, tips usually prolonged and sometimes foliaceous; petals 5; stamens many; carpels many, inserted on bottom, and sometimes also on sides, of hypanthium; fruit an aggregate of bony achenes, enclosed by but free from the fleshy hypanthium.

a. Styles united in a column exserted beyond opening of hypanthium.
   b. Leaflets 3, rarely 5; stipules nearly entire, glandular-ciliate; petals pink, 2-3 cm long.
      R. setigera Michx., Prairie R.
   b. Leaflets 7-9; stipules toothed.
      c. Inflorescence many-flowered; flowers about 2 cm wide; petals usually white; stems climbing or arching; styles glabrous. *R. multiflora Thunb.
      c. Inflorescence few-flowered; flowers 4-5 cm wide; stems trailing; styles pubescent.
         *R. wichuraiana Crep.
a. Styles separate, not or only slightly exserted in a broad mass.
   b. Sepals unlike, some of them pinnatifid; prickles stout, decurved; mouth of hypanthium 1-2 mm wide.
      c. Leaves glabrous, without glands; styles pubescent; fruit ovoid, 1.5-2 cm long, scarlet.
         *R. canina L., Dog R.
      c. Leaves pubescent and glandular beneath; fruit subglobose.
         d. Styles pubescent; sepals persistent in fruit. *R. eglanteria L., Sweet-brier
         d. Styles glabrous; sepals deciduous in fruit. *R. micrantha Sm.
   b. Sepals alike in form; mouth of hypanthium wider than 2 mm.
      c. Pedicels and hypanthium usually glabrous; sepals persistent in fruit.
         d. Upper surface of leaflets dark green and rugose; stems densely prickly; stem and bases of prickles pubescent. *R. rugosa Thunb., Japanese R.
         d. Upper surface of leaflets pale green, smooth; stems glabrous, without prickles or with few near base. R. blanda Ait. Smooth R.
      c. Pedicels and hypanthium stipitate-glandular; sepals deciduous from fruit; flowers pink, about 5 cm wide, stipules narrow, sides parallel.
         d. Prickles stout, recurved; stipules with inrolled edges; leaflets pubescent beneath.
            R. palustris Marsh., Swamp R.
         d. Prickles slender, straight; stipules not inrolled; leaflets glabrous or pubescent beneath.
            R. carolina L.

22. Prunus L., Peach, Plum, Cherry

Shrubs and trees; leaves simple, alternate, blades serrate, pinnately veined, disklike or toothlike glands on petiole; hypanthium usually deciduous in fruit; sepals and petals 5; stamens 15 or more; carpel 1; ovules 2; fruit a 1-seeded drupe.

a. Ovulary and fruit velvety; stone sculptured; leaves lanceolate, conduplicate in bud; flowers sessile or nearly so; petals pink. *P. persica Batsch, Peach
a. Ovulary and fruit glabrous; flowers pediceled; petals usually white.
   b. Flowers 15 or more in racemes that are longer than wide, at end of leafy branches of current year, appearing with or after leaves; leaves folded in bud; petals white; flowers about 1 cm wide; fruit black, globose, 8-10 mm wide.
      c. Shrub; leaves obovate or ovate, glabrous beneath or with tufts of hair in vein axils; glands of petiole usually rounded or disk-shaped; sepals deciduous in fruit. P. virginiana L., Choke C.
      c. Tree; leaves lance-oblong to ovate, tapering to tip, often villous along midvein beneath; glands of petiole often toothlike; sepals persistent in fruit. P. serotina Ehrh., Wild Black C.
   b. Flowers in umbels, corymbs, or short racemes, appearing before or with the young leaves.
      c. Flowers in umbels.
         d. Flower clusters with involucre of persistent inner bud-scales and sometimes some small leaves at base; petals 10-15 mm long; leaves folded in bud.

    e. Young leaves pubescent beneath, conspicuous glands on petiole, teeth not incurved; inner bud scales spreading or reflexed; hypanthium constricted at apex. *P. avium L., Sweet C.

    e. Young leaves glabrous beneath, teeth incurved, glands of teeth near sinus; inner bud scales erect. *P. cerasus L., Sour C.

  d. Flowers without persistent bud scales or bracts at base.

    e. Sepals glabrous on upper side; leaves folded in bud.

      f. Low shrubs; leaf-blades oblanceolate or obovate, lower part of margins entire, glands between the teeth; sepals glandular.

        g. Blades long-tapering to both base and apex. P. pumila L., Sand C.

        g. Blades acute at base, somewhat rounded at apex. P. susquehanae Willd.

      f. Larger shrubs or small trees.

        g. Flowers not less than 2 cm wide; petals white becoming pink; sepals glabrous on lower side, with marginal glands; leaves folded in bud; blades obovate, abruptly pointed, teeth often in 3's, a larger one with a smaller on each side. P. nigra Ait.

        g. Flowers 1.2 cm wide, white; sepals without marginal glands; leaves folded in bud; blades glabrous, lanceolate or ovate, gradually long-pointed at tip. P. pensylvanica L. f., Pin C.

    e. Sepals pubescent on upper side.

      f. Sepals with marginal glands.

        g. Sepals often pubescent on lower side; flowers not more than 1.5 cm wide; petals white; fruit 2-3 cm wide.

          h. Leaves folded in bud, flat, usually oblong-ovate or -obovate, slightly pubescent beneath at maturity. P. hortulana Bailey, Wild Plum

          h. Leaves rolled in bud, somewhat troughed, usually oblong-lanceolate or lanceolate, hairy on midvein at maturity. P. munsoniana Wight & Hedrick, Wild Goose Plum

        g. Sepals glabrous on lower side; flowers at least 2 cm wide; petals white becoming pink; leaves folded in bud, obovate, abruptly pointed, teeth often in 3's, a larger one with a smaller on each side. P. nigra Ait., Canada Plum

      f. Sepals without marginal glands, glabrous on lower side.

        g. Flowers 1.5 cm wide or more, appearing before leaves; leaves folded in bud, obovate or oblong-ovate, sharply serrate, acuminate, teeth without glands. P. americana Marsh., Wild Plum

        g. Flowers not more than 1 cm wide; leaves rolled in bud, lanceolate to oblong-lanceolate, troughed, acute or short-acuminate. *P. angustifolia Marsh., Chickasaw Plum

  c. Flowers in corymbs or short racemes.

    d. Flowers in short racemes of about 6-10; blades ovate to circular, short-acuminate; fruit dark red or black. *P. mahaleb L., Mahaleb C.

    d. Flowers in corymbs or umbel-like clusters; blades lanceolate to ovate, long-tapering to tip; fruit red. P. pensylvanica L. f., Pin C.

LEGUMINOSAE, Pea Family

    Herbs and woody plants; leaves alternate, with pulvini, usually stipuled, usually compound; calyx of 5 (rarely fewer) united sepals, the teeth or lobes sometimes fewer than five; petals 5 (rarely 1 or none); corolla regular, petals united or separate; or corolla somewhat zygomorphic, petals separate below, odd petal enclosed by the others in bud; or corolla zygomorphic and papilionaceous, petals separate below, odd petal (standard) enclosing in bud the 2 lateral petals (wings) and the 2 inner somewhat connivent petals (keel); stamens 5-many, rarely fewer than 5, usually 10, separate, monadelphous, or diadelphous, sometimes alternately long and short; carpel 1, ovulary 1-loculed or rarely 2-loculed; placenta 1, parietal; fruit a legume, sometimes indehiscent, sometimes divided crosswise into 1-seeded segments (loment).

a. Trees and shrubs.

  b. Leaves simple, ovate-cordate, entire; flowers pink- or red-purple, in small clusters along branches of previous years. 5. Cercis

  b. Leaves compound; flowers in spikes, racemes, or panicles.

    c. Shrubs; petal 1, the standard, dark-purple; leaves pinnately compound, gland-dotted. 15. Amorpha

c. Petals more than 1.
d. Flowers bisporangiate, papilionaceous, petals white or pink; leaves odd-pinnate.
e. Flowers in racemes; stamens diadelphous. 18. Robinia
e. Flowers in elongate panicles; stamens separate. 7. Cladrastis
d. Flowers monosporangiate or some bisporangiate, not papilionaceous; petals greenish-yellow or -white; leaves even-pinnate.
e. Flowers in panicles; hypanthium tubular; leaflets (except 2-4 basal ones) again compound; thorns absent. 2. Gymnocladus
e. Flowers in slender racemes; hypanthium cup-shaped; early leaves pinnate, later ones bipinnate; with or without thorns. 3. Gleditsia
a. Herbs, sometimes twining, and woody vines.
b. Corolla regular or only slightly zygomorphic; stamens separate.
c. Leaves bipinnate; corolla regular, greenish-white. 1. Desmanthus
c. Leaves pinnate; corolla slightly zygomorphic, yellow. 4. Cassia
b. Corolla zygomorphic, papilionaceous; odd petal outside in bud.
c. Leaves with terminal tendrils.
d. Style filiform, terete, hairy all around at tip. 24. Vicia
d. Style flattened, hairy on side toward free stamen; usually leaflets longitudinally veined or stipules conspicuous. 25. Lathyrus
c. Leaves without tendrils.
d. Leaves simple. 8. Crotalaria
d. Leaves palmately 5-11-foliolate. 9. Lupinus
d. Leaves 3-foliolate. (See fourth d.)
e. Leaflets serrulate.
f. Flowers in long slender racemes; petals white or yellow. 11. Melilotus
f. Flowers in heads, short spikes, or short racemes.
g. Fruits curved or coiled; petals not united with stamen-tube. 12. Medicago
g. Fruits not curved or coiled; keel and lateral petals somewhat united with stamen-tube. 10. Trifolium
e. Leaflets entire.
f. Stamens separate; flowers showy, in racemes. 6. Baptisia
f. Stamens monadelphous or diadelphous.
g. Petals yellow; flowers in heads, short spikes, or umbels.
h. Stamens monadelphous; stipules adnate to petiole for half their length forming a tube; fruit of 1 or 2 short segments. 23. Stylosanthes
h. Stamens diadelphous; leaflets actually 5, lower pair stipulelike; fruit linear. 13. Lotus
g. Petals not yellow.
h. Fruit a loment of 2-several segments, pubescent, each hair curved or hooked at tip; leaflets with or without stipels; flowers in simple or panicled racemes; stem sometimes prostrate, not twining. 21. Desmodium
h. Fruit not a loment.
i. Leaflets without stipels; stems not twining.
j. Blades gland-dotted; racemes and peduncles axillary, elongate; fruit cross-wrinkled. 14. Psoralea
j. Blades not gland-dotted; fruit not cross-wrinkled.
k. Calyx subtended by small bracts; stipules free from petiole. 22. Lespedeza
k. Calyx without bracts; stipules united with petiole; flowers in dense heads. 10. Trifolium
i. Leaflets with stipels; stems often twining.
j. Woody or half-woody vine to many meters long; leaflets often lobed; flowers red-purple; racemes elongate. *Pueraria lobata (Willd.) Ohwi, Kudzu-vine
j. Herbaceous, stem much shorter.
k. Style bearded lengthwise on upper surface.
l. Keel spirally coiled; flowers in elongate racemes; stem twining (erect in some cultivated species). 27. Phaseolus
l. Keel not spirally coiled; flowers solitary or in heads, umbels, or short racemes.

134

    m. Flowers 4-6 cm long; standard much longer than other petals; erect or twining. 29. Clitoria
    m. Flowers 1.5 cm long or less; standard not much longer than other petals; erect or twining. 28. Strophostyles
   k. Style glabrous except sometimes at base; flowers in racemes.
    l. Pedicels subtended by wide striate bracts with truncate summit; calyx without bracts; racemes short, dense; leaflets acute. 30. Amphicarpa
    l. Pedicels and calyx subtended by small pointed bracts; racemes narrow, interrupted; leaflets obtuse or emarginate. 31. Galactia
 d. Leaflets more than 3; leaves pinnately compound.
  e. Flowers in umbels or heads.
   f. Corolla yellow; leaflets 5, lower pair stipulelike. 13. Lotus
   f. Corolla rose-purple to blue; leaflets many. 20. Coronilla
  e. Flowers in racemes or spikes.
   f. Vines.
    g. Herbaceous; corolla brown-purple; leaflets 5-7. 26. Apios
    g. Woody; corolla lavender-blue; leaflets 7-many. *Wisteria
   f. Not vines.
    g. Blades gland-dotted. 16. Dalea
    g. Blades not gland-dotted.
     h. Standard ovate or circular; fruit flattened, villous; tenth stamen partly free. 17. Tephrosia
     h. Standard narrow; fruit terete; tenth stamen free. 19. Astragalus

1. Desmanthus Willd.

 D. illinoensis (Michx.) MacM. Herbs; leaves even-bipinnate, leaflets very small; stipules small, slender; flowers small, in peduncled axillary heads; corolla greenish-white, regular; stamens 5, separate; fruits up to 3 cm long, often twisted or curved in the dense head.

2. Gymnocladus Lam., Kentucky Coffee Tree

 G. dioica (L.) K. Koch Tree with stout branchlets; leaves large, compound, upper leaflets again compound; leaflets ovate; axillary buds superposed, sunken; pith large, pinkish-brown; imperfectly diecious; flowers monosporangiate, each with vestiges of the other sporophylls, in terminal panicles; hypanthium tubular; corolla regular, greenish-white; stamens 10, separate; fruit large and woody, pulpy between the few large seeds.

3. Gleditsia L.

 G. triacanthos L., Honey-locust. Tree with branched supra-axillary thorns; leaves pinnately compound or the later ones partly or wholly decompound, often crowded on short branchlets, leaflets obscurely serrate; buds superposed, at least the lower at each node beneath base of petiole; flowers monosporangiate or rarely bisporangiate, in spikelike racemes; sepals and petals 3-5; stamens 3-10, separate; fruit large, with sweet pulp between the many seeds. Some cultivated forms thornless.

4. Cassia L.

 Ours herbs; leaves even-pinnate, leaflets oblong, entire; corolla nearly regular or zygomorphic; petals yellow in ours; stamens 5-10, separate, often unequal, sometimes some of them imperfect; anthers opening by 2 apical pores; fruit often septate.

a. Leaflets usually 3 cm long or more, not sensitive to touch; 3 upper anthers imperfect; stipules deciduous.
 b. Leaflets 4-6; fruit 4-seeded, not appearing jointed; gland between leaflets of lowest pair.
  C. tora L., Sickle-pod
 b. Leaflets 8-20; fruit flattened.
  c. Ovulary densely villous; fruit sparsely villous, segments as long as wide; petiole-gland club-shaped, stalked. C. hebecarpa Fern., Wild Senna
  c. Ovulary appressed-pubescent; fruit glabrous or glabrate, segments wider than long; petiole-gland sessile. C. marilandica L., Wild Senna

a. Leaflets 2 cm long or less, somewhat sensitive to touch; anthers all perfect but unequal; stipules persistent; petiole-gland saucer-shaped.
   b. Stamens 10; calyx about 1 cm long; petals 1-2 cm long. C. fasciculata Michx., Partridge-pea
   b. Stamens 5; calyx 3-4 mm long; petals 4-8 mm long. C. nictitans L. Wild Sensitive Plant

5. Cercis L.

C. canadensis L., Redbud. Small tree; leaves simple, entire, ovate-cordate, palmately veined, 2-ranked; flowers pink-purple, zygomorphic, in small clusters along branches of preceding years, appearing before leaves of the season; calyx 5-toothed; petals 5, uppermost smallest, in bud enfolded by the lateral ones; fruit flat, 6-10 cm long, several-seeded.

6. Baptisia Vent., False-indigo

Herbs; leaves palmately 3-foliolate, usually blackening in drying; flowers in racemes; calyx 4-5-toothed; stamens 10, separate; ovulary stipitate in persistent calyx; fruit inflated.

a. Racemes 1-few, several- to many-flowered; flowers 2-3 cm long.
   b. Petals blue or bluish; stipe of fruit not longer than calyx. B. australis (L.) R. Br.
   b. Petals white; stipe of fruit longer than calyx. B. leucantha T. & G.
a. Racemes few-flowered, many, at ends of many branches; flowers 1-1.5 cm long. B. tinctoria (L.) R. Br.

7. Cladrastis Raf., Yellowwood

*C. lutea (Michx. f.) K. Koch Tree with yellow wood; leaves odd-pinnate, oval or ovate leaflets alternate on rachis; lateral buds superposed, enclosed by hollow base of petiole; stipules none; flowers 2-2.5 cm long, in large drooping terminal panicles; corolla white; stamens 10, separate; fruit stalked in the calyx, flat, several-seeded.

8. Crotalaria L., Rattlebox

C. sagittalis L. Leaves simple, oval or the upper ones narrower; stipules inversely sagittate; corolla yellow, shorter than calyx; stamens monadelphous, the sheath split on upper side; anthers of 2 forms; peduncles terminal and axillary, racemes few-flowered; fruit coriaceous, inflated, the several seeds loose at maturity.

9. Lupinus L., Lupine

L. perennis L. Herb; leaves palmately 5-11-foliolate; leaflets oblanceolate; stipules adnate to petiole; flowers 1-1.5 cm long, in a terminal raceme; corolla blue or sometimes pink or white; stamens monadelphous; anthers of 2 forms; fruit flat, pubescent, constricted between the 5-6 seeds.

10. Trifolium L., Clover

Herbs; leaves 3-foliolate, leaflets usually serrulate; stipules united with petioles for part of their length; flowers small, in heads or short dense racemes; claws of petals more or less united with stamen tube; stamens diadelphous; fruit short and straight, often included in persistent calyx, indehiscent or tardily dehiscent, 1-6-seeded.

a. Corolla yellow.
   b. Leaflets all sessile or nearly so; standard striate-sulcate in age. *T. agrarium L., Yellow Hop C.
   b. Terminal leaflet stalked.
      c. Heads of usually 20 or more flowers; standard conspicuously striate-sulcate in age.
         *T. procumbens L., Low Hop C.
      c. Heads usually of not more than 15 flowers; standard not, or scarcely, striate-sulcate in age.
         *T. dubium Sibth.
a. Corolla not yellow.
   b. Inflorescence longer than wide.
      c. Corolla crimson; flowers 12 mm long or more; leaflets obovate. *T. incarnatum L., Crimson C.

  c. Corolla grayish-white or pink; flowers no more than 7 mm long; leaflets oblanceolate or narrower. *T. arvense L., Rabbit-foot C.
 b. Inflorescence globose.
  c. Flowers sessile or nearly so; corolla rose-purple to white; one or two leaves just under the
   head. *T. pratense L., Red C.
  c. Flowers stalked.
   d. Calyx lobes twice as long as tube or more, pubescent, sometimes minutely so.
    e. Leaflets broadly obovate; plants stoloniferous. T. stoloniferum Muhl., Running
     Buffalo C.
    e. Leaflets narrower; plants not stoloniferous. T. reflexum L., Buffalo C.
   d. Calyx teeth longer or shorter than tube but not twice as long; calyx glabrous or nearly so.
    e. Leaf-bearing stems erect; calyx tube shorter than the shorter teeth. *T. hybridum L.,
     Alsike C.
    e. Leaf-bearing stems repent; petioles and scapes erect; calyx tube longer than the shorter
     teeth. *T. repens L., White C.

11. Melilotus Mill., Sweet Clover

  Tall herbs, fragrant in drying; leaves 3-foliolate, leaflets serrulate, terminal one stalked; stipules partly adnate to petiole; flowers small, in elongate peduncled racemes; calyx lobes slender, nearly
equal; stamens diadelphous; fruit ovoid, exceeding the calyx, indehiscent or finally dehiscent, 1-2-
seeded.

a. Corolla white, 4-5 mm long; fruit reticulate. *M. alba Desr., White S.
a. Corolla yellow, 5-6 mm long.
 b. Fruit cross-ribbed, glabrous or glabrate. *M. officinalis (L.) Lam., Yellow S.
 b. Fruit pubescent; rare. *M. altissima Thuill.

12. Medicago L.

  Herbs; leaves pinnately 3-foliolate, leaflets serrulate, terminal one stalked; stipules partly adnate to petiole; flowers in axillary globose or short-cylindric spikes or racemes; stamens diadelphous;
fruit coiled in Ohio species, usually indehiscent; seeds 1-few.

a. Corolla yellow, 2-4 mm long; annual; prostrate or ascending; leaflets obovate, to 1.5 times as long
 as wide. *M. lupulina L., Hop or Black Medick
a. Corolla blue, violet, white, or yellow, 7-12 mm long; perennial; stems erect; leaflets 2 or more
 times as long as wide. *M. sativa L., Alfalfa

13. Lotus L.

  *L. corniculatus L., Bird's-foot Trefoil. Herb with 5-foliolate pinnately-compound leaves, often
with glands in position of stipules, the 2 lower leaflets remote from the others and appearing to be stipules; flowers about 1 cm long, in headlike umbels on axillary peduncles; petals yellow or marked with
red or brown; stamens diadelphous; fruit several-seeded.

14. Psoralea L., Scurf-pea

  Herbs; leaves more or less gland-dotted, in ours pinnately 3-foliolate; flowers in spikes or racemes, corolla blue or purple to white; stamens diadelphous or monadelphous; fruit ovoid, indehiscent.

a. Leaflets ovate-lanceolate, acuminate; peduncles equaling or shorter than subtending leaves; calyx
 minutely hairy or glabrate. P. onobrychis Nutt.
a. Leaflets narrowly lanceolate, not acuminate; peduncles longer than subtending leaves; calyx hirsute.
 P. psoralioides (Walt.) Cory

15. Amorpha L.

  A. fruticosa L., False Indigo. Shrub; leaves odd-pinnate, leaflets many, entire, 2-4 cm long,
gland-dotted; flowers small, in dense panicled racemes; petal 1, the standard, purple, folded around
stamens and style; stamens 10, separate above, monadelphous at base; fruit oblong, gland-dotted, 1-2-
seeded.

16. <u>Dalea</u> Willd.

*<u>D</u>. alopecuroides Willd. Leaves pinnately compound, leaflets many, stipules minute; flowers small, in dense spikes; calyx villous; corolla white or pink, wings and keel adnate to stamen-tube; fruit indehiscent, 1-seeded.

17. <u>Tephrosia</u> Pers.

<u>T</u>. <u>virginiana</u> (L.) Pers., Goat's Rue, Devil's Shoestring. Silky-villous herb; leaves odd-pinnate, the several leaflets entire, oblong to elliptic, with mucronate apex; flowers 1.5-2 cm long, in a terminal raceme; standard yellowish, pubescent on outside; wings and keel pink or rose-purple, coherent; stamens 10, 1 partly free, 9 united; fruit linear, pubescent, several-seeded.

18. <u>Robinia</u> L., Locust

Trees or shrubs; leaves odd-pinnate, leaflets entire; flowers 2-2.5 cm long, in axillary or terminal racemes; corolla white, pink, or rose; stamens diadelphous; fruit flat, several-seeded.

a. Trees; corolla white; twigs and fruit glabrous or puberulent; stipular spines usually present.
   <u>R</u>. pseudoacacia L., Black L.
a. Corolla pink to rose-purple; twigs and fruit bristly or glandular-viscid; stipules setaceous, not spines.
   b. Twigs and peduncles covered with brown bristlelike hairs; shrubs. *<u>R</u>. <u>hispida</u> L., Bristly L., Rose-acacia
   b. Twigs and peduncles glandular-viscid; trees. *<u>R</u>. <u>viscosa</u> Vent., Clammy L.

19. <u>Astragalus</u> L., Milk-vetch

Leaves odd-pinnate, leaflets entire; flowers in spikes or racemes; corolla white, yellowish, or purple, usually long and narrow; stamens 10, diadelphous; fruit sometimes partially or wholly divided lengthwise into 2 locules.

a. Fruit scarcely inflated, 2-loculed, less than 1 cm thick; racemes long-peduncled, densely many-flowered. <u>A</u>. <u>canadensis</u> L.
a. Fruit much inflated, 1-loculed, 1 cm thick or more; racemes short. <u>A</u>. <u>neglectus</u> (T. & G.) Sheld.

20. <u>Coronilla</u> L., Crown-vetch

*<u>C</u>. <u>varia</u> L. Herb; stems ascending; leaves odd-pinnate; leaflets oblong, several-many; flowers in umbels on axillary peduncles; corolla purple or pink; stamens diadelphous; fruit linear, coriaceous, transversely segmented.

21. <u>Desmodium</u> Desv., Tick Trefoil

Herbs; leaves 3-foliolate; stipels usually present; flowers in racemes or panicles, petals white, pink, or purple; stamens 10, remnant of column persistent in fruit; fruit a loment of 2 to several flat 1-seeded indehiscent segments, pubescent with hooked hairs.

a. Stamens monadelphous; mature loment-stipe longer than persistent stamen-column; loment deeply indented on lower margin, straight or concave on upper; woodland plants.
   b. Terminal bladelets round-ovate, abruptly long-acuminate, about as wide as long; leaves crowded at base of long terminal peduncle (rarely scattered); petals pink. <u>D</u>. <u>glutinosum</u> (Muhl.) Wood
   b. Terminal bladelets ovate, obovate, or rhombic, longer than wide, not strongly acuminate.
      c. Petals usually pink; flowers usually in a raceme, on scape arising from base of plant; leaves crowded at top of flowerless stem (rarely scattered). <u>D</u>. <u>nudiflorum</u> DC.
      c. Petals white; flowers few, in short racemes terminal and rarely axillary on leafy slender erect or decumbent stems. <u>D</u>. <u>pauciflorum</u> (Nutt.) DC.
a. Stamens diadelphous; loment-stipe rarely as long as persistent stamen-column; loment usually indented at both margins, sometimes not equally; stem leafy.
   b. Stems prostrate; bladelets almost circular; stipules to 12 mm long, ovate, persistent; flowers about 1 cm long, floral bracts conspicuous. <u>D</u>. <u>rotundifolium</u> DC.
   b. Stems erect or ascending.

c. Petioles 5 mm long or less; bladelets narrow, length 4-10 times width; loment-segments 2-3, oval or circular; flowers 3-6 mm long. D. sessilifolium (Torr.) T. & G.
c. Petioles, except sometimes of upper leaves, more than 5 mm long.
  d. Stipules 3 mm wide or more, slightly cordate or clasping; flowers about 1 cm long; loment-segments 3-6.
    e. Petals usually pink, drying green; flowers in a panicle; loment-segments longer than wide, angled below; bladelets scarcely reticulate, ovate, hairy, often yellowish-green. D. canescens (L.) DC.
    e. Petals white; flowers usually in a raceme; loment-segments oval to circular, rounded below; bladelets strongly reticulate, lance-ovate. D. illinoense Gray
  d. Stipules narrower than above, often early deciduous.
    e. Flowers 3-6 mm long; loment-segments 1-3, almost circular, upper edge often thickened; bladelets 3 cm long or less, length no more than twice width, rounded at apex, or, if to 5 (7) cm long, then firm and strongly reticulate.
      f. Terminal bladelet to 5 cm long or more, lanceolate, firm, strongly reticulate, pilose beneath; rare. D. rigidum (Ell.) DC.
      f. Terminal bladelet 3 cm long or less, half as wide or more, apex blunt.
        g. Bladelets pilose above, less so beneath; petioles mostly 0.5 to 1 cm long, densely long-pilose. D. ciliare (Muhl.) DC.
        g. Bladelets almost glabrous above, somewhat pilose and paler beneath; petioles usually more than 1 cm long, sparsely hairy. D. marilandicum (L.) DC.
    e. Flowers 6-12 mm long; bladelets and loments not as above.
      f. Bladelets velvety-villous beneath, ovate, large, acute to blunt at tip; petioles densely pilose; stipules reddish, hairy beneath, usually reflexed; petals pink, drying green; loment-segments usually angled below. D. viridiflorum (L.) DC.
      f. Bladelets glabrous or hairy, not velvety-villous; stipules not reddish.
        g. Floral bracts conspicuous but deciduous before anthesis; stipules lanceolate, 8-17 mm long; loment-stipe usually shorter than calyx; flowers crowded.
          h. Petioles usually 4 cm long or more; terminal bladelet acuminate, ovate; loment-segments angled below; stipels long. D. cuspidatum (Muhl.) Loud.
          h. Petioles to 2 (3) cm; terminal bladelet obtuse or acute, lanceolate or lance-ovate; loment-segments rounded below. D. canadense (L.) DC.
        g. Floral bracts small; stipules small or early deciduous; loment segments usually 3-5, angled below; flowers 6-10 mm long.
          h. Bladelets glaucous beneath, ovate, mostly glabrous; stem glabrous or nearly so; loment-stipe much longer than calyx. D. laevigatum (Nutt.) DC.
          h. Bladelets not glaucous beneath. D. paniculatum (L.) DC. Two intergrading varieties: var. paniculatum has terminal bladelet 3 - 10 times as long as wide, stem and under surface of bladelets glabrous or hairs mostly appressed; var. dillenii (Darl.) Isely (D. dillenii Darl.) has bladelets 2-3 times as long as wide, stem and under surface of bladelets hairy, some hairs spreading.

22. Lespedeza Michx., Bush-clover

Herbs and shrubs; leaves rather small, 3-foliolate, leaflets without stipels; flowers usually small, solitary or in spikes, heads or racemes; corolla purple, white, or yellowish, or flowers apetalous, in separate clusters or mixed with those with petals, fruits forming in both; stamens diadelphous; fruit flattened, ovate, acuminate, indehiscent, 1-seeded.

a. Stipules wide, striate; lateral veins of leaflets parallel, ending at margin.
  b. Hairs of stem downwardly appressed. *L. striata (Thunb.) H. & A., Common L.
  b. Hairs of stem upwardly appressed. *L. stipulacea Maxim., Korean L.
a. Stipules subulate; lateral veins of leaflets usually anastomosing.
  b. Stems trailing.
    c. Pubescence of stem spreading. L. procumbens Michx.
    c. Pubescence of stem appressed or stem glabrous. L. repens (L.) Bart.
  b. Stems erect or ascending.
    c. Flowers in dense spikes or heads; corolla yellowish-white with purple spot on standard; calyx as long as fruit or longer.

      d. Spikes usually shorter than peduncles; calyx and fruit about equal; length of leaflets less than twice width. L. hirta (L.) Hornem.

      d. Spikes dense, headlike, usually longer than peduncles; calyx longer than fruit; length of leaflets more than twice width. L. capitata Michx.

   c. Flowers not in dense heads or spikes; corolla purple, pink, or white lined with purple; calyx shorter than fruit.

      d. Leaflets cuneate at base, truncate or retuse and mucronate at tip; flowers axillary, solitary or 2-3 together; corolla white with purple veins. *L. cuneata (Dumont) G. Don, Sericea L.

      d. Leaflets not cuneate at base; flowers purple, in several-flowered racemes.

         e. Peduncles mostly longer than subtending leaves.

            f. Stem not or little branched, hairs spreading; racemes closely several-flowered. L. nuttallii Darl.

            f. Stem bushy-branched, glabrous or hairs appressed; racemes loosely few-flowered. L. violacea (L.) Pers.

         e. Flower-clusters sessile or peduncles mostly shorter than subtending leaves.

            f. Leaflets more than twice as long as wide; clusters of petaliferous flowers sessile or nearly so. L. virginica (L.) Britt.

            f. Leaflets less than twice as long as wide; clusters of petaliferous flowers often short-stalked. L. intermedia (Wats.) Britt.

23. Stylosanthes Sw., Pencil-flower

S. biflora (L.) BSP. Low herb; stem branching from base, erect to almost prostrate; leaves pinnately 3-foliolate; stipules adnate to petiole for about half their length and to each other; leaflets with stipels; flowers small, in small leafy heads or short spikes; corolla yellow or orange; keel curved upward; stamens monadelphous; fruit of 1 or 2 segments, the lower one when present empty and stalk-like, the upper 1-seeded, dehiscent.

24. Vicia L., Vetch

Herbs, sometimes somewhat vinelike; leaves pinnately compound, usually with terminal tendrils; stipules half-sagittate; flowers 1 or 2 in axils or in axillary racemes; calyx irregular or almost regular, sometimes gibbous; stamens diadelphous; fruits dehiscent, 2-many-seeded.

a. Flowers solitary or 2-3 together, sessile or short-peduncled.

   b. Pubescent; flowers 18 mm long or more; fruit dark, pubescent, with constrictions; seeds flattened, 5 mm wide. *V. sativa L., Common V.

   b. Glabrous or glabrate; flowers 18 mm long or less; fruit tawny, not constricted; seeds globose, less than 5 mm wide. *V. angustifolia Reichard

a. Flowers in peduncled spikes or racemes or, if solitary, then long-peduncled.

   b. Calyx teeth about equal, less than 1 mm long, about as wide as long; flowers loosely clustered in raceme; corolla white, keel tipped with blue; fruit long-beaked. V. caroliniana Walt., Wood V.

   b. Calyx not regular or the teeth longer than 1 mm, some of them narrow-attenuate; racemes densely flowered.

      c. Racemes of not more than 10 flowers.

         d. Flowers 1.5 cm long or more; stipules several-toothed. V. americana Muhl.

         d. Flowers less than 1 cm long; stipules entire or with one basal lobe.

            e. Flowers 3-4 mm long; racemes 3-6-flowered; fruit tapering to evident beak, pubescent, 2-seeded. *V. hirsuta (L.) S. F. Gray

            e. Flowers 7-8 mm long; racemes 2-3-flowered, or flower solitary; fruit abruptly narrowed to minute beak, glabrous, 3-5-seeded. *V. tetrasperma (L.) Moench

      c. Racemes of more than 10 flowers.

         d. Pedicel attached at end of calyx; blade of standard as long as claw. *V. cracca L.

         d. Pedicel attached to lower side of calyx (calyx with sac or spur extending backward beyond pedicel attachment); blade of standard much shorter than claw.

            e. Stem spreading-pubescent, especially above; lower calyx-teeth long-hairy, 2 mm long or more. *V. villosa Roth, Hairy V.

            e. Stem glabrous, appressed-pubescent, or with a few spreading hairs; lower calyx-teeth rarely long-hairy, 2 mm long or less. *V. dasycarpa Ten.

25. Lathyrus L.

Climbing or trailing herbs, mostly glabrous; leaves with 1-6 pairs of leaflets and (in ours) ending in simple or branched tendrils; stipules usually conspicuous; flowers usually showy, solitary or in racemes from the axils; corolla yellow, purple, or white, or of other colors in cultivated species; fruit dehiscent, 2- to many-seeded.

a. Leaflets 2-6 pairs.
  b. Corolla yellowish-white; stipules asymmetric, semi-cordate, sometimes toothed; racemes shorter than subtending leaves. L. ochroleucus Hook.
  b. Corolla purple.
    c. Stipules symmetric, half to as large as leaflets, with 2 basal lobes. L. japonicus Willd., Beach Pea
    c. Stipules asymmetric, usually smaller, with 1 basal lobe.
      d. Peduncles nearly as long as to longer than subtending leaves; stems winged or wingless; leaflets usually 4-6, sometimes 8. L. palustris L.
      d. Peduncles no more than 2/3 as long as subtending leaves; leaflets 8-12. L. venosus Muhl.
a. Leaflets 1 pair.
  b. Stem wingless.
    c. Stipules symmetric, with 2 basal lobes; petals yellow; rhizomes slender. *L. pratensis L.
    c. Stipules with 1 basal lobe; petals purple or pink; rhizome bearing tubers. *L. tuberosus L.
  b. Stem winged.
    c. Peduncles 4-10-flowered; flowers odorless; corolla pink to white. *L. latifolius L., Everlasting Pea.
    c. Peduncles usually not more than 3-flowered, rarely 4; flowers fragrant; corollas of various colors. *L. odoratus L., Sweet Pea.

26. Apios Medic.

A. americana Medic. Twining herbs; rhizomes with tubers; leaves pinnate, stipules small, leaflets 5-7, ovate to lanceolate, acute or acuminate, stipels obscure; flowers 1 cm or more long, bracted, in axillary peduncled racemes; upper lobes of calyx minute, lowest lobe longest; corolla brown-purple, keel strongly incurved; stamens diadelphous; fruit several-seeded.

27. Phaseolus L., Bean

P. polystachios (L.) BSP. Twining herb; leaves pinnate; stipules small; leaflets 3, ovate, with stipels; racemes longer than the leaves, each flower with 2 bractlets; calyx veiny, its lobes short; corolla purple, 1 cm long or more, keel spirally coiled; stamens diadelphous; fruit linear, 4-5-seeded.

28. Strophostyles Ell., Wild Bean

Twining herbs; leaves pinnate; leaflets 3, with stipels; flowers in peduncled racemes or heads, each flower subtended by 2 small bracts; corolla pink, purple, or white; keel strongly upcurved; stamens diadelphous; fruit with several seeds.

a. Bractlets, calyx-tube and blades densely hairy; flowers usually 5-7 mm long; seeds shining.
  S. leiosperma (T. & G.) Piper
a. Bractlets and calyx-tube glabrous or sparsely hairy; flowers 7-15 mm long; seeds woolly or scurfy.
  b. Bractlets shorter than calyx-tube; leaflets rarely lobed. S. umbellata (Muhl.) Britt.
  b. Bractlets not shorter than calyx-tube; leaflets often lobed. S. helvola (L.) Ell.

29. Clitoria L., Butterfly-pea

C. mariana L. Glabrous twining herb; leaves pinnate; stipules subulate; leaflets 3, ovate to ovate-lanceolate, with stipels; flowers about 5 cm long, axillary, solitary or 2 together on a peduncle, each with 2 bractlets; corolla pale blue; standard much longer than remainder of corolla; keel shorter than wings, coherent with them, incurved; stamens monadelphous below; legume stipitate, several-seeded.

30. Amphicarpa Ell., Hog-peanut

A. bracteata (L.) Férn. Slender twining herb; leaves pinnate; leaflets 3, ovate, with stipels; flowers without corolla or with small corolla present near base of plant in addition to petaliferous flowers in axillary racemes or panicles from upper part of plant; corolla pale purple to white, small, each pedicel bracted; pods from upper flowers usually 3-seeded, those at base of plant or below ground usually 1-seeded.

a. Hairs of stem retrorsely appressed-pubescent; leaflets thinly pubescent; sides of legume glabrous, hairs on suture ascending. var. bracteata.
a. Hairs of stem brownish, spreading; leaflets densely pubescent; legume villous, hairs on suture retrorse. var. comosa (L.) Fern.

31. Galactia P. Br., Milk-pea

G. volubilis (L.) Britt. Stems prostrate or twining; leaves pinnate, stipules small; leaflets 3, ovate, with stipels; flowers small, in axillary few-flowered racemes; calyx 2-bracted, 4-lobed, the 2 upper lobes fused; corolla purple or pink; stamens diadelphous; keel adherent to wings; fruit flattened, densely pubescent, few-seeded.

LINACEAE, Flax Family

Linum L., Flax

Herbs; leaves sessile, entire, narrow, alternate or opposite; flowers hypogynous, perfect, in clusters; corolla regular; sepals, petals, and stamens 5, petals separate, filaments connate at base; carpels 5, united; styles 5 or united at base; capsule partly or completely 10-loculed, a false septum in each carpel; fruit a capsule; seeds 10.

a. Petals blue, 1 cm long or more.
　b. Annual; stigmas linear. *L. usitatissimum L., Common F.
　b. Perennial; stigmas capitate. L. lewisii Pursh
a. Petals yellow, usually less than 1 cm long.
　b. Two dark glands at base of leaf; 3 lines decurrent from leaf-base to beyond next node below; all sepals glandular-ciliate. L. sulcatum Riddell
　b. No glands at base of leaf; outer sepals not glandular.
　　c. Three lines decurrent from leaf-base to beyond next node below; base of stem usually decumbent. L. striatum Walt.
　　c. One prominent line decurrent from center of leaf-base, often becoming faint before next node below; capsule depressed at summit.
　　　d. Inflorescence loosely corymbiform, more or less dichotomously branched; inner sepals not glandular below middle; blades not subulate-pointed. L. virginianum L.
　　　d. Inflorescence stiffly corymbiform. L. medium (Planch) Britt.
　　　　e. Only upper blades subulate-tipped; inner sepals sparingly glandular. var. medium
　　　　e. All but lower blades subulate-tipped; inner sepals glandular-ciliate. var. texanum (Planch.) Fern.

OXALIDACEAE, Wood Sorrel Family

Oxalis L., Wood Sorrel

Low herbs; leaves palmately compound, 3-foliolate, leaflets obcordate, entire; flowers hypogynous, bisporangiate, in cymes or umbels; sometimes cleistogamous flowers developing; perianth regular; sepals 5, separate; petals 5, separate or united at base; stamens 10, alternate ones shorter, usually united at base; carpels 5, united; styles, stigmas, and locules 5; fruit a somewhat elongate capsule.

a. Without aerial leafy stems; leaves all basal.
　b. Petals usually violet; umbels few-flowered; with bulblike base. O. violacea L., Violet O.
　b. Petals white or pink-lined; flowers usually solitary; with slender rhizome. O. montana Raf.
a. With aerial leafy stem; petals yellow.
　b. Petals more than 1 cm long; leaflets 2-5 cm wide, margins usually purple-brown; flowers in cymes; pedicels not deflexed. O. grandis Small
　b. Petals usually less than 1 cm long; leaflets 2 cm wide or less, margins not purple-brown.

c. Stem prostrate, rooting, flowers in simple umbels; fruiting pedicels deflexed.
  *O. corniculata L.
c. Stem erect.
  d. Flowers in umbels; fruiting pedicels deflexed; hairs of fruit appressed. O. dillenii Jacq.
    (O. stricta L.; O. florida Salisb.)
  d. Flowers in cymes or branched umbels; fruiting pedicels not deflexed; hairs of fruit spread-
    ing. O. stricta L. (O. europaea Jord.)

GERANIACEAE, Geranium Family

Herbs; leaves opposite or alternate; flowers hypogynous, bisporangiate, in cymose or umbellate clusters or rarely solitary; perianth regular or slightly zygomorphic; sepals 5, separate; petals 5, separate, alternating with glands; stamens 10 or 5; carpels 5, united, in fruit with long erect slender beak; locules 5; stigmas 5; ovules 1 or 2 per carpel; fruit a 5-seeded capsule.

a. Stamens with anthers 10 (5, in 1 species with palmately divided leaves). Geranium
a. Stamens with anthers 5; leaves pinnately compound. Erodium

Geranium L., Cranesbill

Leaves palmately lobed or divided or rarely compound; carpels dehiscing by coiling outward from base upward, remaining attached at tip.

a. Leaves compound, the 3-5 leaflets pinnately divided; petals rose-purple; carpel-bodies, at maturity,
  separating from styles. G. robertianum L., Herb Robert
a. Leaves palmately deeply divided; carpel-bodies not separating.
  b. Only 1 pair of cauline leaves usually present; flowers 2-3 cm wide; petals pink-purple.
    G. maculatum L., Wild C.
  b. More than 2 cauline leaves present; flowers usually smaller.
    c. Sepals awn-tipped.
      d. Carpel-bodies glabrous; pedicels of fruit much longer than calyx. *G. columbinum L.
      d. Carpel-bodies pubescent; pedicels of fruit slightly longer to shorter than calyx.
        e. Hairs of carpel-bodies about 1 mm long; seeds faintly reticulate. G. carolinianum L.
        e. Hairs of carpel-bodies short, spreading, sometimes glandular; seeds strongly reticu-
          late. *G. dissectum L.
    c. Sepals not awn-tipped.
      d. Stamens 10; carpel-bodies glabrous, wrinkled. *G. molle L.
      d. Stamens 5; carpel-bodies pubescent. *G. pusillum L.

Erodium L'Her., Storksbill

*E. cicutarium (L.) L'Her. Leaves pinnately compound, leaflets cleft or pinnatifid; sepals awned; petals rose-color, one a little smaller; stamens with anthers 5, filaments without anthers 5; carpel-bodies sharp-pointed at base, at dehiscence separating, the styles becoming spirally twisted.

ZYGOPHYLLACEAE, Caltrop Family

Tribulus L., Caltrop

*T. terrestris L., Puncture-weed. Stems prostrate; leaves even-pinnate, opposite, stipuled, one at each node larger; flowers about 1 cm wide, hypogynous, bisporangiate, solitary in axils; perianth regular; sepals and petals 5, stamens 10, petals yellow; carpels 5, united; style 1; ovulary 5-loculed; fruit splitting into 5 segments, each with 2 spines and a band of tubercles.

RUTACEAE, Rue Family

Shrubs and small trees (in ours) with alternate gland-dotted leaves; flowers hypogynous, bisporangiate or monosporangiate; perianth regular; petals separate.

a. Leaves pinnately compound; carpels separate; fruit an aggregate of follicles. Xanthoxylum
a. Leaves 3-foliolate; carpels united; fruit a samara. Ptelea

143

Xanthoxylum Gmel., Prickly-ash

X. americanum Mill. Shrub; leaves pinnately compound; stems and sometimes petioles prickly; diecious; flowers small, in axillary clusters, appearing before the leaves; calyx none; petals 4-5; stamens 4-5 in staminate flowers, on hypogynous disk; carpels 3-5, separate, becoming 1-2-seeded follicles.

Ptelea L., Hop-tree

P. trifoliata L. Shrub or small tree; leaves 3-foliolate, petioled; flowers bisporangiate or monosporangiate, in dense terminal compound cymes; carpellate flowers sometimes with rudimentary stamens; sepals 3-5; petals 3-5, small, greenish-white, stamens as many; ovulary 2-loculed; fruit a flat 2-seeded samara.

SIMAROUBACEAE, Quassia Family

Ailanthus Desf., Tree-of-heaven

*A. altissima (Mill.) Swingle Tree; leaves alternate, pinnately compound, leaflets many, each with 1 or 2 glandular teeth near base; flowers hypogynous, greenish, small, in terminal panicles; flowers staminate, carpellate, or bisporangiate; sepals 5, united at base; petals 5, separate; stamens 10 in staminate, fewer in bisporangiate, none in carpellate flowers; carpels 2-5, styles united, ovularies separate or nearly so, fruit an aggregate of samaras.

POLYGALACEAE, Milkwort Family

Polygala L., Milkwort

Ours herbs; leaves simple; flowers hypogynous, zygomorphic, bisporangiate, in spikes, racemes, or heads; sepals 5, the 3 of the outer circle (the uppermost and the 2 lower) small, the two inner or lateral ones (wings) larger and petaloid; petals 3, united, the lower one concave and often bearing a fringed crest; stamens 6 or 8, filaments united below in a sheath split on upper side and more or less united with petals; anthers 1-loculed, opening by a pore at or near the tip; carpels 2, united; style 1; ovulary 2-loculed; capsule 2-seeded.

a. Flowers about 1.5 cm long, rose-purple to white, pediceled, 1-4 in a short terminal cluster.
  P. paucifolia Willd., Fringed M.
a. Flowers smaller, more numerous, in spikes, racemes, or heads.
  b. Petals united in a cleft tube 7-10 mm long, twice as long as wings or more; flowers pink or rose-purple; glabrous and glaucous. P. incarnata L., Pink M.
  b. Petals shorter, equaling or shorter than the wings.
    c. Leaves, at least the lower, whorled or opposite.
      d. Wings deltoid, long-pointed; racemes dense, thick-cylindric to ellipsoid; leaves mostly whorled. P. cruciata L.
      d. Wings obovate, rounded at tip, less than 3 mm long; racemes slenderly cylindric to conic; lower leaves opposite. P. verticillata L.
        e. Wings shorter than capsule. var. verticillata
        e. Wings about equaling capsule. var. ambigua (Nutt.) Wood
    c. Leaves all alternate.
      d. Racemes spherical or short-cylindric, dense and headlike, 1-2 cm long, often 1 cm thick or more; stems solitary; annual.
        e. Wings ovate or oval, rounded above; flowers rose-purple, green, or white; lobes of aril linear, 3/4 length of seed. P. sanguinea L.
        e. Wings elliptic; flowers pink to rose-purple, petals yellow-tipped; lobes of aril obovate, 1/4 length of seed. P. curtissii Gray
      d. Racemes slender-cylindric; stems tufted, perennial or biennial; aril somewhat shorter than seed.
        e. Perennial with woody rhizome; racemes dense; flowers white; pedicels 1 mm long or less; blades acuminate, lower leaves scalelike; cleistogamous flowers absent.
          P. senega L., Seneca Snakeroot
        e. Biennial; racemes rather loose; flowers rose-purple or white, spreading; some pedicels much more than 1 mm long; blades obtuse; cleistogamous flowers present in basal racemes. P. polygama Walt.

EUPHORBIACEAE, Spurge Family

Ours herbs, often with milky juice; rarely diecious; flowers hypogynous, monosporangiate; calyx and corolla present, or corolla or both absent; perianth, when present, usually regular, usually of separate parts; carpels 2 or 3, united, locules and styles as many, styles usually 2-lobed; stigmas sometimes dissected into many filiform segments; ovules 1 or 2 in each locule; placentae axile; fruit a capsule.

a. Flowers without a perianth, borne in a cyathium; juice milky. 6. Euphorbia
a. Flowers with at least a calyx, not borne in a cyathium; juice not milky.
  b. Leaves opposite, glabrous; styles 2; ovulary 2-loculed. 2. Mercurialis
  b. Leaves alternate; if some leaves are opposite, then leaves stellate-pubescent.
    c. Leaves large, peltate, palmately lobed; stamens branched. 4. Ricinus
    c. Leaves not peltate; stamens not branched.
      d. Ovules 2 in each locule; leaves entire, 2-ranked, 1-2 cm long. 5. Phyllanthus
      d. Ovules 1 in each locule; leaves larger, or not entire, or both.
        e. Pubescence not stellate; carpellate flowers subtended by a several-lobed or -toothed bract. 3. Acalypha
        e. Pubescence stellate; carpellate flowers not subtended by such bract. 1. Croton

1. Croton L.

Our species monecious; stellate-pubescent; leaves mostly alternate; flowers in small crowded clusters terminal on stem and branches; sepals commonly 5; petals the same number or absent; stamens 5-many; hypogynous disk present; capsule 2- or 3-seeded; seed with caruncle.

a. Leaf blades toothed; ovulary 3-loculed; styles 3, each bifid. C. glandulosus L.
a. Leaf blades entire or undulate.
  b. Ovulary 2-loculed; capsule 1-seeded; styles 2, each bifid. C. monanthogynus Michx.
  b. Ovulary 3-loculed; capsule usually 3-seeded; styles 3, each 2-3 times forked. C. capitatus Michx.

2. Mercurialis L., Mercury

*M. annua L. Diecious; leaves opposite, petioled, stipuled; blades lanceolate, serrate; staminate flowers in peduncled interrupted spikes, carpellate in short axillary clusters; sepals 3; petals none; stamens 8-20; ovulary 2-loculed; styles 2; capsule 2-seeded, pubescent.

3. Acalypha L., Three-seeded Mercury

Monecious; leaves alternate, with stipules; flowers monosporangiate, in spikes, the carpellate sometimes in separate ones, sometimes at base of staminate spike; spike subtended by foliaceous bract; calyx of staminate flowers 4-parted, of carpellate, 3-5-parted; petals none; stamens 8-16; styles 3, dissected.

a. Capsule prickly; leaf blades somewhat cordate at base; carpellate and staminate flowers in separate spikes. A. ostryaefolia Riddell, Hornbeam T.
a. Capsule not prickly; blades not cordate at base; the 2 kinds of flowers usually in same spike, carpellate at base.
  b. Leaf blades ovate; petioles almost as long as blades; bracts usually 5-7-cleft. A. rhomboidea Raf.
  b. Leaf blades ovate-lanceolate to linear, longer than petioles; bracts 9-15-lobed.
    c. Bracts usually not glandular, the lobes lanceolate and acute; staminate spikes equaling or surpassing bracts; petiole usually exceeding bract it subtends; stem usually with both spreading and incurved hairs. A. virginica L.
    c. Bracts usually glandular, the lobes ovate or deltoid; staminate spike usually much exceeding bract; petiole usually not exceeding bract it subtends; stem usually with only incurved hairs. A. gracilens Gray

4. Ricinus L., Castor-bean

*R. communis L. Tall, stout, glabrous, glaucous; leaves large, alternate, peltate, palmately-lobed; flowers monosporangiate, in panicles, the staminate below, the carpellate above, sometimes some bisporangiate ones between; sepals 3-5; petals none; stamens many; seeds caruncled, poisonous; capsule spiny.

5. Phyllanthus L.

P. caroliniensis Walt. Glabrous annual; stems erect, usually branched; leaves simple, entire, alternate, stipuled, 2-ranked, 1-2 cm long; flowers minute, monosporangiate, in pairs in the axils, 1 staminate, 1 carpellate; sepals 6, petals none; stamens 3; styles 3, bifid; 2 seeds in each locule.

6. Euphorbia L., Spurge

Ours herbs with milky juice; monecious; flowers monosporangiate, the staminate consisting of a single stamen, the carpellate consisting of a gynecium of 3 united carpels, ovulary 3-loculed, 1 ovule in each locule, styles 3, bifid; inflorescence-unit a cyathium, consisting of several staminate flowers and 1 carpellate flower within a cup-shaped involucre with edge 4- or 5-lobed, the 1-5 glands in the sinuses sometimes with petal-like appendages; pedicel of carpellate flower (gynophore) elongating in fruit and arching over edge of cup; cyathia variously clustered.

a. Glands of cyathium with white or pink petal-like appendages or, if these are absent or reduced, then plant prostrate, mat-forming.
  b. Leaves alternate below inflorescence; plant erect.
    c. Glabrous; leaves green; glands 5. E. corollata L.
    c. Stem and cyathia pubescent; upper leaves and those subtending rays of inflorescence entirely white or conspicuously white-margined; glands 5. *E. marginata Pursh, Snow-on-the-mountain
  b. Leaves all opposite, small, blade oblique at base; plants mostly procumbent.
    c. Plants, including capsules, glabrous.
      d. Leaf-blades entire.
        e. Glands of cyathium minute, not appendaged; blades linear-oblong; stipules separate. E. polygonifolia L.
        e. Glands of cyathium with minute appendages; blades round-ovate; stipules united. E. serpens HBK.
      d. Leaf-blades serrate, sometimes only at tip.
        e. Seed-faces with 3-6 transverse ridges; leaf-blades narrowly oblong; glands of cyathium flat. E. glyptosperma Engelm.
        e. Seed-faces pitted and with short cross ridges; glands of cyathium depressed in middle. E. serpyllifolia Pers.
    c. Stems, at least the younger, pubescent, sometimes only in 1 line; leaves usually pubescent.
      d. Ovulary and capsule pubescent.
        e. Seed-faces with 3-4 cross ridges; blades serrulate, length more than twice width; cyathium split a short distance on 1 side. E. supina Raf.
        e. Seeds almost smooth; blades entire or remotely serrulate, length not more than twice width; cyathium split to base on 1 side. E. humistrata Engelm.
      d. Ovulary and capsule glabrous.
        e. Stem erect or obliquely ascending, glabrous or pubescent in lines; seeds obtusely angled; leaves usually with a red spot. E. maculata L.
        e. Stem prostrate or spreading, hirsute; seeds sharply angled. E. vermiculata Raf.
a. Glands of cyathium without petal-like appendages, sometimes with crescent-shaped horns; plants mostly erect or spreading.
  b. Cyathia in dense terminal clusters, 4-5-lobed, lobes fimbriate; gland usually 1.
    c. Blades linear to ovate, toothed; stem and leaves pubescent; bracts under inflorescence green or whitened at base. E. dentata Michx.
    c. Blades of various shapes, some unlobed, some fiddle-shaped, entire to toothed; stem and leaves glabrous or nearly so except in younger parts; bracts under inflorescence often wholly or partly red. E. heterophylla L.
  b. Cyathia in mostly umbelliform clusters subtended by a whorl of leaves, usually 4-lobed and with 4 glands; leaves mainly alternate below inflorescence.
    c. Leaf-blades entire; glands of cyathium crescent-shaped or 2-horned.
      d. Seeds smooth; stem-leaves linear, those of inflorescence broader. *E. cyparissias L.
      d. Seeds pitted or tubercled.
        e. Blades of inflorescence and stem-blades both linear-lanceolate. *E. exigua L.

     e. Blades of inflorescence ovate, differing in shape from stem-blades; rays of inflorescence usually 3.
        f. Each lobe of capsule crested with 2 wings; seeds with 4 rows of 3-4 pits on outer face, 2 furrows on inner face; stem-leaves petioled, blunt to retuse at apex. *E. peplus L.
        f. Lobes of capsule not winged; stem-leaves sessile or subsessile.
           g. Leaves of inflorescence subulate-tipped; seeds brown beneath a gray coating, several transverse furrows on each side; stem-leaves acute. *E. falcata L.
           g. Leaves of inflorescence wider than long, sometimes connate at base, not subulate-tipped; seeds pitted all over; stem-leaves obovate, blunt to retuse. E. commutata Engelm.
c. Leaf-blades serrulate; glands of cyathium elliptic, oval, or circular.
  d. Ovulary and capsule smooth; inflorescence usually 5-rayed; all blades obovate or oblanceolate. *E. helioscopia L.
  d. Ovulary and capsule warty; stem-blades oblanceolate.
     e. Inflorescence usually 3-rayed; blades glabrous beneath; cyathium glabrous; bifid styles separate to base. E. obtusata Pursh
     e. Inflorescence usually 5-rayed; blades pubescent beneath; cyathium usually pubescent; bifid styles united at base. *E. platyphylla L.

CALLITRICHACEAE, Water Starwort Family

Callitriche L., Water Starwort

Small herbs growing in mud or submersed in water with floating rosettes of leaves at stem tips; leaves opposite; monecious; flowers minute, in leaf axils, hypogynous, monosporangiate; the two kinds usually in different axils; perianth absent; flower subtended by 2 small bracts; stamen 1; carpels 2, united; styles 2, filiform; ovulary 4-loculed by growth of false septum in each carpel; fruit separating into 4 one-seeded portions. Mature fruit is necessary for identification.

a. Fruit wider than long, somewhat stalked, the lobes separated by a deep groove, each with 2 sharp edges; styles shorter than fruit; terrestrial; leaves oblanceolate, 3-nerved, of uniform shape. C. deflexa A. Br. var. austini (Engelm.) Hegelm.
a. Fruit sessile; aquatic or terrestrial; submersed leaves linear, floating leaves obovate to spatulate with rounded summit; leaves and stems with minute scales.
  b. Fruit longer than wide, obovoid or ellipsoid; carpels with sharp margins, separated by wide V-shaped groove on the broad side, by a shallow groove on the edge; styles shorter than fruit. C. palustris L.
  b. Fruit about as long as wide, obovoid to nearly circular, all margins rounded, both grooves shallow; styles as long as or longer than fruit. C. heterophylla Pursh

LIMNANTHACEAE, False-mermaid Family

Floerkea Willd., False-mermaid

F. proserpinacoides Willd. Diffuse delicate glabrous herb; leaves pinnately divided into narrow lobes; flowers small, hypogynous, bisporangiate, solitary in axils, stalked; sepals 3, separate; corolla regular; petals 3, separate, white, about 2 mm long, shorter than sepals; stamens 6; carpels 3 or 2, ovularies separate or nearly so, styles arising from base of ovularies, united at base but separate at top; carpels separating, each 1-seeded, indehiscent.

ANACARDIACEAE, Sumac Family

Trees, shrubs, and woody vines with milky or resinous sap; leaves alternate, simple or compound; flowers small, hypogynous, monosporangiate or bisporangiate, in panicles or in small spikes paniculately clustered; perianth regular; sepals and petals 5, stamens 5, inserted under edge of intrastaminal disk; carpels 3, united; styles and stigmas 3; ovulary 1-loculed with 1 ovule; fruit a drupe.

a. Leaves compound; fruit symmetric. Rhus
a. Leaves simple; fruit asymmetric. Cotinus

Cotinus Duham., Smoke-tree

*C. coggygria Scop. Shrub with obovate or elliptic pinnately-veined leaves; flowers in terminal panicles, monosporangiate or bisporangiate, the pedicels on which no fruits develop becoming long and plumose; fruits glabrous, asymmetric, style lateral.

Rhus L., Sumac

Small trees, shrubs, and vines; leaves 3-foliolate or pinnately compound; flowers greenish-white or yellow, in panicles or in short crowded spikes, bisporangiate and monosporangiate on same plant; style terminal.

a. Leaflets 3.
  b. Vines; flowers in axillary panicles; fruits white or drab; leaves not fragrant; leaflets stalked, stalks of lateral ones short. R. radicans L., Poison Ivy
  b. Erect shrubs; flowers in small solitary or clustered spikes, appearing before or with the young leaves; fruits red; leaves fragrant; terminal leaflet short-stalked, lateral ones sessile. R. aromatica Ait., Fragrant S.
a. Leaflets more than 3; flowers appearing after the leaves.
  b. Rachis of leaf winged; panicles terminal; fruit red, pubescent. R. copallina L., Dwarf or Mountain S.
  b. Rachis of leaf not winged.
    c. Leaflets entire; panicles axillary; fruit white or drab. R. vernix L., Poison S.
    c. Leaflets serrate; panicles terminal; fruit red, pubescent.
      d. Twigs and petioles glabrous and glaucous. R. glabra L., Smooth S.
      d. Twigs and petioles velvety-pubescent. R. typhina L., Staghorn S.

AQUIFOLIACEAE, Holly Family

Trees and shrubs, evergreen or deciduous; leaves alternate, simple, with minute stipules; flowers hypogynous, monosporangiate, each kind bearing vestiges of the other sporophylls, or bisporangiate; mostly diecious, but some bisporangiate flowers present with the monosporangiate ones; sepals 4-6, small, or absent; perianth regular; petals 4-8; stamens as many, sometimes slightly united with petals, alternate with them; carpels 4-several, united, locules as many, 1 or 2 ovules in each locule; style short or none; fruit a drupe.

a. Calyx present, persistent in fruit; petals white or greenish, united at base. Ilex.
a. Calyx none or deciduous; petals yellowish, separate, linear-oblong. Nemopanthus.

Ilex L., Holly

Deciduous or evergreen; staminate flowers usually clustered, the carpellate clustered or solitary; petals and stamens 4-8; ovulary 4-8-loculed, 1 ovule in each locule; fruit a red or yellow drupe.

a. Leaves evergreen, spiny-toothed to almost entire, coriaceous. I. opaca Ait., American H.
a. Leaves deciduous, serrate, oblanceolate, elliptic, or obovate, not coriaceous. I. verticillata (L.) Gray, Winterberry.

Nemopanthus Raf., Mountain-holly

N. mucronatus (L.) Trel. Glabrous shrub; leaves thin, elliptic or oblong, entire or nearly so, 5 cm long or less; calyx minute and deciduous or absent; flowers axillary, slender-pediceled, solitary or clustered; drupe red.

CELASTRACEAE, Staff-tree Family

Small trees, shrubs, and woody vines; leaves simple; flowers hypogynous, bisporangiate or monosporangiate; hypogynous disc present; perianth regular; sepals, petals, and stamens 4-5; carpels 2-5, united, locules of ovulary and lobes of stigma as many; style short or none; fruit a capsule; seeds with arils.

a. Leaves opposite; small trees or erect or prostrate shrubs.
  b. Leaves deciduous; capsule 3-5-loculed; aril orange or red. 1. Euonymus

b. Leaves evergreen; capsule 2-loculed; aril white. 2. Pachystima
a. Leaves alternate; vines. 3. Celastrus

1. Euonymus L., Spindle-tree

Small trees and erect or prostrate shrubs; young stems usually green; leaves opposite, finely serrate, pinnately-veined; flowers bisporangiate, about 1 cm wide or less, axillary, solitary or clustered; sepals, petals, and stamens 4 or 5; sepals united at base, calyx short and flat; petals spreading, rounded; stamens short, arising at edge of flat disk covering gynecium; carpels 3-5; capsule red or purple; seeds 1-4 in each locule.

a. Leaves petioled; flower parts in 4's; fruit smooth.
   b. Petals purple; leaf-blades pubescent beneath. E. atropurpureus Jacq., Wahoo.
   b. Petals white, green, or yellowish; leaf-blades glabrous beneath. *E. europaeus L.
a. Leaves subsessile; flower parts in 5's; fruit tubercled; petals greenish-purple.
   b. Erect shrub; leaves ovate-lanceolate, acuminate. E. americanus L., American Strawberry-
     bush.
   b. Prostrate shrub, stems rooting; leaves obovate, obtuse. E. obovatus Nutt.

2. Pachystima Raf.

P. canbyi Gray   Small evergreen shrub; leaves opposite, coriaceous, linear-oblong, entire or serrulate, 1-2 cm long; flowers small, bisporangiate, solitary or clustered in axils; sepals, petals, and stamens 4; carpels 2; seeds 1 or 2, arils white.

3. Celastrus L., Bittersweet

C. scandens L. Woody vine; leaves alternate, pinnately veined, serrate; diecious, each kind of flower with rudiments of the other sporophylls, or some bisporangiate flowers on the same plant as the monosporangiate ones; panicles terminal; petals greenish-white, small; sepals, petals, and stamens 5; carpels 3; fruit orange; seeds 1 or 2 in each locule, arils scarlet.

STAPHYLEACEAE, Bladdernut Family

Staphylea L., Bladdernut

S. trifolia L. Shrub or small tree; bark of twigs striped; leaves opposite, 3-foliolate, leaflets finely serrate; stipules early deciduous; flowers hypogynous, bisporangiate, in drooping terminal clusters; perianth regular; sepals 5; petals 5, greenish-white, separate, about 1 cm long; stamens 5, inserted outside the hypogynous disk; carpels 3, united; ovulary 3-lobed, 3-loculed; placentae axile; ovules several in each locule; styles 3, separate below, coherent above; fruit a bladdery capsule.

ACERACEAE, Maple Family

Acer L., Maple

Trees and shrubs with opposite simple palmately-lobed (or rarely compound) leaves; flowers monosporangiate or bisporangiate, small, hypogynous or perigynous, in umbels, corymbs, racemes, or panicles; disk often conspicuous; perianth regular; sepals 4-5 or more, separate or united at base; petals none or 4-5 or more, separate; stamens 4-10, usually 8; carpels 2, united; styles united below; stigmas 2; ovulary 2-loculed, ovules 2 in each locule; placentae axile; each carpel becoming winged; fruit 2-seeded, of 2 united samaras (or rarely carpels and samaras 3-4).

1. KEY TO PLANTS IN FLOWER

a. Petals absent.
   b. Diecious; twigs green; disk absent; staminate flowers in corymbs or umbels, carpellate flowers
     in racemes; flowers appearing before or with the leaves. A. negundo L.
   b. Monecious.
     c. Flowers appearing long before the leaves, all sessile or subsessile; ovularies and young fruits
       white-villous. A. saccharinum L.
     c. Flowers appearing with or a little before or after the leaves, in corymbs, pedicels slender and
       drooping. A. saccharum Marsh. and A. nigrum Michx. f.

a. Petals present; flowers pediceled.
  b. Flowers red, appearing much before the leaves, in small clusters. <u>A</u>. <u>rubrum</u> L.
  b. Flowers appearing with or after the leaves.
    c. Flowers in erect corymbs or panicles.
      d. Petals white. *<u>A</u>. <u>tataricum</u> L.
      d. Petals not white.
        e. Inflorescence a slender pubescent panicle; sap not milky; petals greenish, much exceeding sepals. <u>A</u>. <u>spicatum</u> Lam.
        e. Inflorescence a corymb; sap milky; petals yellow or yellow-green.
          f. Corymbs glabrous; lobes of leaves pointed. *<u>A</u>. <u>platanoides</u> L.
          f. Corymbs pubescent; lobes of leaves rounded. *<u>A</u>. <u>campestre</u> L.
    c. Flowers in drooping panicles or racemes; petals yellow or yellow-green.
      d. Flowers in a slender glabrous raceme. <u>A</u>. <u>pensylvanicum</u> L.
      d. Flowers in a pubescent panicle or raceme. *<u>A</u>. <u>pseudoplatanus</u> L.

2. KEY TO PLANTS IN LEAF

a. Leaves pinnately compound; one-year-old stems green. <u>A</u>. <u>negundo</u> L., Box-elder.
a. Leaves simple.
  b. Leaf-blades unlobed or only slightly lobed, ovate, serrate or doubly serrate. *<u>A</u>. <u>tataricum</u> L., Tatarian M.
  b. Leaf-blades lobed.
    c. Sap milky; wings of samaras divaricate, in almost a straight line.
      d. Tips of lobes rounded. *<u>A</u>. <u>campestre</u> L., Hedge M.
      d. Tips of lobes pointed. *<u>A</u>. <u>platanoides</u> L., Norway M.
    c. Sap not milky; wings of samaras from almost parallel to divergent at not much more than a right angle.
      d. Three upper lobes of blade entire or with 1-3 large teeth or small lobes on each side.
        e. Stipules present; sinus at base of blade closed or nearly so; sides of blade drooping. <u>A</u>. <u>nigrum</u> Michx. f., Black M.
        e. Stipules absent; sinus at base of blade open; blade flat. <u>A</u>. <u>saccharum</u> Marsh., Sugar M.
      d. Three upper lobes of blade with more numerous teeth, serrations, or small lobes.
        e. Terminal lobe narrow at base, widest about halfway between base and apex; blades silvery white beneath, 5-lobed, lobes again lobed or toothed; sinus on each side of terminal lobe a gothic arch. <u>A</u>. <u>saccharinum</u> L., Silver M.
        e. Terminal lobe not or little narrowed at base; sinuses not as above.
          f. Blades 3-lobed, middle lobe much longer than the 2 lateral ones. *<u>A</u>. <u>ginnala</u> Maxim.
          f. Blades 3-5-lobed, middle lobe not much longer than the lobe on either side of it.
            g. Blades 3-lobed, margins of lobes finely and closely serrate. <u>A</u>. <u>pensylvanicum</u> L., Moosewood, Striped M.
            g. Margins of lobes coarsely toothed.
              h. Blades 3-lobed or obscurely 5-lobed, teeth of margin sharp-pointed; branchlets gray-pubescent; blades pubescent beneath; shrub or small tree. <u>A</u>. <u>spicatum</u> Lam., Mountain M.
              h. Blades usually 5-lobed, sometimes 3-lobed; teeth not sharp-pointed and/or branchlets glabrous; trees.
                i. Blades glaucous beneath; sinus at each side of terminal lobe about a right angle. <u>A</u>. <u>rubrum</u> L., Red M.
                i. Blades light green beneath; sinus at each side of terminal lobe with narrow base. *<u>A</u>. <u>pseudoplatanus</u> L., Sycamore M.

HIPPOCASTANACEAE, Buckeye Family

<u>Aesculus</u> L., Buckeye, Horse Chestnut

Trees and shrubs with opposite palmately compound leaves; flowers hypogynous, in terminal panicles, bisporangiate and staminate ones usually in same panicle; sepals 5, united, sometimes oblique or gibbous at base; corolla zygomorphic; petals 4 or 5, clawed; extrastaminal disk present; stamens 5-8, separate; carpels 3, united, ovulary 3-loculed, 2 ovules in each locule; style and stigma 1; fruit a leathery capsule, seeds 1-3, smooth, brown, large.

a. Leaflets mostly 7; winter buds glutinous; petals usually 5, white, marked with red and yellow; fruits prickly. *A. hippocastanum L., Horse Chestnut.
a. Leaflets mostly 5; winter buds not glutinous; petals 4.
  b. Stamens long-exserted; petals pale greenish-yellow, all about the same length; fruit prickly. A. glabra L., Ohio B.
  b. Stamens not or little exserted; petals yellow or reddish, the 2 upper longer and erect; fruit not prickly. A. octandra Marsh., Yellow or Sweet B.

BALSAMINACEAE, Jewel-weed Family

Impatiens L., Jewel-weed, Touch-me-not

Succulent herbs; leaves alternate or whorled, simple; flowers hypogynous, bisporangiate, zygomorphic; sepals 3, the lower one saccate, spurred at base; petals 5, the lateral ones united in pairs, each pair appearing as one lobed petal; stamens 5; carpels 5, united, style 1; ovulary 5-loculed; capsule elastically dehiscent.

a. Flowers purple, rose, or white, solitary or few in axils of sharply toothed short-petioled leaves. *I. balsamina L., Garden T.
a. Flowers yellow or orange, solitary or in clusters of a few in axils of crenate-serrate long-petioled leaves.
  b. Flowers orange-yellow with red-brown mottlings; spur about 8 mm long, lying parallel with the sac; sac longer than wide. I. capensis Meerb., Spotted T.
  b. Flowers pale yellow, sometimes sparingly dotted; spur about 5 mm long, lying at right angle to sac; sac almost as wide as long. I. pallida Nutt., Pale T.

RHAMNACEAE, Buckthorn Family

Shrubs and small trees; leaves simple, alternate or rarely opposite, with stipules; monecious or sometimes diecious; flowers bisporangiate or monosporangiate, perigynous or slightly epigynous; stamens and petals attached to disk lining floral cup; sepals and stamens 5 or 4; petals 5 or 4 or none; carpels usually 3 or 4, united, locules as many; ovules 1 in each carpel.

a. Leaves pinnately-veined; flowers solitary or in few-flowered sessile or short-peduncled clusters in or above axils. Rhamnus.
a. Leaves with 3 strong veins from top of petiole; flowers in terminal or axillary corymbs of clustered umbels. Ceanothus.

Rhamnus L., Buckthorn

Shrubs or small trees; leaves alternate or rarely opposite; flowers small, greenish, perigynous, solitary or in small axillary clusters, bisporangiate or monosporangiate, then each kind often with vestiges of the other type of sporophyll; sepals and stamens 4 or 5; petals 4 or 5 or none; carpels and locules 2-4; fruit a drupe with 2-4 stones.

a. Leaves often opposite or nearly so, with usually 3 pairs of lateral veins; winter twigs usually ending in thorns; flowers monosporangiate, appearing with the leaves; petals and sepals 4. *R. cathartica L.
a. Leaves alternate, with more than 3 pairs of lateral veins; winter twigs not ending in thorns.
  b. Flowers bisporangiate, appearing after the leaves; sepals and petals 5; winter buds hairy, without scales.
    c. Leaves serrulate, oblanceolate-elliptic; pedicels pubescent; flower clusters usually short-peduncled. R. caroliniana Walt.
    c. Leaves entire or nearly so, obovate-elliptic; pedicels glabrous or glabrate; flower clusters usually sessile. *R. frangula L.
  b. Flowers monosporangiate or some of them bisporangiate, appearing with the leaves; petals 4 or none; winter buds with scales.
    c. Leaf-blades crenate-serrate, obovate-elliptic; petals none; sepals and stamens 5. R. alnifolia L'Her.
    c. Leaf-blades finely serrulate, lanceolate-elliptic; sepals, stamens, and petals 4. R. lanceolata Pursh

Ceanothus L., Redroot

Shrubs with alternate leaves; flowers bisporangiate, in corymbs of clustered umbels; sepals and stamens 5; petals 5, white; hypanthium adnate to base of ovulary; carpels 3, sunken in disk; fruit 3-lobed, carpels separating when ripe.

a. Peduncles axillary; blades ovate. <u>C</u>. <u>americanus</u> L., New Jersey Tea
a. Peduncles at ends of leafy stems; blades elliptic-lanceolate. <u>C</u>. <u>ovatus</u> Desf.

VITACEAE, Grape Family

Climbing or trailing woody vines; leaves alternate, simple or compound, with or without stipules; flowers small, hypogynous, bisporangiate or monosporangiate; clusters usually opposite leaves; perianth regular; sepals, petals, and stamens 5 or sometimes 4; calyx minute; petals separate but sometimes connate at tip, greenish; stamens opposite petals; carpels usually 2, rarely more, united, locules as many; style short or absent; placentation central; ovules 1-2 per carpel; fruit a berry.

a. Hypogynous disk not evident; leaves, or some of them, palmately compound, leaflets 3-7, if 3, then
    tendrils ending in adhesive disks. 2. <u>Parthenocissus</u>
a. Ovulary surrounded by hypogynous disk; leaves simple or twice compound.
  b. Petals not expanding, falling with tips united as a cap; glands of hypogynous disk alternate with
      stamens. 3. <u>Vitis</u>
  b. Petals expanding, not united; hypogynous disk cup-shaped. 1. <u>Ampelopsis</u>

Ampelopsis Michx.

With or without tendrils; leaves simple, margins dentate or lobed, or leaves compound with dentate leaflets; flowers monosporangiate, each kind with rudimentary structures of the other kind, or some flowers bisporangiate; calyx somewhat lobed; petals separate; berry rather dry.

a. Leaves twice pinnate. <u>A</u>. <u>arborea</u> (L.) Koehne
a. Leaves simple.
  b. Leaves not or only slightly lobed. <u>A</u>. <u>cordata</u> Michx.
  b. Leaves 3-5 lobed. *<u>A</u>. <u>brevipedunculata</u> (Maxim.) Trautv.

Parthenocissus Planch., Virginia Creeper, Woodbine

With tendrils; leaves palmately compound or simple and lobed; berry 1-4-seeded.

a. Leaves 3-foliolate or simple and lobed; tendrils ending in adhesive discs. *<u>P</u>. <u>tricuspidata</u> (Sieb. &
    Zucc.) Planch., Boston Ivy
a. Leaves palmately compound, leaflets usually 5.
  b. Tendrils ending in adhesive discs; inflorescence elongate, paniculate. <u>P</u>. <u>quinquefolia</u> (L.)
    Planch., Virginia Creeper
  b. Tendrils ending in slender tips; inflorescence with 2 nearly equal divergent branches, each branch
    dichotomously forked. <u>P</u>. <u>inserta</u> (Kerner) K. Pritsch

Vitis L., Grape

With coiling tendrils; blades palmately veined, round-ovate, cordate, lobed or only coarsely toothed; flowers fragrant, the staminate with rudiments of carpels, the carpellate with rudiments of stamens; calyx none or a narrow flange; petals coherent at summit, deciduous as a cap; each carpel 2-ovuled; berry juicy.

a. Mature leaf-blades glabrous beneath or pubescent only in the axils, somewhat pubescent when young
    but not cobwebby; not glaucous.
  b. Blades 3-7-lobed, sinus at each side of terminal lobe usually acute; fruit glaucous; diaphragm at
      nodes 2 mm wide or less. <u>V</u>. <u>riparia</u> Michx., Riverside G.
  b. Blades unlobed or with 2 shallow lateral lobes, sinus at each side of terminal lobe usually wide;
      fruit shining; diaphragm at nodes 2 mm wide or more. <u>V</u>. <u>vulpina</u> L., Frost G.
a. Mature leaf-blades pubescent beneath with cobwebby or ordinary hairs or, if nearly glabrous, then
    glaucous; young blades pubescent beneath with cobwebby hairs, these hairs sometimes brown, and
    sometimes also with ordinary hairs.

b. An inflorescence or a tendril (or both) opposite each of 3 or more successive leaves; fruits 1 cm in diameter or more; pubescence woolly, brown or gray. V. labrusca L., Fox G.
b. An inflorescence or a tendril (or both) opposite no more than 2 successive leaves; fruits less than 1 cm in diameter.
   c. Branchlets angled; leaves unlobed or shallowly lobed, not glaucous beneath.
      d. Branchlets and petioles densely short-pubescent. V. cinerea Engelm.
      d. Branchlets and petioles with thin cobwebby pubescence. V. baileyana Munson
   c. Branchlets terete; blades usually 3-5-lobed, sometimes glaucous beneath. V. aestivalis Michx., Summer G.
      d. Lower surface of blades brown-pubescent, not or little glaucous. var. aestivalis
      d. Lower surface of mature blades glabrous or nearly so, quite glaucous. var. argentifolia (Munson) Fern.

TILIACEAE, Linden Family

Tilia L., Linden, Basswood

Trees; leaves alternate, simple, stipuled, 2-ranked, blades ovate, oblique at base, serrate or dentate; flowers hypogynous, bisporangiate, fragrant, cream-color, in cymes, the peduncle adnate for part of its length to a large bract; perianth regular; sepals and petals 5; a staminode (in our species) opposite each petal; stamens many, sometimes united basally in 5 groups; carpels 5 (in ours), united; ovulary 5-loculed, 2 ovules in each locule; style 1; fruit indehiscent, pubescent, 1-2-seeded.

a. Leaf-blades pubescent beneath with dense white (or brownish) stellate hairs; bracts pubescent on lower side. T. heterophylla Vent.
a. Leaf-blades glabrous or velvety- or thinly-pubescent beneath.
   b. Teeth of blade-margin 3-6 mm long, with long narrow tips.
      c. Blades glabrous except for axillary tufts of hairs beneath; bracts glabrous. T. americana L., American L.
      c. Blades minutely velvety-pubescent beneath; lower surface of bract and adnate part of peduncle often pubescent. T. floridana (V. Engler) Small
   b. Teeth of blade-margin shorter; blades thinly pubescent beneath with stellate and simple hairs; bracts glabrous. T. neglecta Spach

MALVACEAE, Mallow Family

Ours herbs or shrubs; often stellate-pubescent; leaves alternate, simple, usually palmately veined; flowers hypogynous, bisporangiate or monosporangiate; sepals 5, separate, or united at base, frequently subtended by a circle of bracts; corolla regular; petals 5, separate or nearly so; stamens many, the filaments united in a sheath around the style or styles, this sheath adnate to the petals at base, filaments separate distally; carpels 5-many, united; styles as many as carpels or style 1 with apical branches as many as carpels; fruit a capsule or separating into the dehiscent or indehiscent 1- to several-seeded carpels.

a. Flowers with circle of bracts below the calyx.
   b. Carpels 5; fruit a 5-loculed capsule. 7. Hibiscus
   b. Carpels more than 5, united in a ring, separating at maturity.
      c. Involucral bracts 6 or more, united at base. 2. Althaea
      c. Involucral bracts 3; petals notched or obcordate. 1. Malva
a. Flowers without bracts below the calyx.
   b. Flowers in terminal panicles; petals white; blades palmately lobed.
      c. Diecious; lobes of blade acute. 3. Napaea
      c. Flowers bisporangiate; upper lobes of blade long-pointed. 5. Sida
   b. Flowers axillary.
      c. Petals pale blue or violet; blades ovate, dentate and somewhat lobed. 4. Anoda
      c. Petals yellow.
         d. Blades lanceolate, serrate; a small spine below base of petiole. 5. Sida
         d. Blades ovate-cordate, entire or nearly so, velvety-pubescent. 6. Abutilon

1. Malva L., Mallow

Herbs; flowers with 3 bracts below the calyx; petals obcordate; stamen-column anther-bearing at the summit; fruit a disk of many carpels arranged in a circle, each carpel 1-seeded and indehiscent.

a. Leaf-blades with shallow rounded lobes; flowers axillary, pedicels shorter than petioles of subtending leaves.
  b. Petals red-purple, 2-2.5 cm long; stem erect. *M. sylvestris L., High M.
  b. Petals white or pink, about 1 cm long; stem prostrate or ascending. *M. neglecta Wallr., Common M., Cheeses
a. Leaf-blades deeply lobed or cleft, segments of upper blades pinnatifid; petals white to pale purple, 2-3 cm long; fruit pubescent. *M. moschata L., Musk M.

2. Althaea L.

*A. rosea Cav., Hollyhock. Tall erect stellate-pubescent herb; blades large, orbicular, shallowly lobed, margins toothed; flowers 7-10 cm wide, in a raceme, petals of various colors; carpels many, 1-seeded.

3. Napaea L., Glade-mallow

N. dioica L. Tall herb; diecious; blades palmately 5-9-lobed; bracts none; calyx 5-toothed; flowers white, in large terminal panicles, staminate with petals 5-9 mm long, with 15-20 or more stamens, carpellate with smaller petals and some rudimentary stamens; carpels about 10, 1-ovuled.

4. Anoda Cav.

*A. cristata (L.) Schlech. Hairy erect herb; leaves ovate, somewhat lobed, margins dentate; flowers about 1 cm wide, long-peduncled, solitary in axils; calyx with ovate-lanceolate acuminate lobes; petals blue to lavender; carpels awned, hirsute.

5. Sida L., Sida

Herbs; involucral bracts lacking in ours; calyx 5-lobed, persistent in fruit; carpels 5-15, 1-ovuled, dehiscent.

a. Leaves palmately lobed, lobes long-pointed; flowers in large terminal panicle; petals white, about 1 cm long; carpels about 10, each with pointed beak. S. hermaphrodita (L.) Rusby, Tall S.
a. Leaves lanceolate, serrate; flowers in small axillary clusters; petals yellow, about 5 mm long; carpels 5, each 2-beaked. S. spinosa L., Prickly S.

6. Abutilon Mill.

*A. theophrasti Medic., Velvet-leaf. Branched annual, velvety stellate-pubescent; leaf blades broadly ovate, cordate, acuminate, entire or nearly so; flowers axillary, 1.5-2.5 cm wide; involucral bracts none; calyx 5-cleft; petals yellow; carpels 10-15, pubescent, conspicuously beaked, each containing 3-9 ovules, dehiscent.

7. Hibiscus L., Rose-mallow

Shrubs and herbs; leaves palmately lobed or dentate; flowers showy, ours 5-15 cm wide; involucral bracts linear; column of stamens bearing anthers at sides; carpels 5; ovulary 5-loculed, 3 or more ovules in each locule; styles separate at apex; stigmas capitate; fruit a capsule.

a. Shrubs. *H. syriacus L., Rose-of-Sharon
a. Herbs.
  b. Leaves deeply lobed or divided into narrow divisions; calyx conspicuously veined, inflated in fruit; corolla pale yellow with purple eye. *H. trionum L., Flower-of-an-hour
  b. Leaves unlobed or with few lobes.
    c. Upper leaf-blades usually hastate; stem, calyx, and lower leaf surface glabrous or nearly so; petals pink with darker base. H. militaris Cav., Halberd-leaved R.
    c. Leaf-blades not hastate; stem pubescent above, calyx and lower leaf surface stellate-pubescent, upper leaf surface glabrous or glabrate.
      d. Style branches decidedly pubescent; peduncle not fused with petiole; petals pink with darker base or rarely white. H. palustris L., Swamp R.
      d. Style branches glabrous or sparsely pubescent; peduncle often fused with petiole; petals usually cream to white with purple or red base. H. moscheutos L., Swamp R.

HYPERICACEAE (GUTTIFERAE), St. John's-wort Family

Shrubs and herbs; leaves opposite, entire, dotted (except in 1 species); flowers hypogynous, bisporangiate, in cymose clusters or solitary; sepals 4-5; corolla regular, petals separate, 4-5; stamens few to many, separate or basally united in 3 or more groups; carpels 2-5, united, styles as many, separate or united at base; locules of ovulary as many as carpels or locule 1, the placentae then parietal and as many as carpels; fruit a many-seeded capsule.

a. Sepals and petals 4, outer pair of sepals larger; styles 2. Ascyrum
a. Sepals and petals 5; styles or stigmas 3-5. Hypericum

Ascyrum L.

A. hypericoides L., St. Andrew's Cross. Small glabrous pale-green shrub; leaves linear to oblanceolate, narrowed to base, rounded at tip, 2-3 cm long; flowers solitary or in small clusters; stamens many; petals yellow, about 1 cm long; ovulary and capsule 1-loculed.

Hypericum L., St. John's-wort

Herbs and shrubs; flowers usually in cymose clusters; carpels 3 or 5 (rarely 4 or 6); ovulary 1-loculed with parietal placentae, or 3-5-loculed with axile placentae.

a. Petals yellow; stamens many to few, sometimes united at base in 3 or 5 clusters.
  b. Styles, carpels, and locules 5.
    c. Large herb; leaves lanceolate or elliptic; petals 2-3 cm long; capsule 2-3 cm long.
      H. pyramidatum Ait., Great S.
    c. Shrub; leaves linear or oblanceolate; petals 2 cm long or less; capsule 1 cm long or less.
      H. kalmianum L.
  b. Styles and carpels 3; locules 3 or 1.
    c. Stamens 15-40 or more; flowers usually more than 7 mm wide.
      d. Shrub; stamens separate; blades linear to narrowly elliptic; twigs 2-edged.
        H. spathulatum (Spach) Steud.
      d. Herbs, sometimes slightly woody at base.
        e. Petals and sometimes other flower-parts black-dotted or -lined; ovulary and capsule 3-loculed, surface marked with oil-vescicles; stamens in 3-5 groups.
          f. Stem much branched; petals black-dotted near margin; sepals little or not black-dotted; blades 2-4 cm long. *H. perforatum L., Common S.
          f. Stem little branched below inflorescence; petals and sepals lined and dotted with black; blades 4-6 cm long. H. punctatum L.
        e. Petals mostly without black dots and lines; ovulary and capsule 1-loculed or incompletely 3-loculed, mostly without oil-vescicles.
          f. Styles separate; stamens not in groups.
            g. Sepals about equaling capsule; petals coppery-yellow; stamens many.
              H. denticulatum Walt.
            g. Sepals shorter than capsule; petals yellow; stamens usually 20 or fewer.
              H. majus (Gray) Britt.
          f. Styles united at base; stigmas minute.
            g. Blades elliptic to oval, usually 1-3 cm long; seeds minutely striate. H. ellipticum Hook.
            g. Blades linear to narrowly oblong, 3-7 cm long; seeds rough-pitted. H. sphaerocarpum Michx.
    c. Stamens 5-12 (rarely -20); flowers 7 mm wide or less.
      d. Blades minute, scalelike, 1-3 mm long; flowers nearly sessile; capsule much longer than sepals. H. gentianoides (L.) BSP.
      d. Blades more than 3 mm long, linear to ovate, 1-8-nerved.
        e. Bracts of ultimate branches of inflorescence broad, like cauline blades but smaller; blades 3-5-nerved, the larger 1-2 cm long. H. boreale (Britt.) Bickn.
        e. Bracts of ultimate branches of inflorescence subulate.
          f. Blades linear to narrowly lanceolate.
            g. Blades 1-nerved, linear, 2 cm long or less; sepals equaling capsule. H. drummondii (Grev. & Hook.) T. & G.
            g. Blades 1-3-nerved, linear to lanceolate, obtuse at tip; capsule red or purple, longer than sepals; stamens sometimes more than 12. H. canadense L.

f. Blades wider or more than 3-nerved or both.
    g. Bracts of inflorescence, except those near ends of ultimate branches, leafy.
        H. mutilum L.
    g. All bracts above lowest inflorescence-branches subulate.
        h. Blades ovate; flowers 3-4 mm wide. H. gymnanthum Engelm. & Gray
        h. Blades lanceolate; flowers 5-10 mm wide. H. majus (Gray) Britt.
a. Petals pink; stamens usually 9, united in 3 groups, a large gland between each 2 groups.
  b. Filaments united only at base; leaves dotted beneath.
    c. Sepals lanceolate, acute, 5-7 mm long; styles 2-3 mm long. H. virginicum L.
    c. Sepals oblong or elliptic, 5 mm long or less; styles usually not more than 1 mm long.
      H. virginicum L. var. fraseri (Spach) Fern.
  b. Filaments united to above the middle.
    c. Leaves sessile or nearly so, rounded or cordate at base, not dotted beneath. H. tubulosum
      Walt.
    c. Leaves tapering to short petioles, dotted beneath. H. tubulosum Walt. var. walteri (Gmel.)
      Lott

ELATINACEAE, Waterwort Family

Elatine L., Waterwort

E. brachysperma Gray   Tiny herb of wet soil or shallow water; leaves opposite, connected by stipules, obovate; flowers hypogynous, bisporangiate, axillary; perianth regular; sepals, petals, stamens, and carpels usually 3; placentae axile; ovules several, short-cylindric, somewhat curved, with areoles in longitudinal rows.

CISTACEAE, Rockrose Family

Herbs and shrubs; leaves simple, alternate or opposite, entire; flowers hypogynous, bisporangiate; sepals 5, the two of the outer series smaller; corolla regular or none; stamens few to many; carpels 3, united; style often short or lacking; stigmas 1 or 3; ovulary 1-loculed with 3 parietal placentae, sometimes 3-loculed as result of growth inward of placentae; fruit a capsule.

a. Leaves scalelike, 1-4 mm long, closely imbricated; petals 5, yellow; style elongate. 2. Hudsonia
a. Leaves not as above.
  b. Petals 5, yellow, or none; style short. 1. Helianthemum
  b. Petals 3, reddish-brown; style none. 3. Lechea

1. Helianthemum Mill., Frostweed

Stellate-pubescent, herbaceous or somewhat shrubby; earlier flowers with petals and with many stamens; later flowers cleistogamous, with smaller petals or none, and with fewer stamens.

a. Petaliferous flowers 2-4 cm wide, solitary or in clusters of 2; outer sepals not more than 2/3 as long as inner; later axillary branches overtopping capsules; seeds papillose. H. canadense (L.) Michx.
a. Petaliferous flowers 1.5-2.5 cm wide, several in a raceme or corymb; outer sepals nearly or quite as long as inner; later axillary branches usually not overtopping capsules; seeds reticulate.
  H. bicknellii Fern.

2. Hudsonia L.

H. tomentosa Nutt.   Low branched hoary-tomentose shrub; leaves scalelike, alternate, appressed, imbricated; flowers small, bright yellow, solitary at ends of short leafy branches; each outer sepal connate with an inner one; petals 5; stamens 10 or more, style slender; capsule few-seeded.

3. Lechea L., Pinweed

Slender branching herbs often woody at base; basal leafy shoots appearing late in season; leaves small; flowers many, minute, in panicles; petals marcescent; stamens usually 5-15; stigmas 3, plumose, sessile; capsule few-seeded, enclosed in persistent calyx.

a. Two outer sepals evidently shorter than three inner; stem appressed-pubescent.
  b. Calyx and fruit narrowly ellipsoid, almost or quite twice as long as wide; capsule equaling or longer than calyx; leaves of basal shoots no more than 3 times as long as wide. L. racemulosa Michx.
  b. Calyx and fruit subglobose; leaves of basal shoots more than 3 times as long as wide.
    c. Plants canescent; panicle branches ascending; cauline leaves somewhat appressed; longer sepals about equaling capsule; seeds ovoid, smooth. L. stricta Leggett
    c. Plants green, not canescent.
      d. Calyx and fruit ellipsoid; seeds 2-sided, flattened, smooth. L. leggettii Britt. & Holl.
      d. Calyx and fruit globose; seeds 3-sided, with grayish-white covering. L. intermedia Leggett
a. Outer sepals about equaling or slightly longer than inner.
  b. Leaves of basal shoots and cauline leaves linear; calyx and fruit subglobose, calyx completely covering capsule; outer sepals equaling or evidently longer than inner; hairs of stem appressed. L. tenuifolia Michx.
  b. Leaves of basal shoots not more than 3 times as long as wide; cauline leaves lanceolate or oblanceolate.
    c. Hairs of stem appressed; calyx and fruit obovoid. L. minor L.
    c. Hairs of stem spreading; calyx and fruit globose. L. villosa Ell.

VIOLACEAE, Violet Family

Herbs; leaves alternate, stipuled, sometimes all basal; flowers hypogynous, bisporangiate; sepals 5, persistent in fruit; corolla zygomorphic; petals 5, lower one spurred or saccate at base; stamens 5, anthers connivent or united, each with a distal projection; carpels 3, united; style and stigma 1; ovulary 1-loculed; placentae 3, parietal; fruit (in ours) a capsule; seeds many.

a. Anthers united; peduncles not or scarcely longer than flowers; sepals not auricled. Hybanthus
a. Anthers not united, sometimes connivent; peduncles much longer than flowers; sepals auricled. Viola

Hybanthus Jacq., Green Violet

H. concolor (T. F. Forst.) Spreng. Stem erect; blades elliptic, acuminate at both ends, petioled, entire or somewhat toothed; flowers greenish, solitary or 2 or 3 together in leaf-axils; peduncles recurved; sepals nearly equal; petals about as long as sepals, about 5 mm long, lowest wider than the others and saccate at base; anthers almost sessile, united in a sheath, a 2-lobed gland at base; style hooked at tip.

Viola L., Violet

With aerial leafy stems or with peduncles and petioles arising from rhizomes or stolons; flowers usually solitary, largest about 3 cm wide; petals not quite equal, the lowest one spurred; two lower stamens with spurs that project into spur of petal, these two stamens the only ones present in cleistogamous flowers, which are characteristic of most species, are apetalous, and appear after the vernal petaliferous flowers. Hybridization often occurs between plants of related species.

a. Plants without aerial leafy stems, petioles and peduncles arising from rhizomes or stolons.
  b. Petals blue or violet; rhizomes thick and fleshy.
    c. Tip of style pointed, bent downward in a hook; leaves and capsules finely pubescent, blades cordate-ovate; with stolons. *V. odorata L., Sweet V.
    c. Tip of style widened, with a conical beak on one side.
      d. None of the leaf-blades divided, all ovate to reniform, cordate at base, margins crenate-serrate.
        e. Blades glabrous or nearly so; lateral petals bearded.
          f. Hairs of lateral petals knobbed at tip; peduncles exceeding leaves; spurred petal glabrous; cleistogamous flowers on long erect peduncles, sepal-auricles lobed. V. cucullata Ait., Marsh Blue V.
          f. Hairs of lateral petals slender or only slightly clavate; peduncles shorter than to not much longer than leaves.
            g. Spurred petal villous; cleistogamous flowers on ascending peduncles. V. affinis LeConte

157

g. Spurred petal glabrous or nearly so.
  h. Petals pale violet to nearly white; cleistogamous flowers on short prostrate peduncles; rare. V. missouriensis Greene
  h. Petals deep purple or blue; cleistogamous flowers on short spreading peduncles. V. papilionacea Pursh
e. Blades pubescent on both sides (except in V. hirsutula)
  f. Petioles and peduncles glabrous; blades pubescent above, glabrous or nearly so beneath; veins usually purple. V. hirsutula Brainerd
  f. Petioles pubescent; blades pubescent about equally on both sides.
    g. Blades short-petioled, circular or almost so, rounded at apex; peduncles up to twice as long as petioles; rare. V. villosa Walt.
    g. Petioles equaling or longer than peduncles; blades acute or obtuse at apex, not rounded; abundant.
      h. None of the blades lobed. V. sororia Willd.
      h. Blades of middle season lobed, earliest and latest ones unlobed. V. triloba Schwein.
d. Leaf-blades lobed or divided, sometimes only at base, or longer than wide.
  e. Blades longer than wide, more or less lobed near base, or unlobed.
    f. Blades and petioles densely pubescent; petioles often half as long as blades; sepals and auricles ciliate. V. fimbriatula Sm.
    f. Blades and petioles minutely pubescent; petioles equaling or longer than blades; sepals and auricles glabrous. V. sagittata Ait.
  e. Blades about as long as wide, the whole blade lobed or divided.
    f. Blades cordate at base, earliest and latest ones not lobed, those of middle of season 3-5-lobed, middle lobe wide; plants pubescent. V. triloba Schwein.
    f. Blades divided nearly to base into 5-9 (or more) narrow lobes.
      g. Plants glabrous; rhizome vertical; stamens orange, tips exsert; petals all glabrous, sometimes the 2 upper ones of darker color; cleistogamous flowers none. V. pedata L., Bird's-foot V.
      g. Plants more or less pubescent; rhizome not vertical.
        h. Pubescent with long hair; lateral petals bearded, spurred petal glabrous. V. palmata L.
        h. Pubescent with short hair; lateral and spurred petals bearded. V. pedatifida G. Don
b. Petals yellow or white.
  c. Petals yellow; style capitate, beakless; blades round-ovate, small at anthesis, becoming 5-12 cm wide by late summer. V. rotundifolia Michx.
  c. Petals white; rhizomes slender; stolons often present.
    d. Blades 1.5-6 or more times as long as wide.
      e. Blades slenderly lanceolate to oblanceolate, long-tapering to petiole. V. lanceolata L.
      e. Blades about twice as long as wide, cordate, truncate, or cuneate above junction with usually winged summit of petiole. V. primulifolia L.
    d. Blades as wide as or wider than long, cordate at base; petiole not winged.
      e. Petals all glabrous; blades cordate, basal lobes sometimes overlapping, upper surface of basal lobes with small scattered hairs; petioles and peduncles reddish; cleistogamous flowers on prostrate peduncles. V. blanda Willd., Sweet White V.
      e. Lateral petals more or less bearded.
        f. Blades glabrous on both surfaces; lateral petals sparsely bearded; cleistogamous flowers on erect peduncles. V. pallens (Banks) Brainerd
        f. Blades more or less pubescent; lateral petals densely bearded; cleistogamous flowers on prostrate peduncles. V. incognita Brainerd
a. Plants with aerial leafy stems; peduncles axillary.
  b. Stipules not large and leaflike.
    c. Petals yellow; cauline leaves usually on only upper part of stem.
      d. Blades unlobed and cuneate at base, or 3-lobed or 3-parted; rare. V. tripartita Ell.
      d. Blades unlobed, base not cuneate.
        e. Blades halberd-shaped or ovate, usually longer than wide; stem slender, glabrous; stipules 5 mm long or less; petals purple on outside. V. hastata Michx.
        e. Blades broadly ovate, not longer than wide; pubescent or glabrate; stipules longer than 5 mm; petals not violet on outside. V. pubescens Ait.

  f. Plant pubescent; basal leaves 1 or 2; stems erect, usually solitary. var. <u>pubescens</u>, Downy Yellow V.

  f. Plant somewhat pubescent to glabrous; basal leaves 1-5 or more; stems often spreading, usually several. var. <u>eriocarpa</u> (Schwein.) Russell (<u>V</u>. <u>pensylvanica</u> Michx.), Smooth Yellow V.

 c. Petals blue, violet, or white; stem leafy throughout; leaf-blades cordate at base.

  d. Petals white or cream, sometimes tinged with blue on back.

   e. Stipules entire; petals sometimes tinged with blue on back; spurred petal yellow at base and dark-lined; stigma capitate. <u>V</u>. <u>canadensis</u> L., Canada V.

   e. Stipules bristle-toothed; petals white or cream with prominent dark veins at base; stigma slender. <u>V</u>. <u>striata</u> Ait., Striped White V.

  d. Petals blue or violet; stipules bristle-toothed; stigma slender.

   e. Spur 0.8 cm long or more, equaling or longer than petal-blades; petals violet with darker veins, the lateral ones glabrous; style not bent at tip. <u>V</u>. <u>rostrata</u> Pursh, Long-spurred V.

   e. Spur shorter, about half as long as petal-blades or less; lateral petals bearded; style bent at tip.

    f. Plants glabrous, not stoloniferous; stem erect. <u>V</u>. <u>conspersa</u> Reichenb., American Dog V.

    f. Plants finely pubescent; trailing stolons present, rooting at tips; stems becoming prostrate; leaves often purple beneath. <u>V</u>. <u>walteri</u> House

 b. Stipules large and leaflike, deeply lobed; summit of style globose, hollow, with orifice on side; leaf-blades, except the lowest, not cordate at base.

  c. Petals yellow, sometimes with bluish tips, about as long as or shorter than sepals.
   *<u>V</u>. <u>arvensis</u> Murr.

  c. Petals longer than sepals.

   d. Petals pale blue to white; middle lobe of stipules entire or nearly so. <u>V</u>. <u>rafinesquii</u> Greene (<u>V</u>. <u>kitaibeliana</u> R. & S. var. <u>rafinesquii</u> Fern.), Wild Pansy

   d. Petals variously colored, yellow, purple, or white, the two upper darker; middle lobe of stipules crenate. *<u>V</u>. <u>tricolor</u> L., Johnny-jump-up

PASSIFLORACEAE, Passion Flower Family

<u>Passiflora</u> L., Passion Flower

  Ours herbaceous vines climbing by tendrils opposite the leaves; leaves alternate, blades lobed or parted; flowers axillary, perigynous, bisporangiate, peduncled; sepals 5; corolla regular; petals 5, separate, a showy double or triple fringed corona attached to the hypanthium within the corolla; stamens 5, united around stalk of ovulary; styles 3; ovulary 1-loculed; placentae 3, parietal; fruit a berry.

a. Petals and sepals white; corona purple; leaf-blades deeply 3-parted, segments serrate.
 <u>P</u>. <u>incarnata</u> L., Maypop, Purple P.

a. Petals, sepals, and corona greenish-yellow; leaf-blades 3-lobed, lobes entire. <u>P</u>. <u>lutea</u> L., Yellow P.

CACTACEAE, Cactus Family

<u>Opuntia</u> Mill., Prickly-pear

  Stems spreading or prostrate, branching in ours, of flattened fleshy segments bearing small scarious early-deciduous spirally-arranged leaves with axillary areoles containing barbed bristles (glochids) and usually 1 or more large spines; flowers epigynous, bisporangiate, solitary, large; perianth regular, yellow in ours, the many intergrading sepals and petals and the many stamens attached to epigynous hypanthium and with it deciduous in one piece; carpels several, united, parietal placentae and stigmas as many; locule 1; fruit fleshy, bearing areoles.

a. Segments spineless or with 1 or rarely 2 spines per areole. <u>O</u>. <u>compressa</u> (Salisb.) Macbr.
 (<u>O</u>. <u>humifusa</u> Raf.)

a. Spines usually 3-8 per areole except those near base of segments. <u>O</u>. <u>tortispina</u> Engelm.

THYMELAEACEAE, Mezereum Family

Shrubs; leaves alternate; flowers perigynous, bisporangiate; sepals present, sometimes minute; corolla none; stamens 8; carpels united; fruit a drupe.

a. Hypanthium and calyx pale yellow, calyx scarcely evident; style and stamens exserted.   1. Dirca
a. Hypanthium and calyx red-purple; style very short; stamens included.   2. Daphne

1. Dirca L., Leatherwood

D. palustris L. Shrub with tough bark and flexible branches; leaves short-petioled, blades obovate, entire; flowers in small axillary clusters, appearing before the leaves; hypanthium tubular-funnelform, yellow, margin (sepals) wavy or shallowly toothed; locule and ovule 1.

2. Daphne L.

*D. mezereum L. Small shrub; blades oblanceolate, tapering to base; flowers in small sessile lateral clusters, appearing before the leaves; fruit red.

ELAEAGNACEAE, Oleaster Family

Shrubs; stems and simple entire leaves covered with silvery or golden-brown scales; flowers perigynous, monosporangiate or bisporangiate, in small lateral clusters; hypanthium of carpellate and bisporangiate flowers tubular; calyx regular, sepals 4; petals none; carpel and ovule 1; fruit covered by persistent base of hypanthium, hence appearing drupelike.

a. Leaves alternate; stamens 4.   1. Elaeagnus
a. Leaves opposite; stamens 8.   2. Shepherdia

1. Elaeagnus L., Oleaster

Flowers bisporangiate or some of them monosporangiate, on twigs of current season; hypanthium tubular, constricted above ovulary.

a. Twigs with brown scales.   *E. multiflora Thunb.
a. Twigs with silvery scales.   *E. angustifolia L., Russian-olive

2. Shepherdia Nutt., Buffalo-berry

S. canadensis L. Diecious; flowers on twigs of previous season; hypanthium of staminate flowers shallow or cup-shaped; fruit yellow to red.

LYTHRACEAE, Loosestrife Family

Our species herbs; leaves opposite, whorled, or alternate, simple, entire; flowers perfect, perigynous, often dimorphic or trimorphic in length of filaments and of style; petals rarely none or sepals and petals 4-7, stamens as many or twice as many; carpels 2-6, united, ovulary completely or incompletely 2-6-loculed; placentation axile; style 1; fruit a capsule.

a. Corolla not regular; hypanthium spurred or saccate on one side; sepals and petals 6.   6. Cuphea
a. Corolla regular or absent; hypanthium regular.
  b. Hypanthium cylindric, longer than wide; sepals and petals 5-7, stamens as many or twice as many.   5. Lythrum
  b. Hypanthium as wide as long, globose in fruit; petals 5, 4, or none.
    c. Tall plants with alternate, whorled, or opposite leaves; petals usually 5, stamens 10, of 2 lengths; flowers trimorphic.   4. Decodon
    c. Low plants with opposite leaves; petals 4 or none; stamens equal; flowers of one form.
      d. No appendages alternating with sepals; hypanthium shorter than capsule.   1. Peplis
      d. Appendages alternating with sepals; hypanthium equaling capsule.
        e. Flowers usually solitary in axils; bases of leaves tapering.   2. Rotala
        e. Flowers usually clustered in axils; bases of at least upper leaves auricled.   3. Ammannia

1. Peplis L., Water-purslane

P. diandra Nutt. Rooting in mud, at margin of or in shallow water; blades opposite, submersed ones thin, elongate, sessile, emersed ones shorter, narrowed to base; flowers small, greenish, solitary in axils; sepals 4, petals none; stamens 4; carpels and locules 2; capsule globular, indehiscent.

2. Rotala L., Tooth-cup

R. ramosior (L.) Koehne   Low herbs of wet soil; stems erect or prostrate; blades opposite, short-petioled, linear-spatulate, 1-3 cm long; flowers small, solitary and sessile in axils; sepals 4, about as long as appendages; petals 4, minute, white or pink, early deciduous; stamens 4; capsule 4-loculed, dehiscent.

3. Ammannia L.

A. coccinea Rottb. Low glabrous herbs of wet places; blades opposite, sessile, narrow, auricled at base; flowers small, axillary, usually clustered; hypanthium 4-angled; sepals 4, small, appendages in sinuses; petals 4, small, early deciduous; stamens 4 or 8; capsule bursting irregularly.

4. Decodon J. F. Gmel., Swamp Loosestrife

D. verticillatus (L.) Ell. Herbaceous to slightly shrubby; growing in swamps; stem angled, elongate, usually arching and rooting at nodes; leaves opposite, whorled, or alternate, lanceolate, short-petioled; flowers showy, trimorphic, in nearly sessile clusters in axils of upper leaves; sepals 4-8, alternating with linear appendages; petals 5, purple-pink, 10-15 mm long; stamens 10; capsule 3-5-loculed, dehiscent.

5. Lythrum L., Loosestrife

Stems 4-angled; leaves opposite, alternate, or whorled; flowers often dimorphic or trimorphic, solitary or in axillary clusters, sometimes in spikes; hypanthium cylindric, striate; sepals 5-7, alternating with appendages; petals 5-7, purple, red-purple, or white; stamens as many or twice as many as petals, inserted lower in hypanthium; style slender; capsule 2-loculed.

a. Leaves opposite or whorled; flowers in clusters in upper axils, clusters in spikelike inflorescence; plants more or less downy; stamens twice petals. *L. salicaria L.
a. Leaves, at least part of them, alternate; flowers solitary or in 2's in axils; stamens not more numerous than petals.
  b. Petals 2-3 mm long; hypanthium not winged. L. hyssopifolia L.
  b. Petals 5 mm long; hypanthium narrowly 12-winged. L. alatum Pursh

6. Cuphea P. Br.

C. petiolata (L.) Koehne, Blue Waxweed. Viscid-hairy; stem erect, branching; leaves opposite, ovate-lanceolate, petioled; flowers solitary or in 2's in axils, sessile or short-stalked; hypanthium tubular, oblique, ribbed, spurred on one side; sepals 6, appendages in sinuses; corolla irregular; petals red-purple, about 1 cm long; stamens usually about 11-12.

MELASTOMACEAE, Meadow Beauty Family

Rhexia L., Meadow Beauty

R. virginica L. Herb; stem 4-angled; blades opposite, sessile, longitudinally ribbed, ovate or elliptic, serrulate; flowers perigynous, bisporangiate, in a terminal cyme; sepals 4; corolla regular, petals 4, separate, 1-2 cm long, red-purple; stamens 8, anthers dehiscing by terminal pores, each with a small spur; carpels and locules 4; style and stigma 1; fruit a capsule.

ONAGRACEAE, Evening-primrose Family

Herbs; leaves alternate or opposite, simple; flowers epigynous, perfect, epigynous hypanthium present or absent; corolla regular or rarely somewhat zygomorphic; sepals and petals usually 4 (2-6); petals separate or rarely none; stamens usually 8 (2-8); carpels usually 4 (2-6), united, locules as many or rarely 1; style 1, sometimes short; stigma 1 or with lobes as many as carpels; fruit a capsule or nutlike.

a. Sepals, petals, and stamens 2. 6. Circaea
a. Sepals and petals 4 to 6 or rarely none.
  b. Epigynous hypanthium very short or absent.
    c. Petals purple, red-purple, pink, or white; calyx and short hypanthium, when present, deciduous; seeds with tufts of hairs. 3. Epilobium
    c. Petals yellow or greenish or none; calyx persistent in fruit; seeds without tufts of hairs.
      d. Bractlets usually present on ovulary or just below flower; stamens 4; length of capsule less than twice width. 2. Ludwigia
      d. Bractlets absent; stamens 8-12; length of capsule twice width or more. 1. Jussiaea
  b. Epigynous hypanthium elongate, slender.
    c. Petals usually yellow, if pink or white then 2-4 cm long; fruit prismatic, cylindric, or clavate; flower regular. 4. Oenothera
    c. Petals white or pink, slender, less than 1 cm long; fruit fusiform, indehiscent; flower somewhat irregular. 5. Gaura

1. Jussiaea L.

Leaves alternate, mostly entire; flowers axillary; sepals 4-6; corolla regular; petals same number as sepals, yellow in our species, 0.5-1 cm long; stamens 8 to 12; carpels 4 or 5, each many-ovuled; style short; stigma 4-5-lobed; capsule elongate.

a. Petals and sepals 4; stem winged; fruit 4-angled, 1-2 cm long. J. decurrens (Walt.) DC.
a. Petals and sepals 6; stem not winged; fruit slenderly cylindric, 2-5 cm long. J. leptocarpa Nutt.

2. Ludwigia L., False Loosestrife

Of wet soil or water; stems erect or prostrate; flowers solitary, sessile or short-stalked, in upper axils; perianth regular; sepals, petals, and stamens 4, or petals none, sepals persistent in fruit; carpels 4; ovulary often square in cross section; bractlets usually present on ovulary or below.

a. Leaves alternate; flowers stalked or sessile.
  b. Petals yellow or none, about as long as sepals; bractlets near summit of flower stalk; capsule about as long as wide, sharply wing-angled. L. alternifolia L., Seedbox
  b. Petals greenish, minute, or wanting; bractlets on ovulary; capsule longer than wide, angles not winged. L. polycarpa Short & Peter
a. Leaves opposite; flowers sessile; ovulary with 4 green longitudinal bands on the sides; petals none; stems floating or prostrate, often rooting at nodes. L. palustris (L.) Ell. var. americana (DC.) Fern. & Griscom

3. Epilobium L., Willow-herb

Leaves alternate or opposite, blades entire or dentate; flowers solitary or in terminal racemes; epigynous hypanthium short or absent; sepals 4; petals 4, purple to pink or white; corolla regular or slightly zygomorphic; stamens 8; carpels 4; style short; stigma entire or 4-cleft; capsule linear, dehiscent; seeds comate.

a. Flowers small-bracted, showy, in terminal racemes, without an epigynous hypanthium; corolla red-purple, not quite regular; petals entire; stigma 4-cleft. E. angustifolium L., Fireweed
a. Flowers axillary, solitary or, if aggregated in spikes or racemes, then leafy-bracted; short epigynous hypanthium present; corolla regular.
  b. Stigma 4-cleft; petals about 1 cm long, red-purple, notched; plant long-villous. *E. hirsutum L.
  b. Stigma entire; petals smaller.
    c. Stem with decurrent lines from leaf-bases; blades denticulate or serrulate, margins not revolute.
      d. Coma red-brown at maturity; blades lanceolate, tapering to base, grayish; seeds without short neck below coma. E. coloratum Biehler
      d. Coma whitish at maturity; blades lanceolate to ovate, rounded at base; seeds with short neck below coma. E. glandulosum Lehm.
    c. Stem without decurrent lines from leaf-bases; blades entire or undulate, margins revolute.

d. Hairs of stem spreading; lateral veins of leaf obvious. E. strictum Muhl.
d. Hairs of stem appressed or incurved; lateral veins scarcely obvious. E. leptophyllum Raf.

4. Oenothera L., Evening-primrose

Leaves of first year often in rosettes; blades linear to lanceolate; flowers solitary in axils or in terminal racemes; epigynous hypanthium long and slender, deciduous in fruit; sepals 4, at first connate, later more or less separate and reflexed; corolla regular; petals 4, usually large; stamens 8; carpels 4; style elongate; stigma entire or 4-lobed or -cleft.

a. Leaves basal, runcinate-pinnatifid. O. triloba Nutt.
a. Cauline leaves present.
 b. Petals pink or white, 3 cm long or more; capsule broadened toward tip. O. speciosa Nutt.
 b. Petals yellow.
  c. Capsule cylindric or somewhat wider toward base, sometimes ridged but not winged; stamens equal.
   d. Stem erect, often tall; flowers in often elongate terminal raceme, bracts smaller than foliage leaves; blades entire to repand-dentate. O. biennis L.
   d. Stem decumbent to erect; flowers few, in axils of leaves scarcely smaller than foliage leaves; blades sinuate to pinnatifid. O. laciniata Hill
  c. Capsule clavate to oblong, often angled or broadly winged; stamens of 2 lengths.
   d. Inflorescence nodding in bud, erect in flower; petals less than 1 cm long; anthers less than 3 mm long; hairs of stem below inflorescence appressed or curved or short. O. perennis L.
   d. Inflorescence erect in bud; petals usually longer.
    e. Capsule and epigynous hypanthium with spreading gland-tipped hairs; fruit ellipsoid or oblong. O. tetragona Roth
    e. Capsule and epigynous hypanthium without gland-tipped hairs; capsule clavate.
     f. Capsule sessile or nearly so; epigynous hypanthium 12-25 mm long; bracts usually foliaceous; anthers 5-9 mm long; tips of sepals villous. O. pilosella Raf.
     f. Capsule stalked; hypanthium 5-15 mm long; bracts usually not foliaceous; anthers 4-6 mm long. O. fruticosa L.

5. Gaura L.

G. biennis L. Rosette blades oblanceolate; cauline blades lanceolate, tapering at both ends, denticulate; epigynous hypanthium elongate; sepals and petals 4, rarely 3; sepals reflexed; petals narrow, turned to one side of flower, white becoming pink; stamens 8; carpels 4; ovulary becoming 1-loculed; fruit fusiform, 4-sided, 1-4-seeded, hard and indehiscent.

6. Circaea L., Enchanter's-nightshade

Leaves opposite, dentate, ovate or oblong, base rounded, truncate, or subcordate; flowers small, in terminal and lateral racemes, epigynous hypanthium short; sepals 2, reflexed; petals 2, white, notched or 2-lobed; stamens and carpels 2; style elongate; fruit bristly, obovoid, indehiscent.

a. Fruit longitudinally-ridged, 2-loculed, 3.5-6 mm wide, including hairs; stigma shallowly 2-lobed; rhizome slender. C. quadrisulcata (Maxim.) Franch. & Sav.
a. Fruit not ridged, 1-3 mm thick, including hairs; stigma 2-cleft.
 b. Rhizome slender; fruit 2-loculed, 1-2-seeded. C. canadensis Hill
 b. Rhizome thickened; fruit 1-loculed, 1-seeded. C. alpina L.

HALORAGACEAE, Water-milfoil Family

Aquatic or marsh herbs; emersed leaves entire, toothed, or pinnatifid, submersed ones pinnatifid with linear or filiform segments; monecious; flowers epigynous, bisporangiate or monosporangiate, sessile, axillary, solitary or few together; perianth regular; carpels 3-4, united, locules and ovules as many.

a. Flowers monosporangiate or bisporangiate, usually the carpellate below, the staminate above, and the bisporangiate in the transition zone; flower parts in 4's; leaves usually opposite or whorled. Myriophyllum
a. Flowers bisporangiate, the parts in 3's; leaves alternate. Proserpinaca

Myriophyllum L., Water-milfoil

Submersed; flowers small, 2-bracted, in our species in axils of emersed leaves or bracts, the whole forming spikes; calyx of 4 sepals deeply parted in the staminate, toothed in the carpellate; petals 4 or none; stamens 4 or 8; carpels 4, separating at maturity, each 1-seeded, indehiscent; styles and stigmas separate.

a. Flowers on emersed stems, the bracts smaller than foliage leaves.
   b. Bracts, at least the lower, longer than flowers and fruits.
      c. Submersed leaves with about 5 pairs of divisions, both alternate and in whorls on same stem; stamens 4; each carpel ridged, the ridges tubercled. M. pinnatum (Walt.) BSP.
      c. Submersed leaves all whorled, with 7 to 14 pairs of divisions.
         d. Carpels in fruit smooth; stamens 8; bracts pinnatifid. M. verticillatum L. var. pectinatum Wallr.
         d. Carpels in fruit rounded or keeled, beaked; stamens 4; bracts serrate to entire; stem stout. M. heterophyllum Michx.
   b. Bracts shorter than flowers and fruits; stamens 8; stems white in drying; carpels in fruit smooth or finely tubercled. M. exalbescens Fern.
a. Flowers in axils of submersed leaves, leaves with 10-25 pairs of divisions; often cultivated in aquaria. *M. brasiliense Camb., Parrot's Feather

Proserpinaca L., Mermaid-weed

P. palustris L. Erect or decumbent; emersed leaves lanceolate, serrate; submersed leaves pinnatifid with linear or filiform divisions; flowers bisporangiate; sepals 3, triangular; petals none; stamens, carpels, short styles, stigmas, and locules 3; fruit 3-seeded, indehiscent.

ARALIACEAE, Ginseng Family

Trees, shrubs, and herbs; leaves alternate or whorled, compound to decompound, or sometimes simple; flowers small, epigynous, bisporangiate or some bisporangiate and some monosporangiate, in solitary or clustered umbels; sepals 5, minute, or represented by only a rim; petals 5, separate; stamens 5; stamens and petals from edge of epigynous disk or short hypanthium; carpels, in ours, 2-5, united; styles, stigmas, and locules as many; ovule 1 in each locule; placentation axile; fruit a berry.

a. Styles 5; leaves alternate or basal, decompound; umbels clustered. Aralia
a. Styles 2 or 3; leaves whorled, compound; umbel solitary. Panax

Aralia L.

Herbs, shrubs, or trees; leaves often very large; petals green or white; styles 5; berry black or dark-purple.

a. Shrub or small tree with prickly stem and petioles; umbels in large panicle. A. spinosa L., Hercules's-club, Angelica Tree
a. Herbs, sometimes slightly woody at base, not prickly, sometimes bristly-hispid.
   b. Solitary leaf and leafless peduncle arising from rhizome; umbels 2-7, often 3. A. nudicaulis L., Wild Sarsaparilla
   b. Stem leafy.
      c. Leaflets cordate to truncate at base, ovate to circular; umbels many, in panicle or raceme; stem not bristly-hispid. A. racemosa L., Spikenard
      c. Leaflets rounded to tapering at base, lanceolate to ovate; umbels several, in loose terminal corymbiform cluster; stem bristly-hispid near base. A. hispida Vent., Bristly Sarsaparilla

Panax L., Ginseng

Herbs; leaves in a single whorl of usually 3, palmately compound; umbel solitary.

a. Leaflets 3-5, sessile; styles usually 3; berry yellow, about 5 mm in diameter. P. trifolium L., Dwarf G.
a. Leaflets 5, with stalks; styles usually 2; berry red, about 1 cm in diameter. P. quinquefolium L., Ginseng

164

UMBELLIFERAE, Parsley Family

Herbs, often large and coarse; internodes of stem often hollow; leaves alternate, usually compound or decompound, petioles usually sheathing; flowers small, bisporangiate or rarely monosporangiate, epigynous, in usually compound umbels, rarely in heads; bracts subtending umbel (involucre) and bractlets subtending umbellet (involucel) usually present; calyx of 5 sepals, at most 2-3 mm long, or only a rim or flange, or absent; corolla regular or rarely zygomorphic, of 5 separate usually inflexed petals; stamens 5, on epigynous disk; carpels 2, united; styles 2, often with enlarged base (stylopodium); stigmas 2; ovulary 2-loculed, 1 ovule suspended from top of each locule; fruit a schizocarp (of 2 mericarps), each carpel with 5 primary ribs and alternating secondary ones, ribs sometimes corky, winged, or bearing prickles; oil tubes usually present in carpel wall; fruit sometimes flattened, dorsally, if parallel to face along which carpels are joined (commissure), laterally, if contrary to this face.

a. Leaves all simple, none of them deeply divided.
  b. Leaves linear, parallel veined; flowers in heads. 3. Eryngium
  b. Leaves ovate to circular, not parallel veined.
    c. Leaves petioled, palmately veined, peltate or with petiole attached at sinus; stems prostrate or floating. 1. Hydrocotyle
    c. Leaves perfoliate, ovate; stems upright. 9. Bupleurum
a. Some or all leaves compound, decompound, or deeply divided.
  b. Ovulary and fruit covered with prickles; umbellets globose, bisporangiate flowers sessile or nearly so, staminate flowers pediceled; leaves palmately compound or palmately divided.
    2. Sanicula
  b. Ovulary and fruit not prickly or umbellets not as above.
    c. Corolla yellow (rarely purple).
      d. Ultimate divisions of leaf-blade filiform or narrowly linear.
        e. Fruit somewhat flattened dorsally, lateral ribs winged. 20. Anethum
        e. Fruit not flattened dorsally, ribs not winged. 19. Foeniculum
      d. Ultimate divisions of leaf broader.
        e. Leaflets entire. 14. Taenidia
        e. Leaflets serrate, crenate, incised, or lobed.
          f. Cauline leaves once pinnately compound; fruit flattened dorsally, lateral ribs winged.
            26. Pastinaca
          f. Cauline leaves pinnately decompound; fruit flattened dorsally, lateral ribs winged.
            25. Levisticum (See third f.)
          f. Cauline leaves ternately compound or decompound (rarely simple); basal leaves simple or ternately compound or decompound.
            g. Central flower of each umbellet sessile; ribs of fruit not winged. 10. Zizia
            g. Central flower of each umbellet pediceled; some or all ribs of fruit winged.
              22. Thaspium
    c. Corolla white, greenish-white, or pink.
      d. Bulblets borne in axils of upper small leaves; lower leaves compound or decompound, leaflets linear. 11. Cicuta
      d. Without bulblets.
        e. Fruit prickly or bristly.
          f. Fruit appressed-bristly, clavate, length more than twice width; umbel few-rayed; umbellet few-flowered. 5. Osmorhiza
          f. Fruit with spreading or hooked prickles.
            g. Involucral bracts pinnately compound, conspicuous; umbel concave in fruit.
            28. Daucus
            g. Involucral bracts small, simple, or none. 6. Torilis
        e. Fruit not prickly or bristly.
          f. Leaves once compound; leaflets entire, serrate, incised or lobed, but not finely dissected (except in submersed plants of Sium).
            g. Plants coarse and hairy; leaflets often lobed; corolla often zygomorphic; fruit flattened dorsally, lateral ribs winged. 27. Heracleum
            g. Plants glabrous.
              h. Leaves palmately compound, leaflets sometimes deeply lobed; umbels few-rayed; umbellets few-flowered; fruit about twice as long as wide.
              12. Cryptotaenia

h. Leaves pinnately compound; umbels and umbellets several- to many-flowered; fruits ellipsoid or ovoid.
  i. Leaflets finely serrate; sepals minute; fruit somewhat flattened laterally, ribs corky; submersed leaves, when present, 2-3 times dissected. 17. Sium
  i. Leaflets entire or somewhat toothed; sepals evident; fruit flattened dorsally, dorsal ribs slender, lateral ones broadly winged. 24. Oxypolis
f. Leaves, or some of them, decompound.
  g. Fruit at least 3 times as long as wide; umbels few-flowered, sometimes simple; bractlets conspicuous. 4. Chaerophyllum
  g. Fruit less than 3 times as long as wide.
    h. Small delicate early-spring plant usually 1-2 dm tall; leaves basal except one subtending the umbel. 7. Erigenia
    h. Plants larger, with more leaves.
      i. Leaflets ovate to lanceolate, entire, toothed, or lobed, not pinnatifid.
        j. Leaflets lanceolate, sharply toothed, main lateral veins ending in sinuses; stem purple; fruit about as long as wide, brown, the rounded ribs paler than the spaces between. 11. Cicuta
        j. Leaflets ovate, main veins ending in teeth (or some of them in sinuses); fruit usually longer than wide.
          k. Fruit flattened dorsally; lateral ribs broadly winged; upper sheaths bladeless or blades shorter than sheaths. 23. Angelica
          k. Fruit not flattened dorsally; ribs not winged or all narrowly winged; upper blades not shorter than their sheaths.
            l. Lower leaves 3-4 times ternate; fruit terete or slightly flattened laterally; ribs with narrow wings. 21. Ligusticum
            l. Lower leaves 1-2 times ternate; ribs of fruit not winged; styles conspicuously reflexed. 15. Aegopodium
      i. Leaflets pinnatifid or ultimate leaflets linear.
        j. Submersed leaves pinnately dissected; emersed leaves pinnately compound, leaflets linear or lanceolate, serrate; leaves of intermediate form often present. 17. Sium
        j. Without submersed and emersed leaves of different forms.
          k. Sepals evident; ribs of fruit inconspicuous; upper leaves ternate, leaflets linear; lower cauline leaves 1-3 times pinnate, leaflets cut into narrow segments. 16. Perideridia
          k. Sepals none or minute; ribs of fruit prominent.
            l. Stem spotted; bractlets lanceolate to ovate, calyxlike; pedicels glabrous; ribs of fruit wavy. 8. Conium
            l. Stem green; bractlets usually none; bases of petioles widely dilated; carpels 5-angled. 13. Carum (See third l.)
            l. Stem green; bractlets linear, turned to one side; pedicels grooved on upper side, edges of grooves scabrous. 18. Aethusa

1. Hydrocotyle L., Water-pennywort

Low glabrous aquatic or swamp plants with prostrate stems and palmately veined, sometimes peltate, leaves; umbels axillary, subsessile or peduncled, simple, or a series of umbels one above the other; corolla white, in ours; sepals none; fruit as wide or wider than high, flattened laterally.

a. Blades peltate; umbels usually simple, many-flowered. H. umbellata L.
a. Blades not peltate, petiole attached in sinus.
  b. Blades 1-5 cm wide, doubly crenate; umbels subsessile or sessile. H. americana L.
  b. Blades 1-7 cm wide, lobed; peduncle shorter than petiole. H. ranunculoides L. f.
  b. Blades 1 cm wide or less, crenate; peduncle equaling petiole. *H. sibthorpioides Lam.

2. Sanicula L., Black Snakeroot

Stem glabrous; leaves palmately 3-7-divided; bracts leaflike; bractlets small; flowers perfect or staminate, the 2 kinds in the same or in separate umbellets; perfect flowers sessile or short-pediceled, staminate flowers usually on longer pedicels; sepals relatively large, persistent in fruit; fruit ovoid to globose, with hooked prickles.

a. Styles spreading and recurved, longer than bristles of fruit; petals white or yellow.
  b. Petals and anthers yellow; sepals of staminate flowers ovate, soft, shorter than petals, 1 mm long or less; fruit stipitate. S. gregaria Bickn.
  b. Petals and anthers greenish-white; sepals of staminate flowers narrow, rigid, about equaling petals, 1 mm long or more. S. marilandica L.
a. Styles hidden among and shorter than bristles of fruit; petals white.
  b. Sepals of perfect flowers linear-lanceolate, acute, in fruit not exceeding bristles; pedicels of staminate flowers 2-3 mm long; fruit globose, stipitate. S. canadensis L.
  b. Sepals of perfect flowers linear, rigid, incurved, in fruit forming a conspicuous beak longer than bristles; pedicels of staminate flowers about 4 mm long; fruit longer than wide, not stipitate. S. trifoliata Bickn.

3. Eryngium L.

E. yuccifolium Michx., Rattlesnake-master. Stem coarse, erect; leaves linear, 1-3 cm wide, parallel-veined, entire or spiny-toothed; flowers in dense heads 1.5-2.5 cm long, each head subtended by bracts, each flower subtended by a bractlet; sepals large; petals white or blue, erect; fruit obovoid, covered with scales.

4. Chaerophyllum L., Chervil

Leaves decompound; umbels simple or compound, few-flowered; bracts none; bractlets ovate, conspicuous; sepals none; corolla white; fruit 3 or more times as long as wide, flattened laterally; ribs rounded, prominent.

a. Upper stem and lower surface of leaves glabrous or nearly so; umbels peduncled; leaf-segments rounded or blunt at apex; pedicels not clavate; fruit widest at about the middle. C. procumbens (L.) Crantz, Spreading C.
a. Upper stem and lower surface of leaves villous; umbels often sessile; leaf-segments acute; pedicels clavate; fruit widest below the middle. C. tainturieri Hook.

5. Osmorhiza Raf., Sweet Cicely

Leaves ternately decompound; umbels few-rayed; umbellets few-flowered; involucre of few narrow bracts; involucel conspicuous; sepals none; petals white or greenish; fruit clavate, elongate, appressed-bristly, slightly flattened laterally, the ribs slender, acute.

a. Stylopodium and style not more than 1.5 mm long, not exceeding petals. O. claytoni (Michx.) Clarke
a. Stylopodium and style 2-4 mm long, exceeding petals; roots anise-scented. O. longistylis (Torr.) DC.

6. Torilis Adans., Hedge-parsley

*T. japonica (Houtt.) DC. Plant hispidulous; leaves pinnately or ternately compound or decompound, ultimate leaflets pinnatifid; involucre and involucel of few slender bracts; pedicels short; sepals small; petals white; fruit flattened laterally, covered with hooked prickles.

7. Erigenia Nutt., Harbinger of Spring

E. bulbosa (Michx.) Nutt. Plant glabrous; stem usually 1-2 dm tall; leaves ternately decompound, ultimate segments linear or spatulate, basal except for the one subtending the inflorescence; bractlets spatulate, somewhat foliaceous; umbel usually 3-rayed, in the axil; pedicels short; sepals none; petals white, not incurved, obovate; fruit flattened laterally, wider than long, notched at top and base; ribs slender.

8. Conium L., Poison-hemlock

*C. maculatum L. Stem spotted; leaves pinnately decompound, ultimate leaflets ovate to lanceolate, pinnatifid; involucre and involucel of lanceolate or ovate bracts; sepals none; petals white; fruit ovoid, somewhat flattened laterally, ribs prominent, wavy, paler than intervals, corky.

9. Bupleurum L.

Glabrous; leaves entire, perfoliate, cauline ones ovate; bracts none; bractlets 5, ovate, surpassing flowers of umbellet; sepals none; petals yellow; fruit ellipsoid, dark-colored.

a. Fruit smooth. *B. rotundifolium L., Thoroughwax
a. Fruit wrinkled or tuberculate. *B. subovatum Link

10. Zizia Koch, Golden Alexanders

Glabrous or nearly so; leaves simple and ovate-cordate, or ternately compound or decompound, margins serrate; involucre none; involucel of linear-lanceolate bractlets; central flower of each umbellet sessile; sepals evident; petals yellow; stylopodium wanting; fruit short-ellipsoid, somewhat flattened laterally, ribs slender, not winged.

a. Basal and cauline leaves ternately compound or decompound. Z. aurea (L.) Koch
a. Basal leaves simple, ovate-cordate; cauline, with 3-5 leaflets. Z. aptera (Gray) Fern.

11. Cicuta L., Water-hemlock

Leaves 1-3 times pinnately compound or upper ones simple; bracts none or few; bractlets present; sepals evident; petals white; fruit ovoid or globose, slightly flattened laterally, ribs rounded, corky, paler than intervals.

a. Leaflets lanceolate; no bulblets in axils. C. maculata L., Spotted W.
a. Leaflets linear; bulblets in axils of upper leaves, which are sometimes simple. C. bulbifera L.

12. Cryptotaenia DC., Honewort

C. canadensis (L.) DC. Glabrous; leaves trifoliolate; leaflets ovate, incised, serrate, doubly serrate, or lobed; umbels few-rayed; umbellets few-flowered; bracts and bractlets few and small or none; sepals none or minute; corolla white; fruit linear-oblong, slightly flattened laterally.

13. Carum L.

*C. carvi L., Caraway. Glabrous; leaves pinnately dissected into linear divisions; bracts and bractlets few and small or none; sepals none; petals white or rarely pink; fruit ellipsoid, slightly flattened laterally, ribs slender, prominent.

14. Taenidia Drude

T. integerrima (L.) Drude, Yellow Pimpernel. Glabrous, somewhat glaucous; leaves 2-3 times compound, ultimate leaflets entire, ovate to lanceolate; bracts and bractlets usually none; sepals small; petals yellow; marginal flowers of umbellet perfect, inner ones staminate; fruit short-oblong, laterally flattened.

15. Aegopodium L.

*A. podagraria L., Goutweed. Glabrous; leaves ternate, the lower usually twice, the upper usually once; leaflets ovate to oblong, serrate; bracts and bractlets usually none; sepals none; petals white; fruit ellipsoid, somewhat flattened laterally; styles and stylopodia conspicuous, styles reflexed.

16. Perideridia Reichenb.

P. americana (Nutt.) Reichenb. Glabrous; leaves 1-3 times pinnately compound, ultimate segments or leaflets narrow; sepals evident; petals white or pink; bracts none or few; bractlets narrow; fruit oblong, flattened laterally.

17. Sium L., Water-parsnip

S. suave Walt. Glabrous; of wet soil or shallow water; submersed leaves, when present, 2-3 times pinnately dissected, segments linear; emersed leaves pinnately compound, leaflets linear to lanceolate, serrate; bracts and bractlets several to many, ovate to narrow; sepals minute; petals white; fruit short-ellipsoid, ribs prominent, corky.

18. Aethusa L., Fool's-parsley

*A. cynapium L. Leaves 2-3 times compound, ultimate leaflets pinnatifid; bracts none or 1; bractlets narrow, turned to one side of umbellet; sepals none; petals white; fruit ovoid, slightly compressed dorsally; ribs prominent, corky, wider and paler than intervals.

19. Foeniculum Mill., Fennel

*F. vulgare Mill. Glabrous, aromatic; leaves 3-4 times dissected into linear or filiform segments; bracts and bractlets none; sepals none; petals yellow; fruit glabrous, ellipsoid, terete or slightly compressed laterally, ribs slender.

20. Anethum L., Dill

*A. graveolens L. Glabrous; leaves finely dissected, segments filiform; bracts and bractlets none; sepals none; petals yellow; fruit ellipsoid, flattened dorsally, lateral ribs winged, dorsal ribs sharp and slender.

21. Ligusticum L.

L. canadense (L.) Britt. Glabrous; roots aromatic; leaves 1-4 times compound, leaflets ovate or broadly oblong, coarsely serrate; involucre usually none; involucel of narrow bractlets; sepals minute; petals white; fruit ellipsoid, scarcely flattened, ribs narrowly winged.

22. Thaspium Nutt., Meadow-parsnip

Leaves compound or decompound or basal ones simple, ovate-cordate; bracts none or few; bractlets several; all flowers of umbellet pediceled; sepals evident; petals yellow or greenish-yellow or rarely purple; fruits ovoid or ellipsoid, terete or slightly compressed dorsally, some or all ribs winged.

a. Basal leaves simple, ovate-cordate, or once compound with 3 or 5 leaflets, or twice compound; cauline leaves usually ternate; leaf-margin white, thickened; nodes glabrous or nearly so. T. trifoliatum (L.) Gray
a. Basal leaves once or twice ternate; cauline leaves twice compound, ultimate leaflets serrate or incised; nodes of stem pubescent with short stiff hairs. T. barbinode (Michx.) Nutt.

23. Angelica L.

Tall, stout; basal leaves decompound, long-petioled; stem-leaves reduced upward, petioles of upper ones conspicuously sheathing and sometimes bladeless; bracts none or inconspicuous; bractlets few to many; umbels large; sepals small or none; petals white or greenish; fruit ellipsoid or as wide as long, dorsally compressed, primary ribs prominent, lateral ribs broadly winged.

a. Stem pubescent above; sheaths of upper leaves tapering at top; fruit thinly pubescent; umbels densely pubescent. A. venenosa (Greenway) Fern.
a. Stem to 3 m high and 3 cm thick, glabrous or nearly so above; sheaths of upper leaves rounded or auricled at top; fruit and spherical umbel glabrous or nearly so. A. atropurpurea L.

24. Oxypolis Raf.

O. rigidior (L.) Raf., Cowbane. Glabrous; of wet places; leaves pinnately compound, leaflets lanceolate or linear, entire or toothed; bracts and bractlets few and narrow or none; sepals evident; petals white; stylopodium thick; fruit ellipsoid, flattened dorsally; dorsal ribs slender, lateral ribs broadly winged.

25. Levisticum Koch, Lovage

*L. officinale Koch   Glabrous or nearly so; leaves pinnately decompound, leaflets cuneate and entire at base, coarsely toothed toward apex; bracts and bractlets narrow, reflexed; sepals small; petals yellow; fruit ellipsoid, somewhat flattened dorsally; ribs winged, the lateral ones broadly.

26. Pastinaca L., Parsnip

*P. sativa L., Wild P.   Stem stout, grooved; leaves pinnately compound, leaflets ovate, toothed, incised, or lobed; bracts and bractlets mostly lacking; sepals none; petals yellow; fruit ellipsoid, much flattened dorsally, lateral ribs broadly winged.

27. Heracleum L., Cow-parsnip

H. lanatum Michx. (H. maximum Bartr.).   Stout; pubescent; leaves large, ternate, leaflets ovate or broader, sometimes lobed; bracts deciduous; bractlets narrow; sepals minute; petals white, obcordate, usually larger on outer than on inner side of flower, and on peripheral than on central flowers; fruit obovate, flattened dorsally, lateral ribs broadly winged, dorsal and intermediate ribs separated by 4 dark lines.

28. Daucus L., Carrot

*D. carota L., Wild C., Queen Anne's Lace.   Leaves pinnately decompound, leaflets linear or lanceolate; umbels concave in fruit; bracts leaflike, once pinnate, leaflets filiform; bractlets linear or pinnate; sepals none; corolla white or pinkish, of marginal flowers larger and somewhat irregular, of central flower(s) of umbel sometimes purple; fruit ellipsoid, flattened dorsally, primary ribs short-bristly, secondary ribs winged and bearing a row of barbed prickles.

CORNACEAE, Dogwood Family

Cornus L., Dogwood, Cornel

Shrubs, small trees, or herbs; leaves opposite or rarely alternate, simple, entire; flowers epigynous, bisporangiate, small, in compound cymes or in umbelliform or headlike clusters sometimes subtended by an involucre of bracts; corolla regular; sepals and petals 4, separate; epigynous disk present; stamens 4; carpels 2, united; locules 2, 1 ovule in each; style 1; fruit a drupe.

a. Flowers in heads or umbels subtended by bracts; fruit red.
  b. Bracts greenish-yellow, about as long as pedicels of the yellow flowers. *C. mas L., Cornelian Cherry
  b. Bracts white, rarely pink, much exceeding the greenish-yellow flowers.
    c. Flowering stems herbaceous, 3 dm high or less, leaves clustered at summit; flower cluster solitary. C. canadensis L., Bunchberry
    c. Small trees; flower clusters usually many. C. florida L., Flowering D.
a. Flowers in compound cymes; petals white; fruit white to blue; shrubs.
  b. Leaves alternate, sometimes appearing almost whorled on flowering stems; fruit blue.
    C. alternifolia L. f.
  b. Leaves opposite.
    c. Leaf-blades glabrous or minutely appressed-pubescent; fruit white or bluish.
      d. Sepals 1-2 mm long; style dilated at tip; pith of 1- or 2-year-old stems brown; blades lanceolate; fruit blue. C. obliqua Raf.
      d. Sepals not more than 0.5 mm long.
        e. Blades green on both sides; lateral veins 4-5 pairs; fruit pale blue. C. stricta Lam.
        e. Blades whitened beneath.
          f. Branchlets gray or brownish; cymes somewhat racemiform; fruit white; pedicels red; lateral veins 3-4 pairs. C. racemosa Lam.
          f. Branchlets red; cymes not racemiform; fruit white or gray-blue; pedicels not red; lateral veins 5-7 pairs. C. stolonifera Michx. Red-osier D.
    c. Leaf-blades pubescent beneath with larger hairs, at least some of which are spreading.
      d. Hairs brown, at least on veins; sepals 1-2 mm long; style dilated at tip; fruit white or bluish; pith brown. C. amomum Mill.
      d. Hairs pale; sepals 0.5 mm long.

e. Blades broadly ovate, abruptly acuminate, lateral veins 7-9 pairs; branches green; fruit blue; pith white. C. rugosa Lam.

e. Blades ovate, scabrous above, lateral veins 3-4 pairs; fruit white; pedicels red; pith brown or white. C. drummondii Meyer

NYSSACEAE, Sour Gum Family

Nyssa L., Sour Gum

N. sylvatica Marsh. Trees; leaves alternate, simple, blades obovate, entire or with a few large teeth; mostly diecious, but some perfect flowers may be present; flowers small, greenish, the staminate in peduncled umbel-like clusters, the carpellate epigynous, in few-flowered peduncled heads; sepals 5, minute; petals 5, small or none; stamens 5 or more, at edge of disk; ovulary 1-loculed; ovule 1; fruit a drupe.

PYROLACEAE, Pyrola Family

Low evergreen herbs or herbs without chlorophyll; flowers hypogynous, perfect, solitary or in terminal clusters; sepals, petals, carpels, and locules usually 5; sepals separate or united at base; petals separate, waxy, corolla regular; stamens twice the petals; carpels united; style 1, stigma broad, peltate; placentae axile; fruit a capsule.

a. Plants green; anthers opening by basal pores, but later bent back against filaments, pores then appearing apical; pollen in tetrads.
  b. Flowers in corymbs or umbels; style none or very short; stem leafy. 1. Chimaphila
  b. Flowers in racemes; style conspicuous; leaves basal. 2. Pyrola
a. Plants without chlorophyll; anthers opening by slits; pollen grains single. 3. Monotropa

1. Chimaphila Pursh, Wintergreen

Nearly herbaceous; blades thick, shining, alternate or almost whorled; sepals 5; petals 5, pink or white; stamens 10; carpels and locules 5, stigma with crenate border.

a. Leaf-blades oblanceolate, bright green; filaments ciliate. C. umbellata (L.) Bart., Pipsissewa, Prince's-pine
a. Leaf-blades lanceolate, mottled with white; filaments villous. C. maculata (L.) Pursh, Spotted W.

2. Pyrola L., Wintergreen

Evergreen herbs; leaves basal; sepals 5; petals 5, cream to white, 1 cm long or less; stamens 10; stigma 5-lobed or -rayed.

a. Style straight; petals erect; raceme secund. P. secunda L.
a. Style pointing downward, tip upturned; petals spreading; raceme not secund.
  b. Leaf-blades longer than wide, crenulate; sepals little if any longer than wide. P. elliptica Nutt.
  b. Leaf-blades circular, entire; sepals longer than wide. P. rotundifolia L.

Monotropa L.

Without chlorophyll; stem with alternate scales or bracts; sepals 2-5 or none; petals 4-5; stamens 8-10; ovulary 4-5-loculed.

a. Plants white; flower solitary. M. uniflora L., Indian-pipe
a. Plants yellow or reddish; flowers in racemes. M. hypopithys L., Pinesap

ERICACEAE, Heath Family

Woody plants, mostly shrubs; leaves alternate or rarely some of them opposite or whorled, deciduous or evergreen; flowers hypogynous or epigynous, bisporangiate or rarely monosporangiate; calyx lobes 4 or 5, rarely sepals separate; corolla regular or slightly zygomorphic, petals 4 or 5, usually united; stamens usually as many or twice as many as petals, anthers sometimes awned, usually opening by terminal pores which are often at ends of slender tubes; carpels 4-10, united, locules as many or rarely twice as many by false septation; style and stigma 1; placentae axile; fruit a capsule, berry, or drupe.

171

a. Flower hypogynous.
  b. Petals separate; leaves evergreen, revolute-margined, densely woolly beneath. 1. <u>Ledum</u>
  b. Petals more or less united.
    c. Corolla funnelform, salverform, campanulate, or bowl-shaped, lobes at least 1/3 as long as
       tube <u>or</u>, if lobes are shorter, then anthers inserted in pockets of corolla during early anthe-
       sis.
      d. Prostrate, pubescent, somewhat shrubby; corolla salverform with slender tube; anthers
         opening lengthwise. 8. <u>Epigaea</u>
      d. Upright shrubs; anthers opening by terminal pores.
        e. Corolla campanulate or funnelform, without pockets. 2. <u>Rhododendron</u>
        e. Corolla bowl-shaped, with 10 pockets in which anthers are inserted at early anthesis.
           3. <u>Kalmia</u>
    c. Corolla urn-shaped, cylindric, or globose, somewhat narrowed at tip, teeth or lobes short.
      d. Deciduous small trees and shrubs.
        e. Corolla ovoid; anthers opening by short clefts; flowers in large panicle terminal on
           branches of current season. 6. <u>Oxydendrum</u>
        e. Corolla globose; anthers opening by terminal pores; flowers in terminal panicle of small
           clusters from buds on stems of previous season. 5. <u>Lyonia</u>
      d. Evergreen shrubs.
        e. Lower leaf-surface covered with small scales; anthers awnless. 7. <u>Chamaedaphne</u>
        e. Leaf-surface without scales; anthers awned or appendaged.
          f. Leaves few, clustered near top of short upright stems, obovate to elliptic, obscurely
             serrate, aromatic; flowers axillary. 9. <u>Gaultheria</u>
          f. Leaves many, not clustered, entire; flowers in terminal clusters.
            g. Small bog shrubs; blades linear or nearly so, white beneath, margins revolute;
               flowers in small umbels or corymbs. 4. <u>Andromeda</u>
            g. Prostrate shrubs; blades oblanceolate, green, margins not revolute; flowers in
               short racemes. 10. <u>Arctostaphylos</u>
a. Flower wholly or partly epigynous.
  b. Corolla 5-toothed or -lobed; upright shrubs.
    c. Corolla campanulate; leaves not resin-dotted. 12. <u>Vaccinium</u>
    c. Corolla urn-shaped, cylindric, or globose, teeth short.
      d. Blades resin-dotted. 11. <u>Gaylussacia</u>
      d. Blades not resin-dotted. 12. <u>Vaccinium</u>
  b. Corolla with 4 lobes or divisions; stems slender, trailing.
    c. Corolla-lobes shorter than tube; calyx subtended by 2 small bracts. 9. <u>Gaultheria</u>
    c. Corolla divided almost to base; 2 small bracts remote from flower. 12. <u>Vaccinium</u>

1. <u>Ledum</u> L. , Labrador-tea

<u>L</u>. <u>groenlandicum</u> Oeder   Bog shrub; blades thick, evergreen, nearly sessile, lanceolate, rusty-
woolly beneath, margins revolute; flowers in terminal corymbs; corolla white, about 1 cm wide, stamens 5-7.

2. <u>Rhododendron</u> L. , Rhododendron, Azalea

Shrubs or small trees; buds scaly; leaves alternate; flowers showy, in racemes or umbels; corolla regular to somewhat zygomorphic; stamens 5-10; style elongate; fruit a capsule.

a. Blades thick, evergreen; corolla rose to white, yellow-spotted. <u>R</u>. <u>maximum</u> L. , Great R.
a. Blades not thick, deciduous, with marginal cilia.
  b. Corolla yellow, orange, or red-orange. <u>R</u>. <u>calendulaceum</u> (Michx.) Torr. , Flame A.
  b. Corolla white or pink.
    c. Ovulary hairy, nonglandular; blades usually glabrous beneath except on midvein; flowers hardly
       fragrant. <u>R</u>. <u>nudiflorum</u> (L.) Torr. , Pinxter-flower
    c. Ovulary stipitate-glandular; blades pubescent beneath; flowers fragrant. <u>R</u>. <u>roseum</u> (Loisel.)
       Rehd.

3. <u>Kalmia</u> L. , Laurel

<u>K</u>. <u>latifolia</u> L. , Mountain L. , Calico-bush. Shrub; winter buds naked; blades evergreen, coria-
ceous, entire, mostly alternate; flowers showy, in terminal corymbs; corolla 2-2. 5 cm wide, rose to

white; filaments at early anthesis inserted in 10 deeply-colored pockets of corolla, later springing erect; style elongate; fruit a capsule.

4. Andromeda L., Bog-rosemary

A. glaucophylla Link   Low bog shrub; blades evergreen, linear or oblong, entire, margins becoming revolute, white-pubescent beneath when young; flowers in umbel-like clusters; corolla urceolate, pink or white, 5-6 cm long; stamens 10, pollen-sacs awned; fruit a capsule.

5. Lyonia Nutt.

L. ligustrina (L.) DC.   Deciduous shrub; blades entire to serrulate, obovate to elliptic; flowers small, many, in small clusters lateral on stems of previous season; corolla globose, whitish; capsule globose.   Rare in Ohio.

6. Oxydendrum DC., Sorrel-tree, Sourwood

O. arboreum (L.) DC.   Small tree; leaves alternate, deciduous, petioled; blades elliptic, finely serrate, acuminate, stiff hairs on midvein beneath; flowers small, in terminal panicled racemes, fruiting panicles persistent in winter; corolla ovoid, white; stamens 10, anthers opening by short clefts; fruit a capsule.

7. Chamaedaphne Moench, Leather-leaf

C. calyculata (L.) Moench   Erect bog shrub; leaves alternate, evergreen; blades minutely crenulate, lower surface scurfy; flowers solitary in axils of smaller upper leaves, in terminal leafy raceme; corolla white, 6-7 mm long, cylindric-urceolate; fruit a capsule.

8. Epigaea L., Trailing Arbutus, Mayflower

E. repens L.   Prostrate shrub; sometimes diecious; blades evergreen, ovate or oblong, cordate or rounded at base, entire, petioles and usually blades pubescent; flowers bisporangiate or monosporangiate, fragrant, in small clusters, 2 bracts beneath calyx; corolla salverform, 1.5 cm long or less, pink to white; anthers dehiscent lengthwise, stamens 10; capsule fleshy.

9. Gaultheria L.

Small evergreen plants, scarcely woody; flowers solitary, in or near axils, 2 small bracts below calyx; stamens twice the petals; corolla usually white.

a. Flowering stem low, erect; leaves few, near summit, 2-5 cm long; flower parts in 5's; corolla urceolate, about 1 cm long; fruit red. G. procumbens L., Teaberry, Creeping Wintergreen
a. Stem prostrate; leaves many, 0.5-1 cm long; flower parts in 4's; corolla campanulate; fruit white.
   G. hispidula (L.) Bigel., Creeping Snowberry

10. Arctostaphylos Adans., Bearberry

A. uva-ursi (L.) Spreng.   Trailing mat-forming shrub; blades coriaceous, evergreen, 1-3 cm long, entire, oblanceolate; flowers bracted in small dense terminal racemes; corolla urceolate, white or pink, about 5 mm long; stamens 10, anthers awned; fruit a drupe.

11. Gaylussacia HBK, Huckleberry

G. baccata (Wang.) K. Koch   Shrub; blades oval or oblong, usually 2-5 cm long, resin-dotted; flowers in short racemes lateral on stems of previous season; corolla cylindric-urceolate, red-tinged, less than 1 cm long; drupe black.

12. Vaccinium L., Blueberry, Deerberry, Cranberry

Shrubs; leaves alternate, deciduous or evergreen; stamens 8-10, anthers opening by terminal pores; berry 4-5-loculed or sometimes 8-10-loculed as result of false septation.

a. Corolla 5-lobed or 5-toothed; upright deciduous shrubs.
  b. Corolla campanulate; stamens and style exserted; twigs not blistered, speckled, or granulate.
    V. stamineum L., Deerberry
  b. Corolla urceolate, cylindric, or globose; stamens and style included; young twigs blistered, speckled, or granulate. Blueberries
    c. Tall shrubs, 1-5 m; leaf-blades acute at both ends.
      d. Blades glabrous or nearly so beneath, entire or serrate; corolla 6-10 mm long; berry blue or black. V. corymbosum L., High-bush B.
      d. Blades quite pubescent beneath; berry dull black. V. atrococcum (Gray) Heller
    c. Low shrubs, 1 m tall or less.
      d. Blades obovate, oval, or elliptic, mostly 1.5-3.5 cm wide, 1/2-2/3 as wide as long, pale or glaucous beneath.
        e. Blades entire to slightly serrulate, obtuse to acute, usually more than half as wide as long. V. vacillans Torr.
        e. Blades serrulate, acute, about half as wide as long. V. pallidum Ait.
      d. Blades narrowly elliptic, mostly 0.5-1.5 cm wide, usually more than twice as long as wide.
        e. Blades entire, green on both sides, lower surface and branchlets densely pubescent.
          V. myrtilloides Michx.
        e. Blades sharply serrulate, glabrous on both sides; branchlets glabrous or minutely pubescent. V. angustifolium Ait.
a. Divisions of corolla 4, linear, reflexed; evergreen; stems slender, trailing; leaves small, entire; berry red. Cranberries
  b. Flowers from uppermost axils; small bracts of pedicel at or below middle; blades whitened beneath. V. oxycoccos L., Small C.
  b. Flowers from lower axils; small bracts of pedicel above middle; blades little whitened beneath. V. macrocarpon Ait., Large C.

PRIMULACEAE, Primrose Family

Herbs; leaves opposite, whorled, alternate, or all basal, simple or rarely pinnately dissected; flowers hypogynous or rarely epigynous; calyx usually 5-lobed (4-7); corolla regular; petals usually 5 (4-7), more or less united at base; stamens as many as petals and opposite them; staminodes sometimes present; carpels 5, united; locule 1; placenta central, free; style 1; fruit a capsule.

a. Aquatic; leaves pinnatifid; peduncles inflated, hollow. 7. Hottonia
a. Not aquatic; leaves simple, blades entire; peduncles not inflated.
  b. Ovulary partly inferior; flowers in terminal racemes or panicles. 6. Samolus
  b. Ovulary superior.
    c. Petals reflexed; flowers in an umbel on a scape; leaves basal. 1. Dodecatheon
    c. Petals spreading or erect; flowers not in an umbel; stem leafy.
      d. Corolla yellow. 2. Lysimachia
      d. Corolla scarlet, blue, pink, or white.
        e. Leaves in a whorl of 5-10 at top of stem. 3. Trientalis
        e. Leaves opposite; flowers solitary in axils. 4. Anagallis
        e. Leaves alternate; flowers solitary in axils. 5. Centunculus

1. Dodecatheon L., Shooting Star

D. meadia L. Leaves all basal, oblanceolate, narrowed to margined petioles; flowers bracted, in an umbel; petals 1-2 cm long, lavender, pink, or white, united only a little at base; stamens 5, exserted, anthers connivent forming a cone; filaments short.

2. Lysimachia L., Loosestrife

Leaves whorled, opposite, or alternate, sometimes punctate; flower parts in 5's or 6's; petals yellow, sometimes dotted or lined with black or red; corolla campanulate to rotate.

a. Staminodes absent; leaf-blades usually dotted.
  b. Petals and sepals not lined or dotted; filaments united; flowers axillary; leaves whorled; plant pubescent. *L. punctata L.
  b. Petals or sepals or both lined or dotted with black or rarely red.

c. Stems prostrate; leaves opposite, circular, 1-3 cm long; flowers axillary, solitary, 2-3 cm wide; filaments united at base. *L. nummularia L., Moneywort

c. Stems erect; leaves lanceolate.
  d. Flowers in short axillary racemes or heads; filaments separate; leaves opposite. L. thyrsiflora L.
  d. Flowers solitary and axillary or in terminal racemes; filaments united at base.
    e. Flowers all axillary; leaves whorled, 4-5 at a node. L. quadrifolia L.
    e. Flowers in terminal racemes.
      f. Leaves usually opposite; bracts small. L. terrestris (L.) BSP.
      f. Leaves opposite or whorled; at least the lowest flowers of raceme in axils of foliage leaves. L. producta (Gray) Fern.

a. Staminodes alternating with the stamens; stamens separate; leaf-blades, petals, and sepals not dotted or lined.
  b. Blades ovate to lance-ovate, base rounded or subcordate; petioles ciliate. L. ciliata L.
  b. Blades linear to lanceolate, tapering to base, sessile or subsessile.
    c. Blades linear, often revolute-margined, lateral veins obscure. L. quadriflora Sims
    c. Blades lanceolate, not revolute-margined, lateral veins evident. L. lanceolata Walt.

3. Trientalis L., Star-flower

T. borealis Raf. Plant small; leaves lanceolate, they and 1 or a few peduncled flowers whorled at top of stem; corolla pink or white, rotate, 1-1.5 cm wide; sepals, petals, and stamens 5-7.

4. Anagallis L., Pimpernel

*A. arvensis L., Scarlet P. Leaves opposite or whorled, ovate, 1-2 cm long; flowers axillary, peduncled, small; flower parts in 5's; petals almost separate, scarlet, white, or blue; capsule circumscissile.

5. Centunculus L., Chaffweed

C. minimus L. Plant very small, erect; leaves alternate, oblong, up to 1 cm long; flowers minute, solitary and subsessile in axils; flower parts mostly in 4's; corolla pink; capsule subglobose, circumscissile.

6. Samolus L., Water-pimpernel

S. parviflorus Raf. Leaves alternate or basal; flowers small, somewhat epigynous, in terminal panicled racemes; flower parts in 5's; petals white; staminodes alternate with the stamens.

7. Hottonia L., Featherfoil

H. inflata Ell. Aquatic; stems submersed and floating; leaves pinnatifid; flowers small, pediceled, in whorls at several successive constricted nodes of inflated peduncles, these peduncles clustered in an umbel, partly emersed; flower parts in 5's; corolla white, shorter than calyx.

EBENACEAE, Ebony Family

Diospyros L., Persimmon

D. virginiana L. Tree; leaves alternate, simple, blades entire; flowers monosporangiate or bisporangiate, hypogynous; generally imperfectly diecious; carpellate flowers axillary, solitary; staminate flowers usually smaller, sometimes in small clusters; sepals 4, united at base; corolla urceolate to campanulate, 4-lobed, pale yellow; stamens usually 16 in staminate flowers, 8 staminodes in carpellate flowers; carpels united, locules 8; styles 4, each 2-lobed; ovules 1-2 in each locule; fruit a berry.

STYRACACEAE, Storax Family

Shrubs or small trees; pubescence often stellate; leaves alternate, simple; flowers in our species epigynous, bisporangiate; calyx entire or lobed; corolla regular; stamens twice corolla lobes or more, their bases united; carpels united, locules as many or ovulary with septa at base only; style 1.

175

a. Sepals and petals 4; ovulary inferior; fruit winged. <u>Halesia</u>
a. Sepals and petals 5; ovulary partly inferior; fruit not winged. <u>Styrax</u>

<u>Halesia</u> L. , Silverbell

   <u>H</u>. <u>carolina</u> L. Flowers in lateral umbel-like clusters, appearing with or before the leaves; pedicels drooping; calyx small, 4-toothed; corolla bell-shaped, 4-lobed, white, 2-2.5 cm long; stamens 8-16, inserted near base of corolla-tube; fruit 4-loculed, 4-winged, dry, indehiscent.

<u>Styrax</u> L. , Snowbell

   <u>S</u>. <u>grandifolia</u> Ait. Leaf-blades tomentose beneath, dentate to entire; flowers in racemes, rachis, pedicels, and calyxes stellate-pubescent; corolla bell-shaped, 5-parted, white.

OLEACEAE, Olive Family

   Trees and shrubs; leaves opposite; flowers hypogynous, bisporangiate or monosporangiate; calyx small or none; corolla regular, when present; petals 4, more or less united, or none; stamens 2; carpels 2, united; locules 2; style 1; fruit a drupe, capsule, or samara.

a. Trees; petals none; calyx minute or absent; leaves compound. 1. <u>Fraxinus</u>
a. Shrubs and small trees; petals 4.
   b. Corolla yellow, lobes longer than tube; flowers in small lateral clusters, appearing before
      leaves. *<u>Forsythia</u> <u>suspensa</u> (Thunb.) Vahl
   b. Corolla not yellow; flowers in panicles.
     c. Petals nearly separate, narrowly linear, white. 3. <u>Chionanthus</u>
     c. Corolla tube about as long as lobes or longer.
       d. Corolla lilac or white; anthers included in corolla-tube. 2. <u>Syringa</u>
       d. Corolla white; anthers not included in corolla-tube. 4. <u>Ligustrum</u>

1. <u>Fraxinus</u> L. , Ash

   Trees; leaves pinnately compound; flowers monosporangiate or bisporangiate, in crowded or open panicles or racemes, appearing before or with the leaves; calyx absent or small; petals none; fruit a samara with 1, or rarely 2-3, seeds.

a. Calyx only a ring, or deciduous in fruit, or absent; fruit oblong, flat, body scarcely thicker than
   wing; lateral leaflets sessile or twigs 4-sided.
   b. Flowers bisporangiate; calyx deciduous; twigs 4-sided; leaflets stalked. <u>F</u>. <u>quadrangulata</u>
     Michx. , Blue A.
   b. Diecious or imperfectly so; calyx minute or none; twigs terete; lateral leaflets sessile. <u>F</u>. <u>nigra</u>
     Marsh. , Black A.
a. Calyx evident in fruit; diecious or imperfectly so; body of fruit terete or nearly so, thicker than
   wing; twigs not 4-sided; leaflets stalked.
   b. Wing little decurrent on the ellipsoid or short cylindric body of fruit; blades whitened beneath;
     stalk of lateral leaflets not winged to base. <u>F</u>. <u>americana</u> L.
     c. Branchlets, petioles, and leaf-rachises glabrous; leaflets glabrous or sparsely pubescent be-
       neath. var. <u>americana</u>, White A.
     c. Branchlets, petioles, leaf rachises, and lower surface of leaflets pubescent. var. <u>biltmoreana</u>
       (Beadle) J. Wright, Biltmore A.
   b. Wing decurrent on narrowly cylindric body of fruit; stalks of lateral leaflets winged to base or
     nearly so; leaf-rachis and petiole grooved on upper side. <u>F</u>. <u>pennsylvanica</u> Marsh.
     c. Branchlets pubescent; petioles, rachises, and lower surface of leaflets tawny-pubescent.
       var. <u>pennsylvanica</u>, Red A.
     c. Branchlets, petioles, and rachises glabrous; lower surface of leaflets glabrous or nearly so.
       var. <u>subintegerrima</u> (Vahl) Fern. , Green A.

2. <u>Syringa</u> L. , Lilac

   *<u>S</u>. <u>vulgaris</u> L. Shrub; blades ovate, base truncate to cordate; flowers lilac or white, about 1 cm wide, fragrant, in panicles; fruit a capsule.

3. Chionanthus L., Fringe-tree

C. virginicus L. Low tree or shrub; blades entire, about 1 dm long; panicles drooping, from buds on last year's stem-segments, appearing with the leaves; calyx minute; petals white, linear, 1.5-3 cm long; drupes dark blue.

4. Ligustrum L., Privet

*L. vulgare L. Shrub; blades 3-6 cm long, ovate-lanceolate; flowers small, in terminal panicles, appearing after leaves; petals white; berry black.

GENTIANACEAE, Gentian Family

Herbs; glabrous or nearly so; flowers perfect, hypogynous; petals 4-12, in ours usually 4 or 5, united; stamens as many, on the corolla; carpels 2, united; ovulary 1-loculed, placentae 2, parietal; style 1; fruit a capsule.

a. Leaves, at least those below the inflorescence, scalelike.
  b. Sepals 2; bracts subtending flowers not scalelike. 5. Obolaria
  b. Sepals 4; all leaves scalelike. 4. Bartonia
a. Leaves not scalelike.
  b. Leaves 3-foliolate or plants aquatic and leaves floating.
    c. Leaves 3-foliolate, alternate. 6. Menyanthes
    c. Leaves simple, floating; plants aquatic. 7. Nymphoides
  b. Leaves simple, not floating, opposite or whorled.
    c. Lower stem-leaves in whorls of 3-5; a fringed gland on each petal. 3. Swertia
    c. Stem-leaves opposite; petals without fringed glands.
      d. Corolla rotate, usually pink with green center, tube short. 1. Sabatia
      d. Corolla tubular, funnelform, or campanulate, tube equaling to much longer than lobes or teeth. 2. Gentiana

1. Sabatia Adans., Marsh-pink

S. angularis (L.) Pursh Stem 4-angled, usually branched; stem-blades clasping; flowers in cymes; calyx lobes slender; corolla rotate, pink with green eye, 2-3 cm wide, tube short, lobes usually 5; style 2-cleft.

2. Gentiana L., Gentian

Leaves sessile or nearly so, entire; flowers showy; calyx 4-5-lobed; corolla 4-5 lobed, sometimes with toothed or fimbriate plaits in the sinuses; anthers separate or cohering; style short; stigma 2-lobed.

a. Corolla without plaits between the lobes.
  b. Corolla 2-6 cm long, lobes fringed; flowers solitary.
    c. Corolla-lobes fringed at summit; blades ovate to lanceolate. G. crinita Froel., Fringed G.
    c. Corolla-lobes toothed at summit, fringed at sides; blades usually lance-linear. G. procera Holm
  b. Corolla 1-2 cm long, lobes not fringed; flowers in dense cymes. G. quinquefolia L.
a. Corolla with erose, toothed, or fringed plaits between the lobes.
  b. Plaits evidently shorter than corolla lobes.
    c. Corolla blue, lobes spreading; blades scabrous-margined; calyx-margin scabrous or ciliate. G. puberula Michx.
    c. Corolla greenish, white, or yellow; blades and calyx smooth-margined.
      d. Base of blades narrowed; seeds wingless; bracteal leaves usually overtopping corollas. G. villosa L., Sampson's Snakeroot
      d. Base of blades wide and clasping; seeds winged; corollas usually overtopping bracteal leaves. G. flavida Gray, Yellow G.
  b. Plaits a little longer than or not more than 1 mm shorter than corolla-lobes; calyx-lobes ciliate-margined; corolla blue to white.
    c. Corolla lobes narrower and a little shorter than the distally dilated plaits, adnate to the plaits to the summit. G. andrewsii Griseb., Closed G.

    c. Corolla lobes wider than the plaits, free from them at summit.
       d. Corolla closed, the lobes equaling or shorter than plaits. <u>G</u>. <u>clausa</u> Raf.
       d. Corolla slightly open, lobes longer than plaits. <u>G</u>. <u>saponaria</u> L.

3. <u>Swertia</u> L., Columbo

    <u>S</u>. <u>caroliniensis</u> (Walt.) Kuntze, American C. Tall; basal leaves very large; lower stem-leaves whorled; flowers in larger terminal cluster of cymes; flower-parts in 4's; corolla 2-3 cm wide, greenish-yellow dotted with purple, each division with a large fringe-bordered gland below the middle; style elongate.

4. <u>Bartonia</u> Muhl.

    <u>B</u>. <u>virginica</u> (L.) BSP. Small, erect; leaves scalelike, chiefly opposite; flowers small, in raceme or panicle, the parts in 4's; corolla 3-4 mm long, yellowish, lobed about halfway to base; seeds covering whole inner capsule-wall.

5. <u>Obolaria</u> L., Pennywort

    <u>O</u>. <u>virginica</u> L. Small; lower leaves scalelike; upper leaves 1 cm long or more, obovate, purplish, 1-3 flowers in each axil, the whole a crowded spike; sepals 2, almost separate; corolla about 1 cm long, whitish, funnelform, 4-lobed; stigma 2-cleft; seeds covering whole inner capsule-wall.

6. <u>Menyanthes</u> L., Buckbean

    <u>M</u>. <u>trifoliata</u> L. Bog and marsh herb; leaves crowded near base of stem, 3-foliolate; raceme scapose; flowers dimorphic as to relative length of style and stamens, the parts in 5's; corolla 1-1.5 cm long, white or pink, lobes fringed on inner surface.

7. <u>Nymphoides</u> Hill, Floating-heart

    *<u>N</u>. <u>peltata</u> (Gmel.) Britt. & Rendle Aquatic; blades floating, nearly circular, cordate; flowers in umbels, the parts in 5's; corolla yellow, 2-3 cm wide.

APOCYNACEAE, Dogbane Family

    Leaves simple, opposite in Ohio species; flowers hypogynous, perfect; sepals 5, united at base; corolla regular, petals 5, united; stamens 5, on the corolla; carpels 2, ovularies separate, styles and stigmas united; fruit an aggregate of 2 slender many-seeded follicles.

a. Stem trailing; flowers axillary, solitary; corolla blue. 1. <u>Vinca</u>
a. Stem erect; flowers in cymes; corolla white or pink. 2. <u>Apocynum</u>

1. <u>Vinca</u> L., Periwinkle

    *<u>V</u>. <u>minor</u> L. Blades entire, glossy, evergreen; corolla salverform, 1.5-3 cm wide; stamens included in corolla-tube; a gland on each side of gynecium, carpels thus appearing to be 4; seeds without hairs.

2. <u>Apocynum</u> L., Dogbane, Indian-hemp

    Herbs with milky juice; flowers small; corolla campanulate, tubular, or urceolate, with 5 appendages within, stamens on base of tube; follicles 0.5-2 dm long; seeds with tuft of long silky hairs at apex.

a. Flowers bent downward; corolla white or pink-striped, 6-10 mm long, lobes spreading or recurved; calyx half length of corolla-tube or less; inflorescence usually overtopping sterile lateral branches below it. <u>A</u>. <u>androsaemifolium</u> L.
a. Flowers erect; corolla white or greenish, 6 mm long or less, lobes ascending; calyx as long as corolla-tube; inflorescence overtopped by sterile lateral branches below it.
   b. Stem-leaves, except sometimes the lowest, petioled. <u>A</u>. <u>cannabinum</u> L., Indian Hemp
   b. Stem-leaves sessile or nearly so. <u>A</u>. <u>sibiricum</u> Jacq.

<u>A</u>. <u>medium</u> Greene, consisting of plants intermediate between <u>A</u>. <u>androsaemifolium</u> and <u>A</u>. <u>cannabinum</u> or <u>A</u>. <u>sibiricum</u>, apparently results from hybridization.

ASCLEPIADACEAE, Milkweed Family

Ours herbs; juice usually milky; blades simple, entire; stipules minute; flowers hypogynous, perfect, clustered; sepals 5, separate or united at base; corolla regular, 5-lobed, with a corona; stamens 5, on corolla-base, united in a tube enclosing ovularies, anthers united with stigma, pollen of each pollen-sac a single waxy mass (pollinium), pollinium from left sac of one anther joined by a connective to pollinium from right sac of adjacent anther; carpels 2, ovularies and styles separate, common stigma large, usually 5-lobed or -crenate; fruit an aggregate of 2 follicles, when both ovularies develop; seeds usually comate.

a. Stems not twining. 1. <u>Asclepias</u>
a. Stems twining.
  b. Corolla white; corona of 5 separate segments, each tipped with 2 awns. 2. <u>Ampelamus</u>
  b. Corolla purple; corona a disk or cup with wavy or lobed margin.
    c. Corolla-lobes about 3 mm long. 3. <u>Cynanchum</u>
    c. Corolla-lobes 7 mm long or more. 4. <u>Gonolobus</u>

1. Asclepias L. (Incl. <u>Asclepiodora</u> Gray), Milkweed

Leaves opposite, alternate, or whorled; flowers in umbels; calyx small, corolla larger, usually reflexed at anthesis; corona of 5 hoods, usually conspicuous, often bearing an exserted horn; follicles ovoid or lance-ovoid; seeds in Ohio species comate.

a. Hoods without horns.
  b. Corolla rotate, greenish; hood purplish, with a crest or keel; fruiting pedicels deflexed; leaves mostly alternate. A. viridis Walt. (<u>Asclepiodora viridis</u> (Walt.) Gray), Spider M.
  b. Corolla reflexed; hood without a crest or keel within.
    c. Leaves many, mostly alternate; hood to 2.5 mm long, reaching base of anther-wings, on column 0.5-1 mm long. <u>A</u>. <u>hirtella</u> (Pennell) Woodson
    c. Leaves mostly opposite, lateral veins horizontal; hood about 4 mm long, reaching middle or tip of anthers, sessile. <u>A</u>. <u>viridiflora</u> Raf., Green M.
a. An incurved horn within each hood.
  b. Leaves alternate below inflorescence, narrowly lanceolate or oblanceolate; juice not milky; corolla red-orange to yellow. <u>A</u>. <u>tuberosa</u> L., Butterfly-weed
  b. Leaves opposite or whorled; juice milky; corolla not as above.
    c. Blades narrowly linear, margins revolute, mostly whorled; petals and hoods white or greenish; horn subulate, much surpassing hood; fruit slender, pedicel erect. <u>A</u>. <u>verticillata</u> L., Whorled M.
    c. Blades wider than above.
      d. Blades sessile or subsessile, base rounded or cordate; lateral veins horizontal; pedicels deflexed in fruit.
        e. Blades with ruffled margin, clasping, upper pair usually shorter than peduncles; hood shorter than horn; corolla greenish-purple. <u>A</u>. <u>amplexicaulis</u> Sm.
        e. Blades with plane margin, subsessile, upper pair usually longer than peduncles; hood surpassing horn and stigma; corolla purple to white. <u>A</u>. <u>sullivantii</u> Engelm.
      d. Blades evidently petioled, rarely cordate at base.
        e. Corolla purple-red, deep purple, or rose-pink; fruit downy or minutely hairy.
          f. Blades lanceolate, lateral veins ascending; corolla deep pink or purple-red; hood 2-3 mm high; fruiting pedicel erect; of low ground. <u>A</u>. <u>incarnata</u> L., Swamp M.
          f. Blades elliptic to ovate, lateral veins horizontal; corolla dark purple; hood 5-7 mm high, longer than flattened horn; fruiting pedicel deflexed. <u>A</u>. <u>purpurascens</u> L., Purple M.
        e. Corolla white, pale pink, dull purple, or greenish.
          f. Leaves most often 8, tapering to apex and base, whorled at middle node, sometimes all opposite; corolla pink; hood white, 3-5 mm long; fruiting pedicels erect. <u>A</u>. <u>quadrifolia</u> Jacq.
          f. Leaves all opposite; fruiting pedicels deflexed.
            g. Lateral veins of blades horizontal.

h. Blades densely soft-hairy beneath; corolla pale purple to green; hood with a tooth at each lateral margin; fruit usually with soft conical processes. A. syriaca L., Common M.

h. Blades glabrous or sparsely hairy beneath; corolla white or pink; hood subglobose, horn flattened, the pointed tip almost horizontal. A. variegata L.

g. Lateral veins ascending; blades tapering to base and apex; corolla white or purplish, hood white or pink, 2-toothed. A. exaltata L.

2. Ampelamus Raf., Sandvine

A. albidus (Nutt.) Britt. Twining; leaves opposite, blades lance-ovate or ovate, cordate; flowers in pedunculed axillary clusters; corolla white, about 6 mm long, corona about as long; fruit smooth; seeds comate.

3. Cynanchum L., Swallow-wort

*C. nigrum (L.) Pers. Twining; leaves opposite or rarely whorled; blades ovate or lance-ovate; flowers small, in pedunculed axillary clusters; corolla purple, longer than corona and stamens; fruit smooth; seeds comate.

4. Gonolobus Michx., Angle-pod

G. obliquus (Jacq.) Schult. Twining; leaves opposite, blades ovate to circular, deeply cordate; flowers in pedunculed axillary cymes; corolla brown-purple, lobes narrow, much longer than corona and stamens; seeds comate.

CONVOLVULACEAE, Morning-glory Family

Herbs, usually trailing or twining, sometimes parasites without chlorophyll, sometimes with milky juice; leaves alternate, simple (rarely deeply divided), sometimes scalelike; flowers hypogynous, bisporangiate; sepals 5 or 4, usually separate; corolla regular, with 5 or 4 lobes or plaits; stamens 5 or 4, on corolla tube; hypogynous disk usually present; carpels 2 or 3, united, locules as many or twice as many as result of false septum in each carpel; fruit a capsule.

a. Plants green, rooting in soil; stems with leaves.
  b. Stigma capitate or with 2-3 lobes, the lobes globular; calyx not subtended by bracts. 1. Ipomoea
  b. Stigmas 2, oblong or linear; calyx subtended by 2 large bracts, or 2 small bracts on flower-stalk some distance below flower. 2. Convolvulus
a. Plants nongreen, often yellow, not rooting in soil after seedling stage; leaves scalelike. 3. Cuscuta

1. Ipomoea L., Morning-glory

Leaves cordate-ovate, entire, lobed, or pinnately parted; flowers large, solitary or in small clusters on axillary peduncles; outer sepals usually larger than inner; sepals, petals, and stamens 5; corolla-margin entire or shallowly lobed; style 1; locules 2-4.

a. Stamens and style exserted; corolla salverform, red.
  b. Blades pinnately parted, segments narrow; corolla-limb star-shaped. *I. quamoclit L., Cypress-vine
  b. Blades entire or lobed; corolla-limb with 5 shallow rounded lobes. *I. coccinea L., Red M.
a. Stamens and style included; corolla funnelform to campanulate, pink, purple, blue, or white.
  b. Locules of ovulary 3; stigma 3-lobed.
    c. Blades usually entire; sepals acute, 10-15 mm long; corolla of various colors. *I. purpurea (L.) Roth, Common M.
    c. Blades usually 3-lobed; sepals abruptly narrowed, tapering to long slender tips, 15-25 mm long; corolla blue changing to rose. *I. hederacea (L.) Jacq., Ivy-leaf M.
  b. Locules of ovulary 2; stigma unlobed or 2-lobed.
    c. Sepals glabrous, obtuse; corolla about 6 cm long or more, white, red-purple in the tube; perennial with very large root. I. pandurata (L.) G. F. W. Meyer, Wild Sweet Potato
    c. Sepals sparsely long-hairy, bristle-tipped; corolla 2 cm long or less, white, pink, or purple; annual; capsule pubescent. I. lacunosa L.

2. Convolvulus L., Bindweed

Trailing, twining, or erect; leaves entire, hastate, sagittate, or cordate at base; flowers large, solitary or few on axillary peduncles; calyx usually subtended by 2 bracts; corolla pink or white, funnelform, margin nearly entire; locules 2-4; style 1; stigmas 2.

a. Calyx not subtended by bracts; peduncle with 2 small bracts; corolla about 2 cm long.   *C. arvensis L., Small or Field B.
a. Calyx subtended by and enclosed by 2 bracts.
  b. Corolla double, pink.   C. japonicus Thunb., Japanese B.
  b. Corolla single.
    c. Stem usually not twining, less than 6 dm long; petioles less than half as long as blades; flowers few, from lower axils; corolla white.   C. spithamaeus L.
    c. Stems twining and climbing, often much longer than 6 dm; petioles often more than half as long as blades; flowers many; corolla white or pink.   C. sepium L., Hedge B.

3. Cuscuta L., Dodder, Love-vine

Parasitic twining vines, slender, often yellow; leaves alternate, scalelike; flowers small, in clusters; sepals 4 or 5, separate or united; petals 4 or 5, united; stamens 4 or 5, on corolla-tube; each with a fringed appendage; ovulary 2-loculed; styles 2; seeds usually 4.

a. Stigma slender; capsule circumscissile.
  b. Flowers yellow; corolla-lobes ascending; style included.   *C. epilinum Weihe, Flax D.
  b. Flowers pink or white; corolla-lobes spreading; style exserted.   *C. epithymum Murr., Clover D.
a. Stigma capitate; capsule indehiscent or breaking open irregularly.
  b. Sepals separate, subtended by 1 or more calyxlike bracts; flowers sessile, in dense clusters.
    c. Bracts oblong, tips recurved or spreading, about as long as calyx; inflorescence twisted, ropelike.   C. glomerata Choisy
    c. Bracts broad, appressed, obtuse; inflorescence dense but not ropelike.   C. compacta Juss.
  b. Sepals united at base, not subtended by bracts.
    c. Sepals and petals mostly 4.
      d. Corolla-lobes obtuse or rounded; calyx-lobes obtuse, shorter than corolla-tube; stamens hardly projecting beyond sinuses of corolla; styles about as long as capsule.   C. cephalanthi Engelm.
      d. Corolla-lobes lanceolate, acute, erect or tips inflexed; calyx-lobes equaling or exceeding corolla-tube; stamens exserted from corolla-tube; styles shorter than capsule.
        e. Calyx-lobes obtuse.   C. polygonorum Engelm.
        e. Calyx-lobes acute.   C. coryli Engelm.
    c. Sepals and petals mostly 5.
      d. Corolla-lobes obtuse, spreading, reflexed in age; calyx-lobes rounded, overlapping below, reaching middle of corolla-tube; capsule narrowed at top, somewhat longer than styles; fringes of scales reaching sinuses or a little beyond.   C. gronovii Willd.
      d. Corolla-lobes acute, spreading, tips often inflexed; calyx-lobes rounded, reaching top of corolla-tube; stamens exserted; capsule rounded or depressed at top.
        e. Flowers 2 mm long or less; seeds 1 mm long; scales reaching about middle of corolla-tube.   C. pentagona Engelm.
        e. Flowers 2 mm long or more; seeds 1.5 mm long; scales reaching sinuses of corolla.   C. campestris Yuncker

POLEMONIACEAE, Phlox Family

Our species herbs; leaves alternate or opposite, simple or compound; flowers hypogynous, perfect, often in cymes; calyx 5-lobed; corolla regular, 5-lobed, salverform or campanulate; stamens 5, on corolla-tube; disk present; carpels 3, united; ovulary 3-loculed; placentation axile; style 1; stigmas 3; fruit a capsule enclosed by the persistent, sometimes accrescent, calyx.

a. Cauline leaves alternate.
  b. Leaves simple; corolla tubular-funnelform.   2. Collomia
  b. Leaves pinnately compound; corolla campanulate.   1. Polemonium
a. Cauline leaves opposite, simple, entire; corolla salverform.   3. Phlox

1. Polemonium L., Greek Valerian, Jacob's Ladder

P. reptans L. Tufted, erect or diffuse, glabrous or pubescent; leaves pinnately compound, basal long-petioled, leaflets entire, lanceolate or oblong; flowers pediceled; corolla blue, 10-15 mm long, about twice as long as calyx, 5-lobed; stamens inserted on corolla tube; capsule ovoid.

2. Collomia Nutt.

C. linearis Nutt. Upper stem pubescent; leaves alternate, entire, linear to lanceolate; flowers clustered in axils of crowded upper leaves; corolla blue-purple to white, about 1 cm long, tubular-funnelform, lobes short; stamens and style included in corolla-tube; seeds 1 in each locule.

3. Phlox L., Phlox

Leaves entire, mostly opposite, bracts sometimes alternate; calyx tubular, 5-ribbed, often scarious between ribs; corolla salverform, tube long and slender, the 5 lobes often emarginate; stamens inserted in corolla tube at different levels, wholly included or slightly exserted; style short or elongate; ovules 1-4 in each locule.

a. Stems prostrate and matted; leaves rigid, linear to subulate, crowded, 1/2 to 2 cm long, often with fascicles of leaves in axils; cymes few-flowered; corolla rose-purple or pink, 1-2 cm wide, lobes deeply notched. P. subulata L.
a. Stems erect or ascending; leaves flat, not rigid and subulate, sometimes linear, longer than 2 cm.
  b. Style reaching beyond middle of corolla-tube and exceeding fruiting calyx.
    c. Lateral veins of leaf-blades anastomosing near the ciliate margin; pedicels short; inflorescence pyramidal; anthers cream-color. P. paniculata L.
    c. Lateral veins of leaf-blades not anastomosing, margins not ciliate.
      d. Rosette leaves evergreen; basal stolons present at anthesis, blades spatulate and petioled; flowering stems villous above, nodes few and distant; flowers few, in a single cyme; calyx glandular-pubescent. P. stolonifera Sims
      d. Rosette leaves often not evergreen; basal flowerless shoots present or absent, leaves not spatulate; inflorescence usually more compound.
        e. Inflorescence cylindrical, much longer than wide; stem dotted with red or purple. P. maculata L.
        e. Inflorescence a flat or rounded cluster of one or a few cymes.
          f. Nodes usually not more than 3-4 below the inflorescence; pedicel usually shorter than calyx; corolla usually 2.5-3 cm wide. P. ovata L.
          f. Nodes usually more than 4; corolla usually less than 2.5 cm wide. P. glaberrima L. (incl. P. carolina L.)
  b. Style not reaching beyond middle of corolla-tube, shorter than fruiting calyx.
    c. Prostrate basal shoots present; corolla usually blue or blue-purple, the lobes usually emarginate; leaf-blades acute or obtuse. P. divaricata L., Common Blue P.
    c. Without prostrate basal shoots; corolla usually red-purple, lobes not emarginate, tube usually hairy; leaf-blades sharp-pointed. P. pilosa L.

HYDROPHYLLACEAE, Waterleaf Family

Herbs, usually pubescent; leaves entire, lobed, or compound, alternate; flowers hypogynous, bisporangiate, solitary or in cymes; sepals 5, united at base or separate; corolla regular, of 5 petals united to middle or a little above, usually appendaged in the tube; stamens 5, inserted on corolla-tube; carpels 2, united; styles 2 or partly united; ovulary 1-loculed; placentae 2, large, sometimes nearly meeting, ovulary then appearing 2-loculed; fruit a capsule.

a. Flowers solitary, peduncled; stamens included. 2. Ellisia
a. Flowers in clusters; stamens usually more or less exsert.
  b. Style bifid to about middle or below; leaves pinnately lobed; corolla campanulate to rotate; flowers in helicoid cymes which become elongate and straight, racemelike. 3. Phacelia
  b. Style bifid only at tip; leaves pinnately or palmately lobed; corolla tubular to campanulate; flower clusters several times forked, often compact, not elongate. 1. Hydrophyllum

1. Hydrophyllum L., Waterleaf

Leaves rather large, palmately lobed or pinnately lobed or divided, petioled; flowers in terminal and lateral cymes; corolla tubular or campanulate, white, blue, or blue-purple; stamens alternating with 5 linear appendages in corolla-tube, exsert, filaments usually villous; capsule globose, pubescent; placentae greatly dilated, enclosing ovules.

a. Cauline blades mostly as wide as long, palmately lobed or shallowly pinnately lobed.
    b. Calyx with reflexed appendages in sinuses; basal leaves pinnately divided; stamens only slightly exsert; plant hairy. H. appendiculatum Michx., Appendaged W.
    b. Calyx without reflexed appendages but sometimes with minute teeth in sinuses; basal leaves sometimes with 2 or 3 small leaflets on petiole below the large rounded blade. H. canadense L.
a. Cauline blades longer than wide, pinnately divided nearly to midvein; stamens long exsert.
    b. Plant rough-hairy; peduncle usually shorter than petiole; leaf-segments blunt. H. macrophyllum Nutt., Large-leaf W.
    b. Plant glabrous or sparingly hairy; peduncle usually longer than petiole; leaf-segments acute. H. virginianum L., Virginia W.

2. Ellisia L.

E. nyctelea L. Branched; leaves pinnately parted; flowers solitary, peduncles arising just above axils or opposite the leaves; corolla white, campanulate-tubular, about as long as calyx, somewhat less than 1 cm long, lobes shorter than tube; stamens included. Rare.

3. Phacelia Juss.

Leaves pinnately lobed to pinnately compound; helicoid cymes lengthening and straightening after anthesis; corolla broadly campanulate, blue or white, 0.5 to somewhat more than 1 cm long, lobes sometimes fringed.

a. Corolla-lobes fringed; filaments villous, about as long as corolla. P. purshii Buckl.
a. Corolla-lobes not fringed, sometimes minutely toothed.
    b. Filaments glabrous, shorter than corolla tube. P. ranunculacea (Nutt.) Constance
    b. Filaments villous; plants stipitate-glandular.
        c. Filaments long exsert; corolla with conspicuous scales in the tube. P. bipinnatifida Michx.
        c. Filaments about as long as corolla; corolla without scales in the tube. P. dubia (L.) Trel.

BORAGINACEAE, Borage Family

Our species herbs, usually rough-pubescent, rarely glabrous; leaves alternate, simple, usually entire; flowers hypogynous, perfect, in cymes which are usually helicoid; calyx of 5 united sepals, lobed, sometimes irregular; corolla tubular, campanulate, salverform, or rotate, 5-lobed, usually regular, tube often bearing, and sometimes closed by, appendages of scales, folds, or crests; stamens 5, inserted on corolla-tube, alternate with lobes; hypogynous disk sometimes present; carpels 2, united; style 1, terminal or arising from base of carpels; stigma capitate or bilobed; ovulary 2-loculed at first, 4-loculed at maturity, a false partition in each carpel; 2 ovules in each carpel, 1 in each of the 4 locules; fruit separating into four 1-seeded nutlets or rarely into two 2-seeded nutlets.

a. Ovulary not deeply lobed; style arising from top of ovulary. 1. Heliotropium
a. Ovulary deeply 4-lobed; style arising from base of ovulary.
    b. Corolla almost unlobed; plant entirely glabrous (in Ohio species); nutlets laterally attached. 8. Mertensia
    b. Corolla lobed; plants pubescent.
        c. Stamens much exserted; corolla zygomorphic. 2. Echium
        c. Stamens included or little exserted; corolla regular.
            d. Nutlets attached by their sides, bearing barbed prickles (glochidia); corolla funnelform to salverform.
                e. Inflorescence mostly without bracts; nutlets attached near their summit, bases spreading or divergent. 6. Cynoglossum
                e. Inflorescence with bracts, bracts sometimes between the pedicels; nutlets erect, attached by middle of, or along whole length of, inner margin.

f. Nutlets attached along most of inner margin; glochidia in 2 marginal rows on external face of nutlet; fruiting pedicels erect; bracts subtending pedicels; blades linear to lanceolate. 9. Lappula

f. Nutlets attached by center of inner face, scar of attachment oval; glochidia all over external face of nutlet; fruiting pedicels reflexed; bracts often between pedicels; blades mostly broader than above. 10. Hackelia

d. Nutlets attached by their bases, smooth or rough, but without barbed prickles.

e. Flowers not subtended by bracts.

f. Plants large and coarse; blades large, acute or acuminate at apex; corolla tubular-campanulate; nutlet with toothed ring at base. 3. Symphytum

f. Plants small; blades usually obtuse at apex; corolla salverform or broadly funnel-form; nutlet without ring at base, flattened. 7. Myosotis

e. Flowers subtended by bracts, bracts often leaflike.

f. Corolla dull white, tubular, teeth or shallow lobes acute and erect; style conspicuously exsert. 4. Onosmodium

f. Corolla white, yellow, or orange, campanulate or salverform; style not or little exsert. 5. Lithospermum

1. Heliotropium L., Heliotrope

*H. indicum L. Blades ovate, decurrent on petiole; flowers small, not bracted; corolla blue, not appendaged in throat; stamens included in corolla-tube; style terminal; ovulary not deeply lobed, in fruit the 2 carpels separating, each 1- or 2-seeded.

2. Echium L.

*E. vulgare L. Leaves lanceolate or oblanceolate, cauline sessile, basal tapering to winged petioles; many short cymes in axils of leaflike bracts; corolla blue or white, zygomorphic, tubular or funnelform, lobes unequal, appendages lacking; stamens unequal, long-exsert; style villous, exsert; nutlets ovoid, rough.

3. Symphytum L., Comfrey

Coarse perennials; blades ovate or lance-ovate, petioles usually winged; corolla tubular-campanulate, limb little wider than tube, teeth or lobes short; appendages and stamens attached at same level; nutlets with toothed ring at base.

a. Stem conspicuously winged; corolla blue or yellowish. *S. officinale L.
a. Stem not or scarcely winged; corolla blue. *S. asperum Lepechin

4. Onosmodium Michx., False Gromwell

O. hispidissimum Mackenz. Leaves sessile, strongly ribbed, the cauline ovate; flowers bracted, helicoid cymes terminal on the branches; corolla whitish, tubular, erect lobes shorter than tube, not appendaged; stamens included; style exserted, bifid at tip; nutlets ovoid, short-necked at base.

5. Lithospermum L., Gromwell, Puccoon

Leaves linear to lance-ovate, sessile; flowers solitary and axillary or in leafy-bracted cymes; corolla funnelform or salverform; stamens usually included in corolla tube; style 2-lobed; nutlets smooth or pitted, erect, often shining.

a. Corolla bright yellow or orange, appendaged, tube evidently longer than calyx, limb 1 cm wide or more.

b. Pubescence soft and dense; corolla 10-15 mm wide, tube glabrous within at base. L. canescens (Michx.) Lehm., Hoary P.

b. Pubescence of stiff or spreading hairs; corolla 15-25 mm wide, tube pubescent within at base. L. caroliniense (Walt.) MacM.

a. Corolla white, greenish-white, or pale yellow, appendaged or not, limb less than 1 cm wide.

b. Leaves linear-lanceolate, without evident lateral veins; corolla white, without appendages in throat; nutlets gray or brownish, wrinkled. *L. arvense L., Corn G.

b. Leaves lanceolate to ovate, with prominent lateral veins; corolla white or pale yellow, append-aged; nutlets white, shining. L. latifolium Michx.

6. Cynoglossum L.

Leaves lanceolate or wider, lower petioled, upper sessile or sometimes clasping; flowers bract-less; corolla funnelform or salverform, appendaged, tube short; stamens included in corolla tube; nut-lets attached near their apex, covered with barbed or hooked prickles.

a. Corolla purple-red (rarely white); style overtopping nutlets; outer face of nutlets flat; stem leafy to top. *C. officinale L., Hound's-tongue
a. Corolla blue to white; style shorter than nutlets; outer face of nutlets rounded; stem not leafy above, peduncle long.
b. Corolla 1 cm wide or more, lobes overlapping; nutlets 7 mm long or more. C. virginianum L., Wild Comfrey
b. Corolla less than 1 cm wide, lobes not overlapping; nutlets 5 mm long or less. C. boreale Fern., Northern Comfrey

7. Myosotis L., Forget-me-not

Low herbs with sessile lanceolate or oblanceolate leaves; flowers very small, not bracted; co-rolla salverform or funnelform, appendaged in the throat; stamens included in corolla tube, filaments short; nutlets flattened.

a. Calyx-hairs all straight, appressed.
b. Style equaling or overtopping nutlets; stem stoloniferous; calyx shorter than corolla tube; corolla blue, 6 mm wide or more. *M. scorpioides L.
b. Style shorter than nutlets; stem not stoloniferous; calyx about as long as corolla tube; corolla blue, 6 mm wide or less. M. laxa Lehm.
a. Calyx-hairs, or some of them, spreading and hooked or glandular.
b. Calyx-lobes equal.
c. Pedicels in fruit equaling or longer than calyx; corolla blue or white, 2-3 mm wide, rarely more; style shorter than nutlets. *M. arvensis (L.) Hill
c. Pedicels in fruit shorter than calyx.
d. Inflorescence more than half length of plant; style shorter than nutlets; some lower flowers in axils of leaves; corolla blue. *M. micrantha Pall. (M. stricta Link)
d. Inflorescence no more than half length of plant; style sometimes longer than nutlets; few, if any, flowers in axils of leaves; corolla yellow, changing to blue. *M. discolor Pers. (M. versicolor (Pers.) Sm.)
b. Calyx-lobes unequal; corolla white, 1-2 mm wide.
c. Pedicels erect or nearly so. M. verna Nutt.
c. Pedicels ascending or spreading. M. macrosperma Engelm.

8. Mertensia Roth, Bluebell

M. virginica (L.) Pers. Glabrous; blades elliptic, obovate, or somewhat narrower, basal with long petioles, upper often sessile; calyx much shorter than corolla-tube; corolla not or scarcely crested, pink when young changing to lavender-blue, rarely white, trumpet-shaped, tube 1-2 cm long, limb shorter and almost entire; style elongate; stigma minute; nutlets rough.

9. Lappula Moench, Stickseed

*L. echinata Gilib. Leaves oblanceolate or narrower; flowers short-pediceled, small-bracted, in helicoid cymes; calyx lobes narrow; corolla blue, 2-3 mm wide, appendaged; stamens and style in-cluded; nutlets 1/4 of a sphere, outer convex surface verrucose and bearing near each margin 2 rows of barbed prickles, attachment scar extending whole length of inner margin; fruiting pedicels erect.

10. Hackelia Opiz, Stickseed

H. virginiana (L.) I. M. Johnston Basal blades round-ovate, sometimes cordate, slender-petioled; cauline blades ovate or narrower, tapering to tip and base, sessile; flowers short-pediceled, in small-bracted helicoid cymes; corolla pale blue or white, about 2 mm wide, appendaged; each nutlet one fourth of a sphere, convex outer side verrucose and covered with barbed prickles, inner angle bearing an oval scar of attachment at about middle; pedicels reflexed in fruit.

VERBENACEAE, Vervain Family

Our species herbs; leaves usually opposite, simple, sometimes incised or lobed; flowers hypogynous, perfect; corolla 4-5 lobed, nearly regular, somewhat 2-lipped; stamens 4, on the corolla, 2 longer; carpels 2, united; locules as many or twice as many; style 1; fruit splitting at maturity into as many 1-seeded nutlets as locules.

a. Flowers in single or panicled spikes; style-tip 2-lobed; locules 4. Verbena
a. Flowers in small heads on axillary peduncles; stigma capitate; locules 2. Lippia

Verbena L., Vervain, Verbena

Leaf-blades serrate, incised, or pinnatifid; flowers bracted, in spikes; calyx 5-toothed or -lobed; corolla salverform or funnelform, 5-lobed; stigma 2-lobed; ovulary 4-loculed, splitting into 4 one-seeded nutlets at maturity.

a. Corolla 2-3 cm long, blue, purple, or white, tube about twice as long as calyx; calyx about 1 cm long. V. canadensis (L.) Britt., Rose V.
a. Corolla much shorter than 2 cm; fruiting calyx not more than 0.5 cm long.
   b. Bracts of inflorescence longer than flowers; blades lobed or incised; plants often mat-forming.
      V. bracteata Lag. & Rodr.
   b. Bracts of inflorescence shorter than flowers; plants erect.
      c. Corolla white; fruits remote in very slender spikes. V. urticifolia L., White V.
      c. Corolla blue or purple, rarely white; fruits imbricated in the spikes.
         d. Plants densely and softly hairy; blades broadly obovate or rounded, sessile or nearly so.
            V. stricta Vent.
         d. Plants glabrous or somewhat rough-hairy.
            e. Blades oblanceolate or linear, tapering to sessile or subsessile base, mostly obtuse at tip. V. simplex Lehm.
            e. Blades lanceolate or lance-ovate, sometimes hastately lobed, petioled, acuminate.
               V. hastata L., Blue V.

Lippia L., Fog Fruit

L. lanceolata Michx. On moist or wet soil; flowering branches erect from horizontal stems; blades lanceolate, acute at tip and base; flowers very small, bracted, in heads or short spikes on long axillary peduncles; corolla pink-purple or white, slightly 2-lipped; ovulary 2-loculed, splitting into 2 one-seeded nutlets.

LABIATAE, Mint Family

Usually aromatic herbs or rarely shrubs; stems square; leaves opposite; flowers hypogynous, perfect, in whorls (or rarely solitary) in axils of foliage leaves; or in axils of bracts, then whorls or solitary flowers in spikes or racemes; or whorls 1-few at or near summit of stem; calyx usually 5-lobed or -toothed, sometimes 2-lipped; corolla zygomorphic, rarely almost regular, of 5 petals, usually 5-lobed or sometimes apparently 4-lobed, usually 2-lipped, upper lip usually 2-lobed, lower lip usually 3-lobed; stamens 4 or 2, on the corolla, when 4, usually of 2 lengths, when 2, the other pair sometimes present as vestiges; pollen-sacs parallel or their bases divergent, in extreme condition the sacs lying end to end in a straight line perpendicular to filament; carpels 2, united, ovulary 4-lobed, 1 ovule in each lobe; style 1, bifid at tip; fruit enclosed in persistent calyx, separating into 4 one-seeded nutlets; in genera Nos. 1-4, style terminal, nutlets laterally attached; in the other genera, style basal, nutlets basally attached.

Suggestion. To find whether inner (upper) or outer (lower) pair of stamens is the longer, cut corolla lengthwise along median line of lower lip, then separate cut edges exposing stamens. The two stamens in the middle are the inner (upper) pair; the two on the outside are the outer (lower) pair.

a. Anther-bearing stamens 4.
   b. Calyx with a protuberance on upper side, 2-lipped; corolla 2-lipped. 5. Scutellaria
   b. Calyx without a protuberance on upper side.

c. Corolla nearly equally toothed or lobed, not or scarcely 2-lipped.
  d. Stamens long-exserted, the portion beyond the corolla strongly curved downward; calyx 2-lipped, inverted in fruit, upper lip much the longer in flower. 2. Trichostema
  d. Stamens and calyx not as above.
    e. Leaf-blades 1-3-nerved, nearly entire; pollen-sacs divergent. 1. Isanthus
    e. Leaf-blades serrate or dentate, pinnately veined; pollen-sacs parallel.
      f. Flowers in clusters in axils of leaves or bracts, clusters sometimes in spikes or heads; corolla usually 4-lobed. 30. Mentha
      f. Flowers solitary in axils of bracts, in spikelike racemes; corolla 5-lobed; calyx 2-lipped in fruit. 32. Perilla
c. Corolla appearing 1-lipped, upper lip very short, sometimes split. (See third c.)
  d. Corolla with short 2-lobed upper lip, lower lip large and spreading; calyx nearly regular, lobes equaling tube. 3. Ajuga
  d. Corolla split on upper side, stamens exserted in the cleft, hence all lobes below stamens, corolla thus appearing to have 5-lobed lower lip but no upper lip; calyx somewhat 2-lipped, toothed. 4. Teucrium
c. Corolla more or less 2-lipped.
  d. Stamens included in corolla-tube; calyx regular, spine-tipped teeth becoming outwardly curved; canescent. 6. Marrubium
  d. Stamens exceeding corolla-tube but sometimes not corolla-tip.
    e. Upper lip of corolla straight, not concave or galeate; outer pair of stamens the longer, rarely all nearly equal.
      f. Corolla yellow, lower lip fringed, much longer than upper; stamens long-exserted beyond corolla; flowers in terminal racemes; rare. 31. Collinsonia
      f. Lower lip of corolla not fringed; stamens usually not long-exserted beyond corolla.
        g. Flowers in dense headlike clusters at end of stem or of branches (rarely a few clusters in upper axils).
          h. Heads several to many; bracts in headlike clusters not setaceous; pollen-sacs parallel. 26. Pycnanthemum
          h. Heads solitary or few; bracts in headlike clusters setaceous, long-ciliate. 25. Satureja
        g. Flowers in small axillary clusters, the clusters sometimes in terminal spikes.
          h. Small, diffusely branched, shrub with elliptic or oblong leaves 5-10 mm long; upper clusters usually continuous. 27. Thymus
          h. Herbs; clusters usually not continuous.
            i. Cauline blades ovate, coarsely crenate; plant lemon-scented; calyx 2-lipped. 24. Melissa
            i. Cauline blades linear to oblanceolate, entire or nearly so. 25. Satureja
    e. Upper corolla-lip concave or galeate (or, if rarely not, then inner pair of stamens the longer); corolla definitely 2-lipped.
      f. Calyx with more than 10 nerves, not 2-lipped, at most somewhat oblique; inner pair of stamens the longer.
        g. Stamens well exserted beyond corolla, not ascending under upper corolla-lip; flowers in dense terminal spikes; pollen-sacs parallel. 7. Agastache
        g. Stamens not or little exserted beyond corolla, ascending under upper corolla-lip.
          h. Corolla 2.5-3 cm long; pollen-sacs parallel; flowers single in axils in terminal leafy spike. 8. Meehania
          h. Corolla shorter; pollen-sacs divergent.
            i. Flowers in short dense terminal spike, bracts spiny-toothed; blades sharply serrate or laciniate. 11. Dracocephalum
            i. Flowers in axils of foliage leaves or of bracts that are not spiny-toothed.
              j. Plants erect, canescent; flower-clusters of several to many flowers in upper axils, sometimes continuous; corolla white; blades ovate. 9. Nepeta
              j. Plants prostrate, not canescent; flowers few or solitary in axils; corolla blue; blades round-ovate or reniform. 10. Glechoma
      f. Calyx with 10 nerves or fewer (in Stachys, nerves sometimes more, or difficult to distinguish); stamens ascending under upper lip of corolla, outer pair the longer except in Leonurus.
        g. Calyx strongly 2-lipped; flowers in terminal dense short spikes; bracts round-ovate, cuspidate. 12. Prunella

187

g. Calyx regular or nearly so, not or scarcely 2-lipped.
  h. Flowers in terminal spike, a single flower in axil of each large or small bract; calyx much shorter than corolla.
    i. Blades broadly ovate, petioled, cordate; flowers few, bracts large; pollen-sacs divergent. 14. Synandra
    i. Blades lanceolate; flowers many, bracts small; pollen-sacs nearly parallel. 13. Physostegia
  h. Flowers in clusters in axils of foliage leaves or of bracts, the clusters sometimes in spikes.
    i. Calyx somewhat flaring at summit, teeth about as long as wide, mucronate, no more than 1/4 as long as tube; pollen-sacs divergent; rare. 15. Ballota
    i. Calyx not as above.
      j. Calyx-lobes tipped with terete spines; pollen-sacs parallel or dehiscing crosswise.
        k. Pollen-sacs parallel or nearly so; inner pair of stamens a little the longer; blades sometimes lobed. 16. Leonurus
        k. Pollen-sacs dehiscing crosswise, appearing divergent; blades not lobed; lower lip of corolla with basal protuberances; rare. 17. Galeopsis
      j. Calyx-lobes or teeth acute or aristate but not ending in hard spines (or sometimes with short spines in Stachys); pollen-sacs divergent.
        k. Flower clusters (except sometimes the lowest) in axils of bracts; inflorescence a terminal spike. 19. Stachys
        k. Flower clusters subtended by ordinary or scarcely reduced foliage leaves, but sometimes crowded near tip of stem. 18. Lamium
a. Anther-bearing stamens 2.
  b. Corolla yellow, lower lip fringed, much longer than upper; stamens long-exsert; flowers in terminal panicle. 31. Collinsonia
  b. Flowers not as above.
    c. The 2 pollen-sacs of an anther far apart, anther-connective elongate, 1 sac often vestigial; corolla and calyx 2-lipped. 20. Salvia
    c. The 2 pollen-sacs of an anther the usual distance apart.
      d. Calyx nearly equally toothed, not 2-lipped.
        e. Corolla strongly 2-lipped, upper lip narrow; flowers large, in large terminal and sometimes subterminal headlike clusters. 21. Monarda
        e. Corolla almost regular or not strongly 2-lipped, 4-lobed; flowers less than 1 cm long; clusters axillary or both axillary and terminal.
          f. Corolla white; flower clusters small, in many axils; blades usually cuneate at base. 29. Lycopus
          f. Corolla purplish or white; clusters terminal and in upper axils; blades rounded at base. 28. Cunila
      d. Calyx 2-lipped.
        e. Flowers in dense terminal and sometimes also subterminal clusters; calyx not hairy in throat. 22. Blephilia
        e. Flowers in small clusters in many axils; calyx hairy in throat. 23. Hedeoma

1. Isanthus Michx., False Pennyroyal

I. brachiatus (L.) BSP. Viscid-puberulent; leaves lanceolate, nearly entire, 1-3-nerved; flowers 1-3 in axillary clusters; calyx regular, 10-nerved; corolla pale blue, about 5 mm long, nearly regular; stamens 4, outer pair a little the longer; pollen-sacs divergent.

2. Trichostema L., Blue-curls

Somewhat glandular-pubescent; flowers usually solitary, terminal, becoming lateral by growth of branches; calyx 2-lipped, inverted in fruit; corolla nearly regular, blue, 0.5-1 cm long, tube longer than calyx; stamens 4, exserted, filaments long and curved, outer pair the longer; pollen-sacs divergent.

a. Leaf-blades oblong to lanceolate, with lateral veins. T. dichotomum L.
a. Leaf-blades linear, with only a midvein. T. setaceum Houtt.

188

3. Ajuga L., Bugleweed

Leaf-blades obovate, dentate or lobed; flowers in whorls of 4-6 in axils of leaflike bracts, the whole a terminal spike; calyx nearly regular, 5-lobed; corolla blue, upper lip short, 2-lobed, lower lip large and spreading; stamens 4, exserted beyond upper lip, outer pair the longer; pollen-sacs divergent.

a. Stoloniferous; stem glabrous or slightly pubescent. *A. reptans L.
a. Not stoloniferous; stem villous. *A. genevensis L.

4. Teucrium L., Germander

Calyx nearly regular or somewhat 2-lipped; corolla 1-1.5 cm long, cleft above, stamens exserted in cleft, hence all lobes below exserted stamens and appearing as one lower lip; stamens 4, outer pair the longer; pollen-sacs divergent.

a. Leaves deeply divided; flowers whorled in upper axils; corolla red-purple. *T. botrys L.
a. Leaves dentate or serrate.
  b. Corolla greenish-yellow; flowers 1 or 2 in axils of bracts in slender secund racemes; calyx glabrous. *T. scorodonia L.
  b. Corolla pink-purple or whitish; flowers several in axils of bracts in dense, not secund, racemes; calyx pubescent. T. canadense L. (Incl. T. occidentale Gray)

5. Scutellaria L., Skullcap

Nonaromatic; flowers solitary in axils or in solitary or panicled racemes; calyx gibbous, 2-lipped, upper lip bearing a crest or protuberance; corolla much longer than calyx, 2-lipped, tube sometimes curved upward; stamens 4, outer pair the longer.

a. Flowers solitary in axils of ordinary leaves.
  b. Blades lanceolate, 2-4 times as long as wide; flowers 1.5 cm long. S. epilobiifolia A. Hamilton, Common or Marsh S.
  b. Blades ovate, mostly less than twice as long as wide; flowers 1 cm long or less.
    c. Rhizome slender; blades 2-5 cm long; nutlets tan or yellow. S. nervosa Pursh
    c. Rhizome a chain of small tubers; blades 0.5-2 cm long; nutlets brown. S. parvula Michx.
a. Flowers in terminal or axillary racemes; bracts smaller than foliage leaves.
  b. Corolla 5-8 mm long; racemes axillary and terminal, usually many; blades ovate to lanceolate. S. lateriflora L., Mad-dog S.
  b. Corolla more than 1 cm long.
    c. Middle and upper cauline blades entire, short-petioled, lanceolate. S. integrifolia L.
    c. Blades dentate, crenate, or serrate.
      d. Blades lanceolate, almost sessile; corolla 1.5 cm long. S. epilobiifolia A. Hamilton
      d. Blades ovate, obviously petioled.
        e. Principal blades cordate at base; lateral lobes of corolla almost equaling short upper lip.
          f. Blades 2-4 cm long; stem below inflorescence glabrous or hairs minute, curved upward, nonglandular. S. saxatilis Ridd.
          f. Blades 8 cm long or more; stem with spreading glandular hairs. S. ovata Hill
        e. Principal blades not cordate or, if slightly so, wings of blade extending short distance down petiole ending in a point; lateral lobes of corolla much shorter than galeate upper lip.
          f. Middle and upper blades blunt at apex, cuneate at base, lower third of margin entire, upper margin crenate; stem usually with spreading long multicellular hairs; calyx glandular-pubescent. S. elliptica Muhl.
          f. Blades large, acute at apex, not strongly cuneate at base, less than 1/3 of margin entire; stem minutely pubescent with curved hairs; calyx pubescent, not glandular.
            g. Calyx sparsely pubescent; racemes usually 1, lowest 2 flowers usually in axils of foliage leaves. S. serrata Andr., Showy S.
            g. Calyx canescent; racemes usually more than 3; usually all flowers in axils of bracts. S. incana Biehler

6. Marrubium L., Horehound

*M. vulgare L. White-woolly; branched; with stellate hairs; blades ovate, crenate, lower petioled; flowers in dense axillary clusters; calyx tubular, its 10 spine-tipped teeth curved outward at maturity; corolla about 6 mm long, white, 2-lipped; stamens 4, included in corolla tube, lower pair the longer.

7. Agastache Clayt., Giant Hyssop

Tall; leaves petioled, ovate, serrate; flowers in usually dense terminal spikes; calyx 5-toothed, 15-nerved, oblique; corolla 2-lipped, middle lobe of lower lip crenate; stamens 4, exserted, upper pair the longer; pollen-sacs nearly parallel.

a. Calyx about 6 mm long, lobes less than 2 mm long; corolla greenish-yellow; blades densely short-pubescent beneath. A. nepetoides (L.) Ktze.
a. Calyx 7 mm long or more, lobes 2-3 mm long; corolla pink or purplish; blades glabrous or villous beneath. A. scrophulariaefolia (Willd.) Ktze.

8. Meehania Britt.

M. cordata (Nutt.) Britt. Low, pubescent, spreading or trailing; blades ovate-cordate, long-petioled, crenate; flowers few, bracted, in terminal spike; calyx 15-nerved, obliquely 5-toothed; corolla blue, 2.5-3 cm long, much longer than calyx, 2-lipped, tube expanded above on lower side; stamens 4, included, upper pair the longer; pollen-sacs parallel.

9. Nepeta L., Catnip

*N. cataria L. Canescent, much branched; blades petioled, ovate, crenate-dentate; flowers in rather dense terminal and axillary clusters; calyx 15-nerved, unequally 5-toothed; corolla whitish dotted with purple, about 1 cm long, 2-lipped; stamens 4, ascending under upper corolla-lip, upper pair the longer; pollen-sacs divergent.

10. Glechoma L., Ground-ivy

*G. hederacea L. Stem prostrate; blades reniform or rounded, crenate, petioled; flowers short-stalked, 1-few in axils; calyx nearly regular, 15-nerved; corolla blue, 1-2 cm long, tube long-exserted from calyx, limb 2-lipped; stamens 4, ascending under upper lip, inner pair the longer; pollen-sacs divergent.

11. Dracocephalum L., Dragonhead

D. parviflorum Nutt. Blades lanceolate or wider, serrate; flowers in dense terminal cluster, bracts spiny-toothed; calyx 15-nerved, somewhat 2-lipped; corolla light blue or purplish, about 1.5 cm long, 2-lipped, tube elongate; stamens 4, ascending under upper lip, inner pair the longer; pollen-sacs divergent.

12. Prunella L., Self-heal, Carpenter-weed

P. vulgaris L. Blades ovate to lanceolate, entire or nearly so, petioled; inflorescence a close spike or head, flowers in axils of wide bracts; calyx strongly 2-lipped, lips closed in fruit, upper lip broad and truncate, 3-lobed or -toothed, lower lip of 2 narrow lobes; corolla blue, pink, or white, 1-1.5 cm long, 2-lipped; stamens 4, outer pair the longer; pollen-sacs divergent.

13. Physostegia Benth., False Dragonhead

P. virginiana (L.) Benth. Stem smooth; blades lanceolate, sharply serrate; flowers showy, in a terminal spike, 4-ranked; calyx membranous, nearly regular; corolla 2-3 cm long, lavender-pink or white, much longer than calyx, 2-lipped; stamens 4, outer pair the longer, pollen-sacs parallel.

14. Synandra Nutt.

S. hispidula (Michx.) Britt. Pubescent; blades broadly cordate-ovate, long-petioled, crenate-serrate; flowers sessile, leafy-bracted, in terminal spike; calyx thin, lobes unequal, sometimes 4-lobed; corolla greenish-yellow or -white, 3 cm long or more, 2-lipped, tube slender below, widened above; stamens 4, outer pair the longer, ascending under upper lip, filaments hairy; pollen-sacs divergent.

15. Ballota L., Black Horehound

*B. nigra L. Erect; blades ovate, toothed, petioled; flower-clusters axillary; calyx regular, 5-toothed, 10-nerved, the border spreading; corolla pubescent, 2-lipped; stamens 4, outer pair the longer, pollen-sacs divergent.

16. Leonurus L., Motherwort

Blades toothed or lobed; flowers small, in dense axillary clusters subtended by leaflike bracts; calyx with 5 nearly equal spine-tipped teeth, the 2 lower somewhat deflexed; corolla white or pink, 2-lipped; stamens 4, ascending under concave upper lip; pollen-sacs parallel.

a. Lower blades palmately-cleft; corolla pink, upper lip long-pubescent; calyx 5-nerved.
   *L. cardiaca L.
a. Lower blades toothed; corolla whitish, upper lip short-pubescent; calyx 10-nerved.
   *L. marrubiastrum L.

17. Galeopsis L., Hemp-nettle

Flowers in clusters in upper axils, the clusters sometimes in spikes; calyx with 5 nearly equal spine-tipped teeth; corolla 2-lipped, white, red-purple, or variegated, lower middle lobe with 2 basal protuberances; stamens 4, ascending under concave entire upper lip, outer pair the longer; anthers dehiscing crosswise.

a. Stem with fine short recurved hairs; blades entire or toothed. *G. ladanum L.
a. Stem with coarse straight or slightly recurved hairs; blades toothed. *G. tetrahit L.

18. Lamium L., Henbit, Dead-nettle

Blades reniform to ovate, lower petioled, upper sometimes sessile; flowers in axillary clusters, these sometimes aggregated in terminal spikes; calyx 5-nerved, nearly equally lobed, lobes tapering to slender points; corolla pink- or red-purple to white, 2-lipped, lower middle lobe clawed at base; stamens 4, outer pair the longer; pollen-sacs divergent.

a. Upper leaves sessile or clasping, round-ovate to reniform; corolla without a ring of hairs within; flowers sometimes cleistogamous, the corolla then small, tubular, with hairy round tip, failing to open. *L. amplexicaule L., Common H.
a. All leaves petioled; corolla with a ring of hairs within.
   b. Upper leaves often purple-red; upper internodes very short, hence flowers in dense terminal leafy-bracted spike; corolla 1.5 cm long or less. *L. purpureum L., Red H.
   b. Leaves green, often blotched with white, internodes not short; corolla 2 cm long or more.
      *L. maculatum L., Spotted H.

19. Stachys L., Hedge-nettle

Flowers in whorls in axils of bracts, the whorls in terminal dense or interrupted spikes; calyx nearly regular; corolla rose-purple or yellow, 2-lipped, upper lip concave; stamens 4, ascending under upper lip, outer pair the longer; pollen-sacs divergent.

a. Calyx, stem, and under surface of leaves white-woolly. *S. germanica L.
a. Calyx, stem, and under surface of leaves pubescent or glabrous, not white-woolly.
   b. Stem pubescent on sides and usually also on angles, hairs or some of them often bristlelike, often bulbous-based; calyx pubescent.
      c. Blades sessile or petioles less than 1 cm long, lanceolate to narrowly oblong, softly pubescent on both sides, rounded to subcordate at base; calyx lobes about equaling tube. S. palustris L.

191

c. Blades with petioles 1-3 cm long, ovate, obovate, or lanceolate, cordate, appressed strigose above, pubescent below at least on veins; stem, especially upper part, with stipitate glands mixed with longer hair; calyx lobes shorter than tube. S. riddellii House

b. Stem glabrous or nearly so on sides, glabrous or pubescent on angles.
  c. Blades subsessile or with petioles less than 1 cm long, lanceolate to narrowly ovate.
    d. Blades essentially glabrous on both sides; calyx glabrous or sparsely hairy; bracts sparsely ciliate. S. hyssopifolia Michx. var. ambigua Gray
    d. Blades often hairy on veins beneath and with stiff appressed ones on upper surface; bristle-like hairs of stem-angles deflexed; bracts bristly ciliate. S. tenuifolia Willd. var. hispida (Pursh) Fern.
  c. Petioles 1-3 cm long; calyx and both surfaces of blades glabrous or nearly so; stem glabrous or some short thick hairs on angles; margin of bracts not ciliate above base. S. tenuifolia Willd. var. tenuifolia

20. Salvia L., Sage

Leaves ovate to linear; flowers in whorls in axils of bracts, the clusters in terminal spikes or racemes; calyx 2-lipped; corolla 2-lipped, lower lip sometimes much the longer; anther-bearing stamens 2; anther-connectives long, the half-anther on lower end absent or vestigial except in S. lyrata.

a. Corolla scarlet, about 4 cm long; calyx scarlet. *S. splendens Ker
a. Corolla blue or violet, less than 4 cm long.
  b. Rosette blades lyrate-pinnatifid to repand; cauline leaves 1-2 pairs; both pollen-sacs bearing pollen; corolla violet. S. lyrata L., Lyre-leaf S.
  b. Stem leafy; lower pollen-sacs rudimentary; corolla blue.
    c. Blades rugose, canescent beneath, crenate; upper calyx-lip 3-toothed.
      d. Cauline blades petioled, canescent above; calyx-teeth aristate. *S. officinalis L.
      d. Cauline blades sessile, glabrate above; upper calyx-teeth minute. *S. nemorosa L.
    c. Blades not rugose, not canescent, glabrate or minutely pubescent.
      d. Calyx 5-lobed; blades ovate, decurrent on petiole. S. urticifolia L.
      d. Calyx 3-lobed; blades lanceolate or linear.
        e. Flowers 6 or more at a node; corolla-tube not longer than calyx. S. azurea Lam. var. grandiflora Benth.
        e. Flowers 2-3 at a node; corolla-tube longer than calyx. S. reflexa Hornem.

21. Monarda L., Horse Mint, Wild Bergamot

Blades ovate or lanceolate, in our species petioled; flowers rather large, in terminal and sometimes also in axillary dense clusters subtended by foliaceous bracts; calyx nearly regular, 13-15-nerved; corolla strongly 2-lipped; anther-bearing stamens 2, exserted or included under narrow erect or arched upper lip; pollen-sacs divergent.

a. Corolla red; bracts usually tinged with red; calyx-throat glabrous or nearly so. M. didyma L., Oswego-tea, Bee-balm
a. Corolla lavender, pink, rose, white or yellow.
  b. Stamens not exserted beyond upper lip; corolla yellowish, dotted with purple; upper lip arched, the tip not villous; calyx-throat villous; axillary clusters often present. M. punctata L.
  b. Stamens exserted beyond upper lip, which is slender, acute, almost straight, sometimes villous at tip.
    c. Corolla deep rose-color, upper lip not villous at tip; bracts and calyx usually purple, calyx glandular-puberulent. M. media Willd.
    c. Corolla lavender, yellowish, or white.
      d. Corolla white or yellowish-white, dark-spotted, upper lip not villous at tip; bracts green or white. M. clinopodia L.
      d. Corolla lavender or deeper purple, upper lip villous at tip; bracts green, gray, or pink-tinged; calyx densely villous in throat. M. fistulosa L. Var. fistulosa has blades villous beneath and hairs of stem spreading; var. mollis (L.) Benth. has blades short-hairy or glabrescent beneath and hairs of stem decurved.

22. Blephilia Raf.

Hairy; blades lanceolate or ovate; flowers in terminal and subterminal globose clusters, the clusters sometimes continuous in spikes; calyx 2-lipped, upper lip much the longer; corolla white or bluish, with purple spots, 2-lipped, 1-1.5 cm long; stamens 2, ascending under the erect concave entire upper lip, exserted; pollen-sacs divergent.

a. Blades of flowering stem narrowed to base, sessile or nearly so; teeth of lower calyx-lip reaching base of teeth of upper lip or beyond. B. ciliata (L.) Benth.
a. Blades of flowering stem rounded or cordate at base, petioled; teeth of lower calyx-lip not reaching so far as base of teeth of upper lip. B. hirsuta (Pursh) Benth.

23. Hedeoma Pers., Pennyroyal

Small annuals; blades entire or nearly so; flowers in axillary clusters; calyx 13-ribbed, somewhat 2-lipped, hairy in throat, gibbous on lower side near base; corolla blue, 3-5 mm long, 2-lipped; anther-bearing stamens 2; pollen-sacs divergent.

a. Blades lanceolate or ovate, sometimes petioled, crenulate; teeth of upper calyx-lip usually about as wide as long, acute, not ciliate. H. pulegioides (L.) Pers., American P.
a. Blades linear, entire, sessile; teeth of upper calyx-lip usually subulate, ciliate. H. hispida Pursh

24. Melissa L., Balm

*M. officinalis L. Lemon-scented; blades ovate, long-petioled, crenate; flowers in small axillary clusters; calyx 2-lipped, somewhat villous, 13-nerved; corolla pale blue to white, 10-15 mm long, 2-lipped, upper lip concave; stamens 4, ascending under upper lip; pollen-sacs divergent.

25. Satureja L., Savory, Calamint

Calyx 10-13-nerved, 2-lipped or almost regular, naked or hairy in throat; corolla purple, pink, or white, 2-lipped; stamens 4, lower pair the longer; pollen-sacs parallel to divergent.

a. Flower-clusters terminal and sometimes also subterminal, dense, the many setaceous long-ciliate bracts about equaling calyx; blades ovate to oblong. *S. vulgaris (L.) Fritsch
a. Flower-clusters axillary, sometimes not dense, without setaceous long-ciliate bracts; blades of flowering stem linear to oblanceolate.
   b. Internodes of stem glabrous; calyx glabrous on outside, hairy in throat. S. arkansana (Nutt.) Briq.
   b. Internodes minutely pubescent; calyx naked in throat. *S. hortensis L.

26. Pycnanthemum Michx., Mountain-mint

Strongly aromatic; stem branched; flowers small, in dense headlike cymes terminal on main stem or branches or sometimes in axils; clusters subtended by leaflike bracts, many smaller bractlets in the cyme; calyx nearly regular or somewhat 2-lipped, 10-13-nerved; corolla 2-lipped, white or purplish, sometimes dotted with purple; stamens 4, lower pair a little the longer; pollen-sacs parallel.

a. Headlike cymes very dense, cyme-branches and pedicels hidden by bracts.
   b. Blades linear to lanceolate, lateral veins joining midvein at middle or below; leaflike bracts below the cyme glabrous.
      c. Blades linear, lateral veins from lower fourth of midvein; calyx-teeth 1-2 mm long; stem glabrous; bracts within the cyme long-acuminate with conspicuous subulate points.
      P. tenuifolium Schrad.
      c. Blades lanceolate or linear-lanceolate, lateral veins from lower half of midvein; calyx-teeth less than 1 mm long; stem pubescent on angles, usually glabrous on sides; bracts within the cyme acute or short-acuminate, without conspicuous subulate points. P. virginianum (L.) Durand and Jackson
   b. Blades lanceolate, some lateral veins joining midvein above middle; leaflike bracts below the cyme velvety-pubescent on upper surface.
      c. Blades of main stem ovate or lance-ovate, rounded at base, shallowly toothed, length less than 3 times width; cymes relatively few, usually solitary at tips of branches; leaflike bracts white-pubescent. P. muticum (Michx.) Pers.

    c. Blades of main stem lanceolate, entire or shallowly toothed, length more than 3 times width, tapering at base; cymes many.
        d. Inner bracts usually shorter than calyx, densely canescent; whole lower surface of blades pubescent. P. pilosum Nutt.
        d. Inner bracts usually longer than calyx, thinly pubescent, densely short ciliate; lower surface of blades glabrate or finely pubescent on veins. P. verticillatum (Michx.) Pers.
a. Headlike cymes somewhat open, branches and pedicels not concealed; leaves white-pubescent beneath.
    b. Corolla white with purple spots, lower lip 4-5 mm long; lower calyx-teeth 1. 5 mm long or more, upper 1/3-1/2 as long; inner bracts and calyx-teeth bearing many long flexuous hairs. P. pycnanthemoides (Leavenw.) Fern.
    b. Corolla pink-purple, lower lip 2-3 mm long; lower calyx-teeth 1. 5 mm long or less, upper 1/2-2/3 as long; long hairs present or absent. P. incanum (L.) Michx.

27. Thymus L., Thyme

*T. serpyllum L., Wild T. Prostrate, much-branched, somewhat woody; blades entire, usually elliptic to oblong, 1 cm long or less; calyx 13-nerved, 2-lipped, villous in throat; corolla purple, 2-lipped, about 5 mm long; flowers in axillary clusters, upper clusters often continuous; stamens 4.

28. Cunila L., Dittany

C. origanoides (L.) Britt. Branched; blades ovate, few-toothed or entire, 2-4 cm long; flowers small, in axillary or terminal small clusters; calyx regular, 10-13-nerved, hairy in throat; corolla rose-purple to white, limb expanded, 2-lipped, 4-lobed; anther-bearing stamens 2.

29. Lycopus L., Water-horehound

Usually with long stolons, often with tubers; blades sessile or petioled, toothed or lobed; flowers small, in dense axillary clusters; calyx nearly equally 4-5-toothed; corolla white, the 4 lobes nearly equal, hairy in throat; anther-bearing stamens 2; pollen sacs parallel.

a. Calyx-lobes acute or acuminate, exceeding the mature nutlets.
    b. Blades sessile, base slightly rounded; internodes short; summit of nutlets wavy. L. asper L.
    b. Blades petioled or tapering to a petiolelike base.
        c. Blades serrate; summit of the 4 nutlets truncate, tubercled. L. rubellus Moench
        c. Lower blades lobed or spreading-toothed; nutlets with smooth edge, not tubercled.
            d. Calyx-lobes 2 mm long or more; bracts usually obvious; blades with spreading teeth or the lower ones lobed. L. europaeus L.
            d. Calyx-lobes 1. 5 mm long; bracts small; lower blades pinnatifid. L. americanus Muhl.
a. Calyx-lobes acute or obtuse, not exceeding the mature nutlets; nutlets tubercled. (Intermediates between the two following species, perhaps hybrids, have been called L. sherardi Steele)
    b. Stem pubescent, hairs appressed; corolla-lobes erect; summit of nutlets truncate, inner angle and outer edge tubercled; plants often with red pigment. L. virginicus L.
    b. Stem glabrous or puberulent; corolla-lobes flaring; summit of nutlets oblique, only outer edge tubercled. L. uniflorus Michx.

30. Mentha L., Mint

Strongly aromatic; blades toothed; flowers small, in axillary clusters or in terminal spikes; calyx nearly regular, 5-toothed or slightly 2-lipped; corolla lavender or white, limb 4-lobed; stamens 4, about equal in length; pollen-sacs parallel.

a. Flowers in terminal spikes continuous at least at tips; sometimes additional whorls of flowers in upper axils.
    b. Leaves sessile or petioles to 3 mm long; spikes (excluding corollas) about 6 mm in diameter.
        c. Plant glabrous or nearly so. *M. spicata L., Spearmint
        c. Plant soft-pubescent; blades rugose-reticulate. *M. rotundifolia (L.) Huds.
    b. Leaves definitely petioled; spikes (excluding corollas) 1 cm thick.
        c. Spikes elongate; calyx-teeth usually hairy; blades lanceolate to ovate-oblong. *M. piperita L., Peppermint
        c. Spikes ellipsoid or capitate; peduncled whorls often present in upper axils; calyx glabrous; blades ovate or round-ovate; lemon-scented. *M. citrata Ehrh.

a. Flowers in separate axillary whorls.
   b. Subtending leaves only 2-3 times as long as whorls of flowers in their axils. *M. cardiaca Baker
   b. Subtending leaves more than 4 times as long as whorls of flowers in their axils. M. arvensis L.

31. Collinsonia L.

Nearly glabrous; blades ovate, serrate, acuminate, large, uppermost usually sessile; flowers slender-pediceled, in terminal panicle or raceme; calyx somewhat 2-lipped; corolla with odor of lemon, 1-2 cm long, yellow, lowest lobe much longer than others, fringed; stamens much exserted, diverging; pollen-sacs divergent.

a. Stamens 2; racemes several, in a panicle. C. canadensis L., Richweed
a. Stamens 4; raceme usually 1. C. verticillata Baldw.

32. Perilla L.

*P. frutescens (L.) Britt. Rather coarse; blades ovate, serrate, purple, long-petioled; flowers small, in axils of bracts in terminal and axillary spikelike racemes; calyx becoming 2-lipped in fruit and distended on lower side; corolla white, about equally 2-lipped; stamens 4; pollen-sacs divergent.

SOLANACEAE, Nightshade Family

Herbs, shrubs, and woody vines; leaves alternate, simple to pinnately compound; flowers hypogynous, perfect, solitary or variously clustered; hypogynous disk often present; calyx 5-lobed or -cleft; corolla-margin almost entire to 5-lobed; stamens usually 5, on the corolla; carpels 2, united; locules usually 2, sometimes 3-5; style 1; fruit a berry or a capsule.

a. Shrubs with long branches or somewhat woody vines; flowers about 1.5 cm wide; berry about 1.5 cm long.
   b. Blades entire; anthers dehiscent longitudinally; corolla salverform or funnelform, red-purple or greenish; flowers 1-few in each axil; berry red-orange. 5. Lycium
   b. Some blades with basal lobes or leaflets; anthers opening by terminal slits or pores; corolla rotate, blue to purple, each petal with 2 greenish spots; flowers in clusters; berry red.
      1. Solanum
a. Herbs, not vines.
   b. Corolla funnelform, salverform, or tubular; fruit a capsule.
      c. Calyx tubular, 4-10 cm long, half or nearly half as long as trumpet-shaped corolla, abscising just above base. 6. Datura
      c. Calyx 2 cm long or less, not abscising just above base.
         d. Calyx 5-lobed; stamens of equal size. Nicotiana
         d. Calyx deeply 5-parted; 1 stamen sometimes smaller. Petunia
   b. Corolla campanulate to rotate; fruit a berry, fleshy or nearly dry.
      c. Corolla rotate, sometimes reflexed; anthers often connivent or connate.
         d. Anthers longitudinally dehiscent, with projections beyond pollen-sacs; plants not prickly.
            2. Lycopersicon
         d. Anthers dehiscent by terminal slits or pores, without projections beyond pollen-sacs; plants sometimes prickly. 1. Solanum
      c. Corolla not rotate; anthers neither connivent nor connate.
         d. Corolla pale blue; calyx 5-parted, segments somewhat sagittate at base; ovulary 3-5-loculed. 4. Nicandra
         d. Corolla yellow or white, sometimes with dark center; calyx 5-lobed, segments not sagittate at base; ovulary 2-loculed. 3. Physalis

1. Solanum L., Nightshade, Potato

Herbs and woody vines; corolla rotate, limb plaited, lobes 5; anthers sometimes connate or connivent around the style, dehiscent by terminal slits or pores; berry globose or ovoid, many-seeded, usually 2-loculed.

a. Plants prickly or stellate-pubescent or both.
   b. Corolla yellow; calyx very prickly; lower anther largest. *S. rostratum Dunal, Buffalo-bur.
   b. Corolla blue, violet, or white; calyx not or little prickly; anthers equal.

    c. Plants silvery-canescent, the hairs many-rayed; blades entire or repand-dentate; prickles few or none; rare. *S. elaeagnifolium Cav.
    c. Plants green, the hairs 4-8-rayed; blades large-toothed or shallowly lobed; prickles usually many; common. S. carolinense L., Horse-nettle
a. Plants not prickly, not stellate-pubescent.
    b. Stem climbing, somewhat woody; some blades lobed; corolla blue or purple; berry red. *S. dulcamara L., Bittersweet N.
    b. Stems not climbing, herbaceous.
        c. Blades pinnately compound or pinnately lobed; corolla blue or white.
            d. Stem wing-angled; leaves pinnately compound. *S. tuberosum L., Potato
            d. Stem not wing-angled; blades pinnately lobed. S. triflorum Nutt.
        c. Blades entire or sinuate-dentate; corolla white; berry black. S. nigrum L., Black N. The European S. nigrum is introduced in Ohio. The native plant, which has a similar appearance, has been distinguished as var. virginicum L. or as S. americanum Mill.

2. Lycopersicon Mill., Tomato

*L. esculentum Mill. Large, hairy, viscid; leaf-blades pinnate; corolla yellow, 1-2 cm wide, rotate; anthers connivent in a cone, each with a projection beyond the pollen-sac; fruit a berry, number of locules variable.

3. Physalis L., Ground-cherry

Herbs; leaf-blades entire to sinuate-dentate; flowers solitary in axils; calyx 5-lobed; corolla rotate to companulate, 5-lobed or -toothed; berry 2-loculed, enclosed in the much-enlarged calyx.

a. Corolla white or pale, slightly lobed; anthers yellow; berry red; fruiting calyx 3-5 cm long, red. *P. alkekengi L., Chinese-lantern
a. Corolla yellow or greenish-yellow, sometimes with dark center; fruiting calyx green, veins sometimes red or purple.
    b. Annual; filaments slender; peduncles in anthesis mostly 1 cm long or less.
        c. Stem glabrous or with short appressed hairs; calyx purple-veined, filled at maturity by the viscid purple berry. *P. ixocarpa Brotero, Tomatillo
        c. Stem densely pubescent or villous; berry yellow at maturity. P. pubescens L.
            d. Plant gray-green because of dense, partly glandular, pubescence; blades with sinuate-dentate margin. var. grisea Waterfall (P. pruinosa L.), Strawberry-tomato.
            d. Plant green, villous, hairs without glands; blades often entire toward base. var. integrifolia (Dunal) Waterfall
    b. Perennial; filaments wide and flat; peduncles in anthesis mostly 1-2 cm long.
        c. Calyx hairy only on the nerves; stem glabrous or hairs ascending; leaves glabrous or sparsely hairy. P. longifolia Nutt. (Incl. P. subglabrata Mackenz. & Bush)
        c. Calyx hairy over whole surface; stem and leaves hairy.
            d. Hairs of stem short, terete, glandular, or long, flattened, jointed, or the 2 kinds mixed; blades rounded or subcordate. P. heterophylla Nees
            d. Hairs of stem not glandular, rather sparse; blades usually tapering to base and decurrent on petiole. P. virginiana Mill.

4. Nicandra Adans., Apple-of-Peru

*N. physalodes (L.) Pers. Leaf-blades sinuate-dentate; flowers axillary, solitary; corolla light blue, 2-2.5 cm wide and long, shallowly 5-lobed; locules of ovulary 3-5; berry nearly dry, enclosed in bladdery calyx.

5. Lycium L., Matrimony Vine

*L. halimifolium Mill. Shrub with long branches, sometimes a vine, sometimes with thorns; blades lanceolate, entire; often tufts of leaves on short stems in axils; flowers axillary, solitary or clustered; corolla purplish; berry orange-red.

6. Datura L., Jimson-weed

Coarse herbs; blades ovate, entire to sinuate-dentate or lobed; calyx tubular, 5-toothed, soon abscising crosswise, base remaining; corolla white to violet, funnelform, toothed or shallowly lobed; ovulary 4-loculed.

a. Glabrous; corolla to 1 dm long; capsule prickly or smooth. *D. stramonium L., Common J.
a. Pubescent; corolla to 2 dm long; capsule prickly. *D. innoxia Mill.

SCROPHULARIACEAE, Figwort Family

Mostly herbs; leaves alternate or opposite, rarely whorled, without stipules; flowers hypogynous, perfect; calyx with 5, or rarely 4, lobes or divisions; corolla zygomorphic or rarely nearly regular, often 2-lipped, sometimes spurred or gibbous, lobes 5 or 4; stamens on corolla-tube, rarely 5, usually 4 or 2; when 4, a fifth sometimes present without an anther or as a vestige; when 2, 1-3 vestiges sometimes present; hypogynous disk usually present; carpels 2, united; ovulary 2-loculed, placentation axile; ovules many, rarely few; style 1; stigma 1 or 2-lobed; fruit a capsule, usually ovoid.

a. Anther-bearing stamens 5; corolla rotate, almost regular. 1. Verbascum
a. Anther-bearing stamens 4; fifth stamen absent, vestigial, or only a filament. (See third a.)
  b. Trees; flowers about 5 cm long, in large terminal panicles; leaves opposite or whorled.
    11. Paulownia
  b. Herbs.
    c. Corolla with slender spur on lower side.
      d. Flowers in terminal racemes; leaves entire, linear to lanceolate. 4. Linaria
      d. Flowers axillary, solitary or rarely in small clusters.
        e. Blades linear to spatulate; plants low, upright, much branched; corolla-throat with palate, but open; capsule asymmetric. 6. Chaenorrhinum
        e. Blades ovate to circular; plants diffuse, or vines; corolla-throat closed by palate; capsule symmetric.
          f. Blades nearly circular, shallowly lobed or coarsely toothed, palmately veined; slender glabrous vine. 2. Cymbalaria
          f. Blades pinnately veined, round- to triangular-ovate, cordate or hastate at base, entire or nearly so; pubescent, diffuse. 3. Kickxia
    c. Corolla without a spur, but sometimes gibbous or saccate.
      d. Corolla nearly regular, scarcely 2-lipped, lobes flat or nearly so, only slightly unequal; vestigial stamen minute or absent.
        e. Leaves alternate; flowers in one-sided raceme; corolla-lobes much shorter than tube.
        17. Digitalis
        e. At least the lower leaves opposite; flowers not in one-sided racemes; corolla-lobes about equaling tube or shorter.
          f. Corolla purple, pink-purple, or rarely white.
            g. Flowers sessile; blades lanceolate.
              h. Corolla dark purple; floral bracts small; blades coarsely toothed. 23. Buchnera
              h. Corolla pink-purple; floral bracts leaflike; blades entire or with 1-2 small basal lobes. 22. Gerardia
            g. Flowers pediceled; blades linear. 22. Gerardia
          f. Corolla yellow.
            g. Anthers not awned at base. 21. Seymeria
            g. Anthers awned at base. 22. Gerardia
      d. Corolla evidently 2-lipped.
        e. Upper lip narrow or folded lengthwise or galeate, enclosing the stamens, which ascend under it and are not or little exserted beyond it.
          f. Pollen-sacs unlike, unequal; floral bracts red, white, or yellow, usually cleft; blades longitudinally veined. 24. Castilleja
          f. Pollen-sacs equal and parallel; bracts not brightly colored; blades pinnately veined.
            g. Corolla white with yellow palate; blades entire or bracts toothed near base; fruit 1-4-seeded. 25. Melampyrum
            g. Corolla yellow or tinged with purple or red; blades toothed to pinnately parted; seeds several to many. 26. Pedicularis
        e. Upper lip not as above; stamens not enclosed in upper lip, but exserted, or included in throat, or enfolded in lower lip.

f. Blades pinnately parted; small, diffuse, pubescent; flowers axillary, corolla about 4 mm long. 15. Conobea
f. Blades serrate to entire.
  g. Flowers axillary, solitary; calyx-teeth much shorter than tube; palate prominent. 12. Mimulus
  g. Flowers in whorls in upper axils or in spikes, racemes, or panicles; calyx-lobes equaling or longer than tube.
    h. Antherless stamen equaling or longer than those with anthers, bearded; flowers in terminal panicles. 10. Penstemon
    h. Antherless stamen absent or smaller than the others.
      i. Flowers in terminal spikes or racemes.
        j. Corolla less than 1 cm long; plant to about 2 dm tall. 13. Mazus
        j. Corolla and plant larger.
          k. Corolla open, without palate, lobes very short or lower median one elongate; leaves alternate. 17. Digitalis
          k. Corolla-throat closed or nearly closed by palate.
            l. Blades entire, opposite or alternate; ovulary and capsule asymmetric; upper lip of corolla the longer. 5. Antirrhinum
            l. Blades serrate, opposite; ovulary and capsule symmetric; lower lip of corolla the longer. 9. Chelone
      i. Flowers clustered or solitary in upper axils, or in panicles.
        j. Corolla brown-purple, tube globose or ellipsoid; upper lip erect. 8. Scrophularia
        j. Corolla blue and white, middle lobe of lower lip folded, enclosing stamens and style. 7. Collinsia
a. Anther-bearing stamens 2; 1-3 others sometimes present as vestiges.
  b. Basal leaves ovate, blades 6-15 cm long, cauline leaves sessile, much smaller; corolla yellow; raceme dense and spikelike. 20. Besseya
  b. Basal and cauline leaves not as above, or basal leaves not present.
    c. Corolla rotate, tube short, lobes 4; capsule usually notched or obcordate, usually flattened; foliage leaves opposite, those subtending flowers alternate; style and calyx persistent. 19. Veronica
    c. Corolla-tube longer than corolla-lobes.
      d. Flowers in terminal spikelike racemes; stamens much exserted; leaves in whorls; corolla 4-lobed; tall. 18. Veronicastrum
      d. Flowers axillary; stamens included or barely exserted; leaves opposite; corolla 2-lipped; plants short.
        e. Calyx subtended by a pair of sepal-like bracts. 14. Gratiola
        e. Calyx not subtended by bracts. 16. Lindernia

1. Verbascum L., Mullein

Tall; ours biennial with winter rosettes; leaves alternate, entire, crenate, or toothed, cauline sessile; flowers in terminal branched or unbranched spikes or racemes; stamens 5, often dimorphic, 3 or all of the filaments villous.

a. Plant woolly with branched hairs; corolla yellow; flowers sessile or nearly so; 3 upper filaments densely villous, other 2 glabrous or somewhat villous.
  b. Cauline blades decurrent to next lower node; flowers 2 cm wide or less. *V. thapsus L., Common M.
  b. Cauline blades decurrent for a short distance or not at all; flowers 2.5 cm wide or more. *V. phlomoides L.
a. Plant not woolly, glandular pubescent in upper part; corolla white or yellow, about 2.5 cm wide; flowers pediceled; all filaments villous. *V. blattaria L., Moth M.

2. Cymbalaria Hill, Kenilworth Ivy

*C. muralis Gaert., Mey., & Scherb. Stem trailing, rooting at nodes; leaves alternate, palmately veined, nearly circular, shallowly lobed; flowers on long slender peduncles solitary in axils; corolla 1 cm long or less, blue with yellow palate, 2-lipped; spur short; stamens 4, included; capsule opening by 2 pores.

3. Kickxia Dum., Cancerwort

Prostrate, glandular-pubescent; leaves alternate or the lower opposite, short-petioled, ovate to circular; flowers small, axillary, long-stalked; corolla 6-8 mm long, 2-lipped, yellow and violet, spur about 5 mm long, palate present; stamens 4; carpels of capsule separately circumscissile.

a. Blades rounded to cordate at base, entire. *K. spuria (L.) Dum.
a. Blades truncate and often hastate at base, lower ones somewhat toothed. *K. elatine (L.) Dum.

4. Linaria Mill.

Erect, usually glabrous; leaves narrow, entire, upper alternate, lower and those of basal shoots often opposite or whorled; flowers bracted in terminal racemes; corolla spurred, 2-lipped; lower lip with palate or ridges; stamens 4, included.

a. Corolla yellow with orange palate, 2-3 cm long; capsule about 1 cm long; seeds flat, winged.
  *L. vulgaris Hill, Butter-and-eggs
a. Corolla blue with 2 paler ridges on lower lip, 1 cm long or less (rarely longer); capsule less than
  5 mm long; seeds wingless. L. canadensis (L.) Dum., Toadflax

5. Antirrhinum L., Snapdragon

*A. majus L., Common S. Leaves opposite or upper alternate, lanceolate, entire; flowers showy, in terminal racemes; corolla variously colored, 2-4 cm long, 2-lipped, palate prominent, tube gibbous at base on lower side; stamens 4; capsule asymmetric, opening by pores.

6. Chaenorrhinum Reichenb.

*C. minus (L.) Lange, Lesser Toadflax. Glandular-pubescent, much-branched; leaves narrow, obtuse, 1-2 cm long, lower opposite, others alternate; flowers solitary in axils, pediceled; corolla 5-8 mm long, blue-purple with yellow palate, 2-lipped, spurred; stamens 4; capsule pubescent, carpels unequal, each opening by terminal pore.

7. Collinsia Nutt.

C. verna Nutt., Blue-eyed Mary. Stem slender; leaves opposite, ovate or lanceolate, lower petioled; flowers few or solitary in upper axils; corolla 1-1.5 cm long, tube gibbous on upper side, deeply 2-lipped, upper lip usually white or pale blue, lower lip bright blue or rarely rose-color, lower middle lobe enclosing the 4 stamens; vestigial stamen glandlike, near base of corolla; seeds usually 4.

8. Scrophularia L., Figwort

Stem tall, erect, 4-angled; leaves opposite, serrate, ovate to lanceolate; flowers in large terminal panicles; calyx 5-lobed; corolla reddish-brown, 1 cm long or less, tube globose or ellipsoid, limb 2-lipped, middle lobe of lower lip reflexed, other lobes erect; stamens 4, usually included; vestigial stamen a flat scale on upper lip of corolla.

a. Vestigial stamen purple, usually longer than wide; flowering in late summer and autumn.
  S. marilandica L.
a. Vestigial stamen yellow-green, usually wider than long; flowering in early summer.
  S. lanceolata Pursh.

9. Chelone L., Turtlehead

C. glabra L. Upright, usually glabrous, of wet places; leaves opposite, serrate; flowers in terminal spikes, each subtended by sepaloid bracts; sepals separate or nearly so; corolla 2-3.5 cm long, 2-lipped, upper lip arched, emarginate, lower lip with densely pubescent palate; filaments with anthers 4, villous, longer than the glabrous antherless one; anthers heart-shaped, woolly.

a. Corolla cream or white or slightly pink at tip; blades lanceolate to ovate. var. glabra
a. Corolla greenish-yellow at tip; blades linear-lanceolate. var. linifolia Coleman
a. Corolla rose-purple at tip; blades broadly lanceolate. var. elatior Raf.

10. Penstemon Mitchell, Beardtongue

Leaves opposite, entire or toothed, upper often clasping; flowers showy, in panicles; corolla more or less 2-lipped, white, pale purple, or blue, 2 cm long or more; anther-bearing stamens 4, not exserted; antherless filament equaling or longer than the others, bearded; style elongate.

a. Corolla abruptly widened distally, consisting of narrower tube and gibbous throat, throat not ridged within, lower lip not much longer than upper; antherless filament slightly bearded.
  b. Corolla 2.5-3.5 cm long, purple outside, nearly white within; anthers glabrous; sepals tapering to spreading tips, margins little or not scarious; stem glabrous or slightly pubescent, the hairs more or less reflexed. P. calycosus Small
  b. Corolla white or faintly purple outside, usually with purple lines within; anthers usually with stiff white hairs on back; sepals caudate-tipped.
    c. Stem glabrous (or with some lines of short pubescence), shining; sepals scarious-margined, 5-8 mm long at anthesis; blades glabrous beneath; corolla 2-3 cm long. P. digitalis Nutt., Foxglove B.
    c. Stem slightly puberulent, dull; sepals little or not scarious-margined, 3-5 mm long at anthesis; blades puberulent beneath, at least on veins; corolla 1.5-2.3 cm long. P. alluviorum Pennell
a. Corolla little widened distally, not gibbous, tube slender, somewhat flattened, throat strongly ridged within, lower lip evidently longer than upper; antherless filament densely bearded.
  b. Corolla with dull purple throat and whitish lobes, tube closed by arching palate; stem with fine, spreading, sometimes gland-tipped, whitish hairs; lower surface of blades glabrate or midvein pubescent. P. hirsutus (L.) Willd.
  b. Corolla purple-lined within, throat not closed by palate; stem closely gray-puberulent or rarely villous.
    c. Corolla 2-3.5 cm long, purple outside; sepals much elongate in fruit; lower leaf-surface, or only midvein, sparsely pubescent with long hairs. P. canescens Britt.
    c. Corolla about 2 cm long, white outside; lower leaf-surface densely pubescent with short hairs. P. pallidus Small

11. Paulownia Sieb. & Zucc., Princess Tree

*P. tomentosa (Thunb.) Steud. Tree; leaves large, opposite or whorled, simple, entire or slightly lobed, viscid-pubescent; flowers in terminal panicles; calyx coriaceous, rusty-pubescent; corolla about 5 cm long, lavender-blue, with some yellow within, unequally 5-lobed, tube much longer than calyx; stamens 4, included; capsule ovoid, 3-4 cm long.

12. Mimulus L., Monkey Flower

Stem 4-angled, angles sometimes winged; leaves opposite, serrate or dentate, lanceolate or wider; flowers solitary in axils; calyx 5-angled, 5-toothed; corolla blue to pink or rarely white, 2-4 cm long, 2-lipped, throat nearly closed by palate; stamens 4; capsule enclosed by persistent calyx.

a. Leaves sessile; peduncles longer than calyx. M. ringens L.
a. Leaves petioled; peduncles mostly shorter than calyx. M. alatus Ait.

13. Mazus Lour.

*M. japonicus (Thunb.) Kuntze   Stem to about 2 dm; basal leaves present, cauline smaller; flowers in terminal raceme; calyx lobed to below middle; corolla blue, about 1 cm long, 2-lipped, lower lip much the larger.

14. Gratiola L., Hedge-hyssop

Low herbs of wet habitats; leaves opposite, sessile, entire or dentate; flowers solitary in axils; sepals nearly separate, subtended by a pair of sepaloid bracts; corolla white, yellow, or purple, 1-1.5 cm long, more or less 2-lipped; anther-bearing stamens 2, sometimes 2 vestiges present.

a. Stem essentially glabrous; leaf-blades tapering to base; pedicels stout, usually less than 1 cm long. G. virginiana L.
a. Stem pubescent, often glandular or viscid; pedicels slender, 1-2 cm long.

b. Leaf-blades broadest at their base. G. viscidula Pennell
b. Leaf-blades broadest above their base. G. neglecta Torr.

15. Conobea Aubl.

C. multifida (Michx.) Benth. Low, branching, pubescent; leaves deeply pinnatifid, segments oblong or linear; flowers small, axillary, pediceled, solitary or in 2's; sepals linear, almost separate; corolla lavender, 2-lipped, lips shorter than tube; stamens 4, of 2 lengths, included.

16. Lindernia All., False Pimpernel

Small, glabrous; leaves opposite, entire or remotely dentate; flowers stalked in the axils; sepals almost separate; corolla violet to white, about 1 cm long or less, 2-lipped; stamens 4, 2 with anthers, 2 (in ours) without.

a. Leaf-blades narrowed to base, longer than peduncles in the axils. L. dubia (L.) Pennell
a. Leaf-blades broadest at rounded base, shorter than peduncles in the axils. L. anagallidea (Michx.)
   Pennell

17. Digitalis L., Foxglove

Tall; leaves alternate; flowers large and showy, in terminal, sometimes 1-sided, racemes; corolla tube somewhat inflated, limb short, or one lobe sometimes much prolonged; stamens 4, included.

a. Corolla purple to white, spotted with dark purple, 4-5 cm long, lobes all short; calyx not woolly;
   blades crenate. *D. purpurea L.
a. Corolla cream-color, 3 cm long or less, lower median lobe much longer than the others; calyx
   woolly, blades entire or nearly so. *D. lanata Ehrh.

18. Veronicastrum Fabricius, Culver's Root

V. virginicum (L.) Farw. Tall, erect; leaves in whorls of 3-6, lanceolate, sharply serrate, acuminate; flowers in panicled dense spikelike racemes; calyx 4-5-parted; corolla almost regular, tube longer than the 4 lobes, usually white; stamens 2, long-exsert; capsule dehiscent by 4 terminal slits.

19. Veronica L., Speedwell

Foliage leaves opposite or rarely whorled, bracteal leaves usually alternate; flowers axillary and solitary or in racemes; calyx usually 4-, rarely 5-, parted; corolla blue, pink, or white, tube short, limb rotate, 4-lobed, slightly zygomorphic; stamens 2, one on each side of upper corolla-lobe, exserted; capsule usually flattened and notched at top; style and sepals persistent.

a. Raceme terminal on main stem, bracts alternate; or flowers in axils of alternate leaves which may
   be like foliage leaves or smaller.
  b. Flower-stalks shorter than subtending bracts.
    c. Blades lanceolate, acuminate, sharply serrate, 4-10 cm long; style twice as long as capsule;
       raceme dense. *V. longifolia L.
    c. Blades 1-3 cm long; style not twice as long as capsule.
      d. Corolla dark blue; leaves ovate, toothed, palmately veined; sepals unequal; capsule pubes-
         cent, deeply notched; style as long as depth of notch or a little more; whole plant pubes-
         cent. *V. arvensis L.
      d. Corolla white or pale blue, sometimes with darker streaks; leaves oblong, glabrous or
         somewhat pubescent.
        e. Corolla white or pale blue; calyx lobes oblanceolate, longer than pedicels; style very
           short. V. peregrina L., Purslane S.
        e. Corolla pale with darker stripes; calyx lobes elliptic, about equaling pedicels; style
           longer than depth of notch of capsule; hairs of stem short, upcurved.
           *V. serpyllifolia L., Thyme-leaf S.
  b. Flower-stalks equaling or longer than subtending bracts, bracts like foliage leaves but sometimes
     smaller; capsule wider than long.

c. Blades 3-5-lobed, evidently petioled; sepals ovate, long-ciliate; capsule thick, not or little notched; style 1 mm long or less. *V. hederaefolia L.
c. Blades crenate-serrate, short-petioled; sepals lance-ovate; capsule flattened, notched; sepals evidently 3-nerved.
  d. Capsule 7 mm wide or more, with wide triangular notch; style longer than depth of notch; pedicels in fruit usually more than 1.3 cm long. *V. persica Poir.
  d. Capsule about 5 mm wide, lobes rounded to notch; style equaling or exceeding depth of notch; pedicels in fruit usually less than 1.3 cm long. *V. agrestis L. (Including V. polita)
a. Racemes axillary or on branches, not terminal on main stem; leaves opposite.
  b. Calyx 5-lobed, lobes unequal; leaf-blades sessile. *V. latifolia L.
  b. Calyx 4-lobed.
    c. Stem pubescent, prostrate or flowering branches erect; blades short, not more than twice as long as wide, toothed; of dry habitats.
      d. Corolla about 1 cm wide; pedicels equaling or longer than calyx; blades ovate, truncate or rounded at base. *V. chamaedrys L.
      d. Corolla about 5 mm wide; pedicels shorter than calyx; blades elliptic, narrowed to base. *V. officinalis L., Common S.
    c. Stem glabrous or somewhat pubescent above; plants of wet habitats.
      d. Blades petioled, ovate or ovate-lanceolate, serrate; capsule thick, a little wider than long, not notched. V. americana (Raf.) Schwein.
      d. Blades of flowering stems sessile, entire or teeth low or remote.
        e. Blades linear to lance-linear, narrowed to base; capsule flat, wider than long, notched, much longer than calyx. V. scutellata L.
        e. Blades wider, upper sometimes clasping; calyx and capsule nearly equal in length.
          f. Blades ovate- to lance-oblong, upper somewhat clasping; capsule about as long as wide, not or scarcely notched; autumnal nonflowering stems with petioled leaves; corolla 4-5 mm wide, pedicels ascending. V. anagallis-aquatica L.
          f. Blades somewhat narrower than above; capsule notched; corolla less than 4 mm wide, pedicels spreading. V. comosa Richter

20. Besseya Rydb.

B. bullii (Eat.) Rydb. Stem unbranched; cauline leaves alternate, ovate, small, sessile; basal leaves much larger, petioled, ovate, cordate at base; flowers in terminal spikelike raceme; corolla yellowish, less than 1 cm long, 2-lipped; capsule somewhat flattened.

21. Seymeria Pursh, Mullein-foxglove

S. macrophylla Nutt. Erect; leaves opposite, the lower large, twice pinnatifid, smaller and less divided upward, uppermost lanceolate and entire; flowers sessile or nearly so, solitary in axils of upper leaves, hence in leafy or bracted spikes; corolla yellow, tube about 1 cm long, pubescent within, slightly longer than the 5 rounded nearly equal lobes; stamens 4, not quite equal, on middle of corolla tube, filaments short, villous; pollen-sacs not awned at base.

22. Gerardia L., False Foxglove, Gerardia

Leaves opposite or upper ones alternate; flowers in axils of upper leaves or bracts, hence in leafy or bracted spikes or racemes; calyx 5-lobed or 5-toothed; corolla large, showy, tube campanulate, lobes nearly equal; stamens 4, filaments usually pubescent; pollen-sacs sometimes awned at base; style elongate.

a. Corolla purple, pink, or rarely white; blades linear or, if wider, then auricled at base.
  b. Blades lanceolate, sessile, at least some with basal auricles; flowers sessile; calyx lobes longer than tube. G. auriculata Michx.
  b. Blades linear, not auricled at base; flowers pediceled.
    c. Pedicels shorter than the calyx.
      d. Corolla usually 2 cm long or less; style 1 cm long or less; sinus between calyx-lobes V-shaped or narrowly U-shaped. G. paupercula (Gray) Britt.
      d. Corolla usually 2 cm long or more; style 1.5 cm long or more; sinus between calyx-lobes broadly U-shaped, almost quadrate. G. purpurea L.

c. Pedicels or most of them longer than the calyx.
    d. Calyx-tube not reticulate-veined; corolla purple-pink; stem glabrous or nearly so.
       G. tenuifolia Vahl
    d. Calyx-tube reticulate-veined; corolla pink; stem narrowly winged and scabrous.
       G. skinneriana Wood
a. Corolla yellow; blades not auricled at base, broader than linear.
  b. Stem, outer surface of calyx, and capsule glabrous.
    c. Stem not glaucous; blades mostly entire, lower rarely lobed; pedicels in anthesis mostly
      shorter than calyx. G. laevigata Raf., Smooth F.
    c. Stem glaucous; lower blades deeply lobed or incised; pedicels in anthesis mostly longer than
      calyx. G. flava L.
  b. Stem, outer surface of calyx; and capsule pubescent or glandular puberulent.
    c. Plant glandless; capsule rusty-pubescent. G. virginica (L.) BSP., Downy F.
    c. Plant with some glandular pubescence; capsule stipitate-glandular. G. pedicularia L.

23. Buchnera L., Bluehearts

B. americana L. Stem rough-pubescent; leaves opposite, sessile, lanceolate, coarsely few-toothed, smaller upward; flowers bracted in terminal peduncled spike; calyx tubular, regularly 5-lobed; corolla nearly regular, salverform, about 2 cm long, deep purple, tube slightly curved, villous within, 5-lobed; stamens 4, included in tube.

24. Castilleja Mutis

C. coccinea (L.) Spreng., Indian Paint Brush. Leaves of basal rosettes mostly entire; stem-leaves alternate, usually 3-5-cleft; flowers in terminal spike, floral bracts usually 3-5-lobed, scarlet, yellow, or white at tip; calyx tubular, 2-3 cm long, deeply 2-lobed; corolla pale yellow, little longer than calyx, 2-lipped; stamens 4, not exserted.

25. Melampyrum L., Cow-wheat

M. lineare Desr. Leaves opposite, linear to ovate-lanceolate, entire, or bracteal ones toothed near base; flowers solitary in upper axils; corolla white and yellow, tube slightly widened upward, longer than lips; stamens 4, enclosed in upper lip; capsule flattened, asymmetric; seeds 2-4.

26. Pedicularis L. Wood Betony, Lousewort

Leaves toothed or pinnatifid, sometimes chiefly basal; calyx usually oblique; corolla strongly 2-lipped, upper lip arched, laterally flattened, enclosing the 4 stamens; lower lip with 2 longitudinal folds or crests; capsule asymmetric, pointed.

a. Flowering in spring; calyx split in front, otherwise nearly entire; leaves alternate; capsule flat-tened, lance-oblong, in fruit twice as long as calyx. P. canadensis L.
a. Flowering in autumn; each half of calyx with a leaflike appendage; leaves partly opposite; capsule ovoid, in fruit not twice as long as calyx. P. lanceolata Michx.

BIGNONIACEAE, Bignonia Family

Trees, shrubs, and vines; leaves opposite or whorled; flowers large, hypogynous, bisporangiate, clustered; calyx of united sepals; corolla showy, campanulate or funnelform, 5-lobed, sometimes 2-lipped; fertile stamens 4 or 2, on the corolla; 1-3 staminodes sometimes present; carpels 2, united; style 1, stigma 2-lobed; ovulary 2-loculed; placentae central; fruit an elongate capsule; seeds flat, winged.

a. Leaves simple; trees. 3. Catalpa
a. Leaves compound; vines.
  b. Leaves of 2 entire leaflets plus a terminal tendril. 2. Bignonia
  b. Leaves pinnately compound, leaflets 5 or more, toothed. 1. Campsis

1. Campsis Lour., Trumpet-creeper

C. radicans (L.) Seem. Climbing; with aerial rootlets; flowers with red-orange corollas 6-8 cm long, short-pediceled, in terminal clusters; corolla tube elongate, limb 5-lobed, slightly 2-lipped; fertile stamens 4; capsule 1 dm long or more.

2. Bignonia L., Cross-vine

B. capreolata L. Glabrous vine, high-climbing; branched tendrils terminating the leaves, the 2 foliaceous leaflets ovate-oblong; often a short branch in the axil bearing 2 small leaves; flowers in axillary clusters; corolla about 5 cm long, orange outside, paler in the throat; capsule linear, 1.5-2 dm long.

3. Catalpa Scop., Catalpa, Indian-bean

Trees; leaf-blades large, ovate, sometimes cordate at base, with prominent glands in axils of veins on lower side; flowers in terminal panicles; corolla white or yellow spotted with violet; anther-bearing stamens usually 2, vestiges 3; capsule slender, beanlike, 2-5 dm long.

a. Corolla 4 cm wide or less, much spotted, all lobes erose; wings of seed narrowed at tip; leaves ill-
   scented when bruised. *C. bignonioides Walt.
a. Corolla 5 cm wide or more, little spotted, lower lobe emarginate; wings of seed not narrowed at tip.
   *C. speciosa Ward., Hardy C.

MARTYNIACEAE, Unicorn-plant Family

Proboscidea Schmidel, Unicorn-plant

P. louisianica (Mill.) Thell. Coarse glandular-pubescent strong-scented herb; leaves opposite or upper alternate, long-petioled, cordate, entire or nearly so; flowers perfect, hypogynous, in terminal racemes; calyx 5-lobed, split to base on lower side; corolla 3-6 cm long, somewhat 2-lipped, 5-lobed; fertile stamens 4, on the corolla; carpels 2, united, style 1; ovulary 1-loculed, placentae 2, parietal, winged; proboscislike style-base of fruit splitting into 2 horns.

OROBANCHACEAE, Broom-rape Family

Low, somewhat fleshy, nongreen herbs parasitic on roots; leaves alternate, scalelike; flowers hypogynous, perfect, in spikes or racemes or solitary; calyx with 2-5 lobes or teeth; corolla of 4 or 5 united petals, 2-lipped; stamens 4, on corolla tube; carpels 2, united; style 1, slender; ovulary 1-loculed, with 2 or 4 parietal placentae; capsule many-seeded.

a. Flowers solitary at ends of leafless stalks; corolla white or lavender. 3. Orobanche
a. Flowers in spikes or spikelike racemes; stems with scalelike leaves.
   b. Stems slender, usually much branched; flowers not or scarcely crowded; scalelike leaves 2-4
      mm long, not overlapping. 1. Epifagus
   b. Stems stout, usually unbranched; flowers crowded in inflorescence.
      c. Calyx lobed, lobes longer than tube; stamens shorter than corolla. 3. Orobanche
      c. Calyx toothed, deeply split on lower side; stamens about as long as corolla. 2. Conopholis

1. Epifagus Nutt., Beech-drops

E. virginiana (L.) Bart. Slender glabrous branched yellow, brown, or purple herbs parasitic on roots of Fagus; leaves scalelike; flowers in panicled spikelike racemes, dimorphic, upper, with open corolla, mostly not seed-forming, lower cleistogamous and seed-forming; calyx 5-lobed; corolla of upper flowers about 1 cm long, cylindric, 4-lobed, of lower flowers caplike.

2. Conopholis Wallr., Squawroot

C. americana (L.) Wallr. Stout glabrous unbranched yellow or light-brown herb parasitic on roots of several woody species; scale leaves ovate, up to 2 cm long, many, overlapping; flowers sessile, bracted, in crowded spike; calyx irregularly toothed, split on lower side; corolla 10-15 mm long, 2-lipped; stamens exserted.

3. Orobanche L., Broom-rape

White, brownish, or purple, glandular-pubescent, parasitic on many species of plants; calyx 5-lobed or 5-cleft; corolla 15-25 mm long, white, lavender, or purple, tube usually curved and longer than the 2-lipped limb.

a. Flowering stem leafless, bearing a single flower; corolla-lobes rounded. O. uniflora L.
a. Flowers crowded in a spike or raceme; corolla-lobes acute. O. ludoviciana Nutt.

LENTIBULARIACEAE, Bladderwort Family

Utricularia L., Bladderwort

Herbs, aquatic or of wet soil; stems floating or prostrate on soil in shallow water, or subterranean in wet soil; leaves alternate, dissected or simple, some or all usually bearing tiny bladders; flowers hypogynous, perfect, in racemes or solitary; calyx deeply 2-lipped; corolla yellow in ours, 2-lipped, of 5 united petals, spurred at base, with prominent palate, upper lip entire or 2-lobed, lower lip entire or 3-lobed; stamens 2, on corolla tube; carpels 2, united; style short or none; stigma 2-lobed; ovulary 1-loculed; placenta central, free.

a. Plants of wet soil; leaves simple, linear-filiform, on delicate underground stems; flowers few, on a
   scape, each with bract and 2 bractlets; capsule covered by calyx. U. cornuta Michx.
a. Plants aquatic; leaves dissected; bractlets absent.
   b. Lips of corolla about equal; leaf-segments narrower after each forking, ultimate ones terete and
      filiform.
      c. Plants floating; leaves much dissected; scapes 4-many-flowered; spur about 2/3 as long as
         lower lip, curved. U. vulgaris L.
      c. Delicate; leaves once or twice forked; scapes 1-3-flowered; spur oblong-conic, much shorter
         than lower lip. U. gibba L.
   b. Lower lip of corolla about twice as long as upper; leaf-segments not much narrowed after each
      forking, ultimate ones flat.
      c. Leaf-segments entire; spur only a sac; palate hardly evident; bladders borne on leaves.
         U. minor L.
      c. Leaf-segments minutely serrate; spur nearly as long as lower lip; palate conspicuous; bladders borne on leafless branches. U. intermedia Hayne

ACANTHACEAE, Acanthus Family

Ours herbs; leaves opposite, simple, entire, without stipules; flowers hypogynous, bisporangiate; sepals 5, narrow, almost separate; corolla almost regular to 2-lipped; hypogynous disk present; stamens 2 or 4, on corolla tube; 1 or 3 staminodes sometimes present; carpels 2, united; style 1; ovulary 2-loculed; seeds on curved projections from the axile placentae; capsule splitting elastically.

a. Corolla 3 cm long or more, funnelform; stamens 4. Ruellia
a. Corolla about 1.5 cm long, zygomorphic, 2-lipped; stamens 2. Justicia

Ruellia L.

Leaves ovate or lance-ovate; flowers large, solitary or in few-flowered clusters, axillary or terminal on branches; corolla lavender-blue, funnelform, limb not quite regular, 5-lobed; stamens 4, lower pair the longer; a staminode usually present; capsule ovoid, somewhat flattened.

a. Calyx-lobes 2-4 mm wide, villous-ciliate, with flat tips, about as long as capsule; plant glabrous or
   glabrate.
   b. Flowers 1-3 on axillary peduncles from 1-3 median nodes, peduncle bearing 2 ovate leafy bracts;
      plant glabrous or glabrate. R. strepens L.
   b. Flowers cleistogamous in glomerules in many axils; corolla smaller, clavate-tubular.
      R. strepens L. forma cleistantha (Gray) McCoy
a. Calyx-lobes usually 1 mm wide or less, with slender tips, longer than capsule; flower clusters sessile or subsessile; bracts lanceolate; plant pubescent to glabrescent.
   b. Leaves sessile or subsessile. R. humilis Nutt.
   b. Leaves petioled. R. caroliniensis (Walt.) Steud.

Justicia L., Water-willow

J. **americana** (L.) Vahl   Growing in colonies in wet soil or in water; stem glabrous; blades lanceolate or narrower, tapering to base, sessile or subsessile, glabrous; flowers in crowded spikes or heads on long peduncles from upper axils; corolla white to pale violet, 2-lipped, lower lip spotted with dark purple; stamens 2, anthers with one pollen sac horizontal; capsule clavate.

PHRYMACEAE, Lopseed Family

Phryma L., Lopseed

P. **leptostachya** L. Herb; leaves opposite, ovate, toothed; flowers hypogynous, perfect, opposite on axis of terminal and usually also axillary spikes; calyx 5-toothed, 2-lipped; corolla rose, pale purple, or white, about 7 mm long, 2-lipped; stamens 4, two of them longer, inserted on corolla; carpels 2, united; style 1, stigma 2-lobed; locule and ovule 1; fruit indehiscent, in reflexed persistent calyx.

PLANTAGINACEAE, Plantain Family

Plantago L., Plantain

Ours herbs; leaves all basal or rarely cauline, usually entire, longitudinally ribbed; flowers bisporangiate or sometimes monosporangiate, hypogynous, sessile; corolla regular, tubular, 4- or rarely 3-lobed, dry-scarious, persistent on capsule; stamens 4 or rarely 2, on corolla; carpels 2, united; ovulary 2-loculed; ovules 1-many in each locule; style and stigma 1; fruit a pyxis.

a. Leaves opposite or whorled on aerial stem; blades linear; flowers in axillary peduncled heads.
  *P. **indica** L.
a. Leaves all in basal rosettes; flowers in spikes or heads on scapes.
  b. Blades linear; plants woolly or more or less villous throughout.
    c. Bracts conspicuous, much longer than flowers. P. **aristata** Michx., Bracted P.
    c. Bracts little, if any, longer than flowers. P. **purshii** R. & S.
  b. Blades lanceolate or wider.
    c. Corolla lobes erect above fruit; scapes long-villous. P **virginica** L.
    c. Corolla lobes spreading; scapes not or rarely villous.
      d. Blades lanceolate, strongly ribbed; bracts with wide hyaline tip and margin; the 2 sepals next to bract united; spike short-cylindric, conic when young. *P. **lanceolata** L., English P.
      d. Blades wider, usually less than twice as long as wide.
        e. Scape hollow; flowers not crowded; midvein of round-ovate bracts and sepals obscure or slender; often some blades cordate at base with veins branching from midvein; rare. P. **cordata** Lam.
        e. Scape solid; flowers crowded, covering axis except at base; bracts and sepals keeled; all veins arising from top of petiole.
          f. Capsule ovoid, circumscissile just below tips of sepals; bracts and sepals ovate, sepals rounded at tip. *P. **major** L.
          f. Capsule lance-ovoid, circumscissile near base; bracts lanceolate; sepals lanceolate, acute. P. **rugelii** Dcne.

RUBIACEAE, Madder Family

Herbs and shrubs; leaves entire, opposite or whorled and bases connected by stipules, or whorled without apparent stipules; flowers epigynous, bisporangiate; calyx minute or none or of usually 4 sepals; corolla rotate, funnelform, or salverform, regular, lobes usually 4; stamens as many as corolla lobes, inserted on corolla tube; carpels 2 (or 4), united; locules as many; ovules 1-several in each locule; style 1 or sometimes cleft; stigmas 2 or rarely 1 or 4; fruit separating into 1-seeded indehiscent carpels, or sometimes a berry or a capsule.

a. Upright shrubs or small trees; flowers in spherical heads; leaves opposite or in whorls of 3.
  7. Cephalanthus
a. Herbs, upright to prostrate.
  b. Prostrate evergreen with small round-ovate petioled opposite leaves; flowers in pairs, ovularies united; stigmas 4; fruit a berry. 6. Mitchella
  b. More or less upright, not evergreen; ovularies single; stigmas 1 or 2.

206

c. Leaves apparently whorled.
  d. Flower-heads subtended by an involucre of about 8 bracts united below; triangular sepals present. 1. Sherardia
  d. Flower-clusters not subtended by an involucre; sepals minute or absent.
    e. Corolla funnelform, tube about equaling lobes. 3. Asperula
    e. Corolla rotate, tube very short. 2. Galium
c. Leaves opposite.
  d. Flowers axillary, sessile; ovulary inferior; stipules dissected into bristles.
    e. Flowers in dense clusters. 4. Spermacoce
    e. Flowers solitary or in clusters of 2-3. 5. Diodia
  d. Flowers terminal and solitary or in terminal cymose clusters; ovulary half to two-thirds inferior. 8. Houstonia

1. Sherardia L., Field Madder

*S. arvensis L. Stem square, slender, prostrate at base; leaves lanceolate, pointed, in whorls of usually 6; flowers small, nearly sessile, in terminal heads, bracts below head about 8, united at base; sepals 4-6, persistent in fruit; corolla pink or blue, about 3 mm long, funnelform, 4-lobed.

2. Galium L., Bedstraw

Herbs with 4-angled stems; leaves whorled; flowers small, in axillary or terminal small or diffuse cymes, these sometimes in panicles; calyx minute or none; petals 4 or 3, slightly united at base; carpels indehiscent in fruit, subglobose, smooth or bristly.

a. Leaves with 3-5 longitudinal nerves, at least at base (lateral pair may be obscure), in whorls of 4; fruit usually bristly or pubescent.
  b. Corolla white; leaves linear or linear-lanceolate; flowers in compact many-flowered panicle; fruit hairy or glabrous. G. boreale L.
  b. Corolla greenish-white, yellowish, or purple; leaves oval, elliptic, or lanceolate; inflorescence few-flowered or loosely flowered.
    c. Flowers and fruits pediceled; leaves oval; plants hairy. G. pilosum Ait.
    c. Flowers and fruits sessile on the few divergent inflorescence-branches; plants somewhat pubescent or glabrous.
      d. Corolla-lobes acute, usually hairy on lower side; leaves obtuse, elliptic, hairy on nerves beneath. G. circaezans Michx., Wild Licorice.
      d. Corolla-lobes acuminate, glabrous, often purple with age; upper leaves lanceolate, acute. G. lanceolatum Torr.
a. Leaves with 1 longitudinal nerve, in whorls usually of 5-8, but sometimes of 4; fruit smooth or bristly; plants often leaning on other plants.
  b. Corolla yellow; fruit usually smooth.
    c. Leaves linear, in whorls of 6-8; flowers in compact terminal panicle. *G. verum L.
    c. Leaves elliptic or oblong, in whorls of 4, soon deflexed; flowers in small axillary cymes. *G. pedemontanum All.
  b. Corolla white.
    c. Fruit bristly; leaves with cuspidate tips.
      d. Stem retrorsely hispid on angles; leaves linear-oblanceolate, in whorls of 6-8, marginal cilia reflexed or divergent; annual. G. aparine L., Common Cleavers
      d. Stem glabrous or sparingly hispid; leaves elliptic, in whorls of 6 on main stem, marginal cilia ascending; perennial; fragrant in drying. G. triflorum Michx., Fragrant B.
    c. Fruit smooth or granular, not bristly.
      d. Leaves with acute or cuspidate tips, in whorls of 6 or more on main stem; corolla mostly 4-lobed.
        e. Leaf- and stem-margins and lower surface of midveins retrorse-scabrous; leaves narrowly oblanceolate; flowers in short cymes. G. asprellum Michx., Rough B.
        e. Leaf-margins, if scabrous, not retrorse-scabrous; stems smooth, pubescent, or sparsely retrorse-scabrous; panicle diffuse.
          f. Leaves mostly in whorls of 8 on main stem, of 6 on branches; stem glabrous to finely pubescent; roadsides and fields. *G. mollugo L.
          f. Leaves in whorls of 6 on main stem, of 4 on branches; stem minutely retrorsely scabrous; plants shining; woods. G. concinnum T. & G., Shining B.

d. Leaves with blunt tips; corolla 3- or 4-lobed.
   e. Corolla lobes mostly 3, obtuse; young internodes of stem retrorsely scabrous, stems much branched and matted; leaves usually scabrous on margins and midvein beneath.
      f. Leaves of main axis in whorls of 4-6, oblanceolate or narrowly elliptic; peduncles 1-, 2-, or 3-flowered (often 3-), peduncles and pedicels smooth and straight. G. <u>tinctorium</u> L.
      f. Leaves of main axis in whorls of 4, linear or linear-spatulate; peduncles mostly 1-flowered (or 3-), peduncles and pedicels slender, arcuate, and often scabrous. G. <u>trifidum</u> L.
   e. Corolla lobes mostly 4, acute; leaves of main axis in whorls of 4 (rarely 5 or 6).
      f. Stem mostly glabrous, nodes pubescent; leaves elliptic-oblong or lanceolate, margin ciliate with fine spreading hairs; fruit about 3 mm wide. G. <u>obtusum</u> Bigel.
      f. Stem glabrous, nodes pubescent; leaves linear-oblanceolate, 1.5 cm long at most, soon deflexed; fruit 1.5 mm wide. G. <u>labradoricum</u> Wieg.

3. <u>Asperula</u> L.

*<u>A</u>. <u>odorata</u> L., Sweet Woodruff. Stem erect; leaves oblanceolate, in whorls of mostly 8, 2-3 cm long; flowers small, in umbel-like terminal cluster; calyx minute or none; corolla white, 4-lobed; fruit with hooked bristles.

4. <u>Spermacoce</u> L.

<u>S</u>. <u>glabra</u> Michx. Stem spreading or decumbent; leaves opposite, elliptic, bases connected by stipular membrane dissected into bristles; flowers small, in dense axillary clusters; sepals 4; corolla white, bearded within, tube short, lobes 4; fruit obovoid, 2-loculed.

5. <u>Diodia</u> L., Buttonweed

Low, branched; leaves opposite; stipules connecting leaf-bases bearing long bristles; flowers small, sessile, 1-3 in each axil; sepals sometimes unequal, sometimes with minute teeth between; corolla pink-purple or white, funnelform or salverform; fruit splitting into 2 indehiscent carpels.

a. Style 2-cleft; sepals 2; fruit with longitudinal furrows; leaves lanceolate. <u>D</u>. <u>virginiana</u> L.
a. Style undivided; sepals 4; fruit not furrowed; leaves linear-lanceolate or linear. <u>D</u>. <u>teres</u> Walt.

6. <u>Mitchella</u> L.

<u>M</u>. <u>repens</u> L., Partridge Berry. Small evergreen herb with prostrate stem; leaves round-ovate, opposite, short-petioled, entire, stipules minute; flowers usually terminal, fragrant, each peduncle bearing 2 flowers with ovularies united, dimorphic as to relative length of stamens and styles; corolla white or pink, funnelform, lobes pubescent on upper side; each ovulary of 4 carpels, stigmas, and locules, 1 style; fruit an 8-seeded scarlet berry.

7. <u>Cephalanthus</u> L.

<u>C</u>. <u>occidentalis</u> L., Buttonbush. Shrub or small tree; leaves opposite or in whorls of 3, petioled, ovate- or lance-oblong, entire, tapering at tip and base, stipules connecting the bases; flowers small, in spherical peduncled clusters terminal or from upper axils; corolla white, funnelform, 4-lobed; style filiform, long-exsert; fruit obconic, angled, splitting into indehiscent carpels.

8. <u>Houstonia</u> L.

Small herbs, usually tufted; leaves opposite, bases connected by stipules; flowers small, partly epigynous, often dimorphic or trimorphic as to relative length of stamens and style, solitary or in terminal cymes; sepals 4, erect; corolla 4-lobed, blue, purple, or white; capsule 2-loculed, dehiscent across the top, few- to several-seeded.

a. Peduncles 1-flowered, filiform; corolla salverform, pale blue or violet-blue with yellow eye; fruit wider than long; stem-leaves few. <u>H</u>. <u>caerulea</u> L., Bluets
a. Flowers in cymes; corolla funnelform.
   b. Leaves 3-5-nerved, rounded to subcordate at base. <u>H</u>. <u>purpurea</u> L.

b. Leaves 1-nerved.
  c. Some of the flowers sessile; capsule obovoid, longer than wide, 2/3 inferior; short axillary fascicles usually present. H. nigricans (Lam.) Fern.
  c. Flowers pediceled; capsule half inferior.
    d. Basal leaves many at anthesis, ciliate; stipules rounded at summit, not fringed or toothed. H. canadensis Willd.
    d. Basal leaves not present at anthesis or not ciliate; stipules often fringed or toothed. H. longifolia Gaertn.

CAPRIFOLIACEAE, Honeysuckle Family

Shrubs and woody vines, rarely herbs; leaves opposite, usually simple, with or without stipules; flowers epigynous, usually perfect, in small clusters or in terminal cymes; sepals, petals, and stamens 5 or 4; calyx sometimes minute; petals united; corolla regular or zygomorphic, sometimes 2-lipped; stamens inserted on corolla; carpels 3-5, united; locules 1-5; style 1, sometimes very short; stigmas separate or united; fruit a berry, drupe, or capsule.

a. Small evergreen somewhat woody plant with trailing stems, upright branches forking bearing 2 flowers; petals 5, stamens 4. 4. Linnaea
a. Shrubs, woody vines, or erect herbs.
  b. Style very short or none; shrubs; flowers in large terminal cymes; corolla white, rotate or broadly campanulate.
    c. Leaves simple. 6. Viburnum
    c. Leaves pinnately compound. 7. Sambucus
  b. Style half as long as corolla or longer; corolla campanulate, tubular, or funnelform.
    c. Sepals linear, separate or almost, not less than 1/3 as long as corolla tube.
      d. Coarse herbs; blades entire; sepals not shorter than corolla tube. 5. Triosteum
      d. Shrubs; blades serrate; sepals 1/3-1/2 as long as corolla tube. 1. Diervilla
    c. Calyx entire or with triangular teeth or lobes, sometimes minute.
      d. Corolla nearly regular, campanulate to funnelform, 3-8 mm long; flowers in small terminal and axillary clusters; shrubs. 3. Symphoricarpos
      d. Corolla zygomorphic to nearly regular, rarely as short as 8 mm, sometimes tubular, sometimes 2-lipped; calyx minute; flowers in terminal and subterminal clusters subtended by connate-perfoliate leaves, or in pairs on axillary peduncles; shrubs and vines. 2. Lonicera

1. Diervilla Mill., Bush-honeysuckle

D. lonicera Mill. Shrub; blades ovate or lanceolate, acuminate, finely serrate; cymes small, terminal or in upper axils; sepals linear; corolla funnelform, nearly regular, 5-lobed, yellow, reddish in age, base somewhat gibbous; style exserted; ovulary 2-loculed; capsule 10-15 mm long, beaked.

2. Lonicera L., Honeysuckle

Woody vines or erect shrubs; blades usually entire; upper 1 or more pairs of leaves connate-perfoliate, subtending flower clusters, or flowers in pairs in axils, ovularies of the pair sometimes basally or completely united; calyx minute; corolla tubular or funnelform, shallowly to deeply 5-lobed, almost regular to 2-lipped; stamens 5; style elongate; ovulary 1-3-loculed; fruit a berry.

a. Upper leaves not connate-perfoliate; flowers in 2's on axillary peduncles.
  b. Corolla almost regular, sometimes saccate at base; erect shrubs.
    c. Ovularies united; fruit blue; rare. L. villosa (Michx.) R. & S.
    c. Ovularies separate; fruit red, rarely yellow.
      d. Corolla yellowish, saccate at base, lobes 1/2-1/3 as long as tube. L. canadensis Marsh., Fly H.
      d. Corolla scarcely saccate, lobes equaling or longer than tube.
        e. Corolla pink to white, glabrous; sepals entire; leaves glabrous beneath. *L. tatarica L., Tartarian H.
        e. Corolla white changing to yellow, pubescent; sepals ciliate; blades pubescent beneath. *L. morrowi Gray
  b. Corolla evidently 2-lipped, one lip narrow, the other 4-lobed or -toothed.

c. Vine; corolla 3-5 cm long, white or yellow, lips and tube about equal; fruit black.
   *L. japonica Thunb., Japanese H.
c. Upright shrubs.
   d. Flowers appearing early, before or with leaves; corolla creamy-white, 1.5 cm long; ovularies sometimes partly united; branchlets solid. *L. fragrantissima Lindl. & Paxt.
   d. Flowers appearing after leaves; fruit red.
      e. Branchlets hollow; ovularies separate; corolla white changing to yellowish.
         f. Peduncles longer than petioles; corolla often pink-tinged, tube gibbous.
            *L. xylosteum L.
         f. Peduncles shorter than petioles; corolla tube not gibbous. *L. maackii Maxim.
      e. Branchlets solid; ovularies partly or wholly united; peduncles longer than petioles; corolla yellow. L. oblongifolia (Goldie) Hook.
a. Uppermost leaves connate-perfoliate; flowers in terminal and subterminal clusters; vines; fruit red.
   b. Corolla tubular, nearly regular, red to orange-yellow, lobes short; stamens and style little exsert. L. sempervirens L., Trumpet H.
   b. Corolla 2-lipped; stamens and style exsert.
      c. Disk of connate blades almost circular, glaucous above and below, apices rounded or emarginate; separate blades obovate to almost circular; corolla pale yellow. L. prolifera (Kirchn.) Rehd.
      c. Disk of connate blades usually longer than wide, green above, apices acute; separate blades oblong to elliptic. L. dioica L.
         d. Blades glabrous beneath. var. dioica
         d. Blades pubescent beneath. var. glaucescens (Rydb.) Butters

3. Symphoricarpos Duham.

Shrubs; leaves oval or circular, entire or somewhat crenate or lobed; flowers small, in small clusters terminal and in upper axils; calyx short, 4-5-toothed; corolla 4-5-lobed, tube gibbous on one side; ovulary 4-loculed, 1 ovule in each of only 2 of them; fruit a berry.

a. Fruit coral-pink, 4-6 mm wide; corolla 3-4 mm long; style pubescent; leaves usually soft-pubescent beneath; branchlets villous or puberulent. S. orbiculatus Moench, Coralberry
a. Fruit white, 6-10 mm wide; corolla 6-8 mm long; style glabrous. S. albus (L.) Blake, Snowberry
   b. Branchlets and lower surface of blades pubescent. var. albus
   b. Branchlets and lower surface of blades glabrous. var. *laevigatus (Fern.) Blake

4. Linnaea Gronov., Twinflower

L. borealis L. var. americana (Forbes) Rehd. Small evergreen with trailing somewhat woody stem; leaves short-petioled, obovate to orbicular, 1-2 cm long, entire to crenate; flower fragrant, nodding, solitary on each of two forks of peduncle terminal on short upright branch; corolla slender-campanulate, nearly regular, shallowly 5-lobed, white tinged with rose, 10-15 mm long; stamens 4; ovulary 3-loculed; fruit dry, 1-seeded. Probably now extinct in Ohio.

5. Triosteum L., Horse-gentian

Herbs; leaves sessile or perfoliate, entire, ovate to oblanceolate; flowers solitary or in clusters of 2-4, axillary, sessile, 2-bracted; sepals linear, persistent, 5; corolla narrowly campanulate, the 5 lobes nearly equal, gibbous on one side; ovulary 3-5-loculed, one ovule in each locule; fruit dry.

a. Sepals with fine, usually even, pubescence on back and margins; flowers 1-3 per axil; corolla puberulent; blades soft-pubescent beneath. T. perfoliatum L.
   b. Middle leaves connate-perfoliate; stem pubescence a mixture of shorter, somewhat glandular, hairs and longer hairs. var. perfoliatum
   b. Middle leaves not connate-perfoliate; stem pubescence of mostly long spreading hairs; usually 1-2 flowers per axil. var. aurantiacum (Bickn.) Wieg. (T. aurantiacum Bickn.)
a. Sepals hispid-ciliate on margin, glabrous or sparingly hispid on back; flowers usually 1 per axil; corolla yellow; stem hispid with mostly long hairs up to 2.5 mm; leaves lanceolate or oblanceolate or wider, hispid-strigose above, not connate. T. angustifolium L.

6. Viburnum L., Arrowwood

Shrubs or small trees; leaves simple, sometimes stipuled; flowers small, in compound umbelliform or corymbiform clusters; sepals 5; corolla rotate or campanulate, deeply 5-lobed, regular, white; stamens exserted; style short or none; stigmas 3; fruit a drupe; in some species outer flowers of cluster have no flower parts except enlarged, often zygomorphic, corollas.

a. Leaf-blades palmately veined and lobed (rarely some unlobed).
   b. Outer corollas not enlarged; petioles without glands; fruit red, becoming blue-black; blades dotted beneath. V. acerifolium L., Maple-leaved A.
   b. Outer corollas enlarged; petioles with glands; fruit red. V. opulus L.
      c. Glands of petiole mostly sessile, tops concave. var. *opulus, Guelder-rose
      c. Glands of petiole clavate, tops convex. var. americanum Ait. (V. trilobum Marsh.), Highbush-cranberry
a. Leaf-blades pinnately veined, not lobed; marginal corollas enlarged only in V. alnifolium.
   b. Blades coarsely dentate, each vein or vein-branch ending in a tooth.
      c. Petioles 7 mm long or less; blades pubescent; linear stipules present. V. rafinesquianum Schultes, Downy A.
      c. Petioles more than 7 mm long.
         d. Blades round-ovate, cordate, pubescent beneath; 3 lowest pairs of lateral veins converging at junction with midvein. V. molle Michx.
         d. Blades broadly ovate, rounded at base or subcordate; 3 lowest pairs of lateral veins not converging.
            e. Petioles glabrous; stone with wide shallow ventral groove. V. recognitum Fern.
            e. Petioles pubescent, at least in the groove; stone with deep ventral groove. The several varieties of V. dentatum L.
   b. Blades serrulate, denticulate, or crenulate.
      c. Blades pubescent beneath with fascicled hairs.
         d. Blades round-ovate, cordate, rusty-tomentose; outer corollas enlarged. V. alnifolium Marsh., Hobblebush
         d. Blades oblong-ovate, gray-tomentose. *V. lantana L., Wayfaring Tree
      c. Blades glabrous, or the pubescence not of fascicled hairs.
         d. Blades crenulate; inflorescence definitely peduncled. V. cassinoides L.
         d. Blades sharply serrulate; inflorescence sessile or peduncle not more than 1 cm long.
            e. Blades long-acuminate; petioles, at least those near the inflorescence, wavy-margined. V. lentago L., Nannyberry
            e. Blades acute or obtuse; petioles winged but not or little wavy-margined.
               f. Midveins beneath and petioles rufous-tomentose; blades glossy above. V. rufidulum Raf., Southern Black-haw
               f. Midveins beneath and petioles glabrous or nearly so; blades dull above. V. prunifolium L., Black-haw

7. Sambucus L., Elderberry

Shrubs with large pith; leaves pinnately compound, stipules and stipels often present, leaflets serrate; calyx minute or rudimentary; corolla rotate, shallowly cup-shaped, regular, white; ovulary 3-5-loculed; style very short; drupe juicy.

a. Pith white; inflorescence usually wider than high, with 5 main rays from base; fruit black. S. canadensis L., Common E.
a. Pith brown; inflorescence usually higher than wide, with 1 main axis from base; fruit red. S. pubens Michx., Red E.

VALERIANACEAE, Valerian Family

Herbs; leaves opposite; flowers small, clustered, epigynous, bisporangiate or monosporangiate; calyx minute or expanding late and appearing pappuslike on the fruit; corolla tubular or funnelform, not quite regular, 5-lobed; stamens usually 3 (1-5), on corolla tube; carpels 3, united; style 1; stigma 1 or 2-3-branched; ovulary 3-loculed, one locule containing 1 ovule, the other 2 empty or failing to develop; fruit indehiscent, dry.

a. Leaves, or some of them, pinnately divided or pinnately compound; fruit 1-loculed, 1-seeded; calyx in fruit a circle of plumose bristles. Valeriana
a. Leaves simple, blades entire or somewhat toothed; fruit 3-loculed, only 1 locule containing a seed; calyx minute or absent. Valerianella

Valeriana L., Valerian

Cauline leaves pinnately compound or pinnately divided; flowers densely clustered, sometimes monosporangiate; calyx minute at anthesis; corolla usually gibbous at base, white or pink; calyx pappuslike in fruit, of several plumose bristles.

a. Corolla tube 1 cm long or more, slender; basal leaves cordate-ovate, simple, or with 1 pair of leaflets; cauline leaves of 3-7 ovate leaflets. V. pauciflora Michx.
a. Corolla tube 5 mm long or less; basal blades entire or, like the cauline, pinnately divided into lanceolate or linear segments.
  b. Basal blades linear-oblanceolate, entire or with 1 or 2 divisions, nearly parallel-veined, thick; cauline blades with a few pinnate divisions. V. ciliata T. & G.
  b. Both basal and cauline blades pinnately divided into several to many segments that are reticulate-veined. *V. officinalis L.

Valerianella Mill., Corn Salad

Stems dichotomously branched; leaves sessile, entire or somewhat toothed near base; flowers small, bracted; corolla almost regular, white or blue; stamens exserted; ovulary 3-loculed, only 1 containing an ovule.

a. Corolla blue; fruit flattened at right angle to septum between seed-containing carpel and empty ones, seed-containing one corky-thickened on back. *V. olitoria (L.) Poll.
a. Corolla white; fruit flattened parallel to septum described above.
  b. Fruit, when viewed from front or back, circular in outline, seed-containing carpel about 1/3 as wide as the 2 empty ones; walls of empty ones breaking leaving a circular opening; rare.
    V. umbilicata (Sulliv.) Wood
  b. Fruit, when viewed from front or back, longer than wide.
    c. Fruit square in cross section, seed-containing carpel only slightly wider than the 2 empty ones, empty ones separated by a shallow groove; common and abundant. V. intermedia Dyal
    c. Fruit triangular in cross section, seed-containing carpel twice as wide as the 2 empty ones, empty ones separated by a narrow groove. V. chenopodifolia (Pursh) DC.

DIPSACACEAE, Teasel Family

Dipsacus L., Teasel

*D. sylvestris Huds. Tall erect herb; stem prickly; blades prickly on midvein beneath, basal ones oblanceolate, crenate, cauline ones lanceolate, entire, sessile or connate-perfoliate; flowers epigynous, perfect, in peduncled dense ovoid or cylindric clusters subtended by an involucre of rigid prickly bracts; rigid bractlets among the flowers longer than the flowers; calyx about 1 mm long, silky; corolla pale purple, not quite regular, tubular, with 4 short lobes; stamens 4, inserted on corolla; carpels 2, only 1 developing; locule and ovule 1; fruit an achene.

*D. fullonum L., Fuller's Teasel, with tips of receptacular bracts recurved, is rare.

CUCURBITACEAE, Gourd Family

Prostrate or climbing vines, usually with tendrils; leaves alternate, palmately veined, usually lobed; ours monoecious; flowers monosporangiate, sometimes with vestiges of the other sporophylls, solitary or clustered; calyx 5-6-lobed; corolla campanulate, rotate, or salverform, regular, 5-6-lobed; stamens 3, more or less united; carpellate flowers epigynous; carpels usually 3, style 1, stigmas 2-3; ovulary 1-loculed with large parietal placentae, or 2-3-loculed with central placentae; fruit usually a pepo. Contains a number of cultivated plants including cucumber, watermelon, squash, pumpkin, and gourd.

a. Corolla yellow, 5-10 cm wide; ovulary and fruit smooth. 1. Cucurbita
a. Corolla white or greenish, 1 cm wide or less; ovulary and fruit prickly.

b. Petals 5; carpellate flowers in peduncled heads. 2. Sicyos
b. Petals 6; carpellate flowers solitary or few in axils. 3. Echinocystis

1. Cucurbita L., Gourd, Squash, Pumpkin

*C. foetidissima HBK., Wild P. Prostrate vine; leaves large, ovate, denticulate and somewhat lobed; flowers usually solitary in axils; corolla yellow, campanulate, 5-10 cm long; fruit subglobose, 5-10 cm wide; rare.

2. Sicyos L., Star-cucumber, Bur-cucumber

S. angulatus L. Climbing; blades shallowly 5-lobed or -angled; flowers small, white or greenish, the staminate in peduncled corymbs or racemes, the carpellate in peduncled heads, both clusters in same axil; calyx and corolla 5-lobed; fruit 1-seeded, indehiscent, dry, about 15 mm long.

3. Echinocystis T. & G., Wild Balsam-apple

E. lobata (Michx.) T. & G. Climbing; blades with 3-7, usually 5, lobes; staminate flowers in rather showy erect panicles or racemes, carpellate flowers solitary or a few together, usually the two kinds in same axil; calyx 6-lobed; petals 6, united a little at base; ovulary 2-loculed; fruit about 5 cm long, ovoid, covered with soft prickles, fleshy, becoming dry, dehiscent by 2 apical pores, 4-seeded.

CAMPANULACEAE, Bellflower Family

Herbs; leaves alternate, simple; flowers epigynous, bisporangiate, often showy, in racemes, spikes, or panicles, or a few together in leaf-axils; sepals 3-5; corolla regular or zygomorphic, of united petals, 5-lobed; stamens 5, separate and on base of corolla, or united in a tube around style; carpels united; ovulary with 2-3 locules; style 1; fruit a many-seeded capsule.

a. Anthers not united; corolla regular or nearly so, not split; ovary 3-loculed.
  b. Leaves palmately veined, usually no more than 2 cm long, round-ovate, cordate-clasping.
    1. Specularia
  b. Leaves pinnately veined, longer, linear to ovate, not cordate-clasping. 2. Campanula
a. Anthers united; corolla zygomorphic, 2-lipped, the tube split on one side nearly to base; ovulary
  2-loculed. 3. Lobelia

1. Specularia Fabricius, Venus's Looking-glass

S. perfoliata (L.) A. DC. Small; stem erect; leaves small, round-ovate, cordate-clasping, palmately veined; flowers sessile, axillary, 1-few at a node, those at upper nodes about 1 cm long, those at lower nodes cleistogamous and smaller; sepals 3-5; corolla blue or purple, nearly rotate; capsule opening by pores.

2. Campanula L., Bellflower

Juice milky; flowers often showy; sepals 5; corolla blue, violet, or white, campanulate, funnelform, or rotate; stigmas and locules 3; capsule opening by lateral pores.

a. Cauline blades linear or lance-linear, entire or sparingly crenulate; flowers few, long-stalked, solitary or in loose panicles; capsule-pores near base.
  b. Basal blades broadly ovate to circular; sepals linear, 4-10 mm long; corolla blue, 1.5 cm long or more. C. rotundifolia L., Harebell
  b. Basal blades lanceolate; sepals narrowly triangular, 4 mm long or less; corolla pale blue or white, 12 mm long or less. C. aparinoides Pursh
    c. Corolla pale blue; pedicels to 1 dm long. var. uliginosa (Rydb.) Gl. (C. uliginosa Rydb.)
    c. Corolla white or nearly so; pedicels usually shorter. var. aparinoides
a. Cauline blades lance-ovate, serrate or dentate; flowers usually many, in spikes or racemes.
  b. Corolla rotate; capsule-pores at summit; blades evenly serrate, acuminate at both ends.
    C. americana L., Tall B.
  b. Corolla campanulate or funnelform; capsule-pores basal; blades irregularly serrate-dentate, not acuminate at base. *C. rapunculoides L.

3. Lobelia L., Lobelia

Juice sometimes milky; flowers often showy, bracted, in racemes, spikes, or panicles; sepals, petals, and stamens 5; corolla-limb 2-lipped, the tube split on one side; stamens united in a tube around the style, some or all anthers hairy; carpels and locules 2; style 1; capsule opening at top.

a. Corolla red, 3-4 cm long, tube with lateral openings. L. cardinalis L., Cardinal L.
a. Corolla blue, purplish, or white.
  b. Corolla tube 1.2 cm long or more, with lateral openings.
    c. Calyx with large deflexed auricles in the sinuses; leaves glabrous or nearly so. L. siphilitica L., Great Blue L.
    c. Calyx without auricles or with very small ones; leaves densely puberulent. L. puberula Michx., Downy L.
  b. Corolla tube 1 cm long or less, without lateral openings.
    c. Cauline leaves linear, basal ones wider; lower lip of corolla not pubescent at base; capsule not inflated. L. kalmii L.
    c. Cauline leaves wider, some of them 1 cm wide or more; lower lip of corolla pubescent at base.
      d. Stem pubescent, usually much branched; capsule inflated. L. inflata L., Indian-tobacco
      d. Stem short-pubescent at base, usually glabrous above, usually not or little branched; capsule not inflated. L. spicata Lam.
        e. Auricles of calyx short, usually 1 mm long or less. var. spicata
        e. Auricles longer, filiform and deflexed. var. leptostachys (A. DC.) Mackenz. & Bush

COMPOSITAE, Composite Family

Ohio species herbs; rarely diecious; leaves alternate, opposite, or whorled; flowers bisporangiate, monosporangiate, or neutral, epigynous, in dense heads on a common receptacle surrounded by an involucre of bracts (phyllaries); pappus in position of calyx, of hairs, bristles, awns, teeth, or scales, or cuplike, or absent; corolla tubular and usually regular, usually 5-lobed, or ligulate and zygomorphic (tubular at base but flat and straplike above); heads discoid (of flowers with only tubular corollas), ligulate (of flowers with only ligulate corollas), or radiate (of a central disk of flowers with tubular corollas surrounded by one or more series of flowers with ligulate corollas, called ray flowers); flowers subtended by bracts (pales), receptacle then chaffy, or flowers bractless, receptacle then naked; stamens 5, inserted on the corolla, filaments separate, anthers usually united around the style or rarely separate or scarcely united; carpels 2, united; ovulary 1-loculed with 1 ovule; style 1, usually 2-branched at apex; stigmas 2; fruit an achene.

There are 3 main leads in the key, corresponding to the 3 types of heads.

a. Heads discoid.
  b. Pappus capillary (of soft hairs or hairlike bristles).
    c. Receptacle chaffy; stem ending in a dense cluster of tiny heads, this cluster subtended by branches which end in similar clusters; plants white-woolly, to 4 dm tall; rare. 14. Filago
    c. Receptacle not chaffy; heads not as above.
      d. Receptacle densely bristly-hairy.
        e. Phyllaries rigid, hooked at tip; stems and leaves not spiny. 49. Arctium
        e. Phyllaries not hooked at tip.
          f. Pappus bristles plumose; plants spiny. 51. Cirsium
          f. Pappus bristles not plumose.
            g. Phyllaries tipped with appendages that are pectinate, lacerate, fringed, toothed, scarious or hyaline, or rarely tipped with long spines; marginal flowers of head sometimes enlarged. 53. Centaurea
            g. Phyllaries linear to ovate, tipped with short spines; wings of stem and leaves spine-toothed. 50. Carduus
      d. Receptacle naked, sometimes pitted (or with some short hairs or bristles in Onopordum).
        e. Stem broadly winged, teeth of wing and of blades spine-tipped; pappus barbellate; plant white-woolly. 52. Onopordum
        e. Stem not winged.
          f. Pappus in 2 series, the outer much the shorter; corollas purple. 1. Vernonia
          f. Pappus-bristles all of approximately the same length.
            g. Pappus-bristles plumose or barbellate; leaves often resin-dotted.
              h. Heads in corymbiform clusters; corollas creamy-white. 4. Kuhnia

       h. Heads in racemes or spikes; corollas usually rose-purple. 5. <u>Liatris</u>
   g. Pappus-bristles neither plumose nor barbellate, sometimes scabrous.
       h. Leaves, except sometimes the upper ones, opposite or whorled. 3. <u>Eupatorium</u>
       h. Leaves alternate, or all basal.
           i. Main phyllaries in a series of about equal length, with sometimes some
               shorter ones at base.
               j. Blades triangular or hastate, serrate or dentate; involucre about 1 cm
                   high. 47. <u>Cacalia</u>
               j. Blades not as above.
                   k. Blades palmately veined; main phyllaries about 5. 47. <u>Cacalia</u>
                   k. Blades not palmately veined; main phyllaries more than 5.
                       l. Involucre about 5 mm high, cylindric. 48. <u>Senecio</u>
                       l. Involucre about 1.5 cm high, usually wider at base. 46. <u>Erechtites</u>
           i. Main phyllaries in several series of unequal length.
               j. Plants white-woolly; phyllaries white, pink, or yellowish.
                   k. Basal rosettes prominent at anthesis, blades oblanceolate to obovate,
                       differing in shape from the few narrow cauline ones; diecious.
                       15. <u>Antennaria</u>
                   k. Basal rosettes absent at anthesis or the blades not much different from
                       cauline ones.
                       l. Monecious; heads alike, outer flowers of head carpellate with fili-.
                           form corollas, a few inner flowers perfect with coarser 5-toothed
                           corollas. 17. <u>Gnaphalium</u>
                       l. Imperfectly diecious; carpellate corollas filiform, staminate ones
                           coarser and lobed; sometimes a few staminate flowers at center
                           of carpellate heads. 16. <u>Anaphalis</u>
               j. Plants not white-woolly.
                   k. Leaf-blades triangular-hastate. 47. <u>Cacalia</u>
                   k. Leaf-blades not triangular-hastate.
                       l. Phyllaries linear; cauline blades 1 cm wide or less. 12. <u>Aster</u>
                       l. Phyllaries lanceolate or wider, usually glandular; cauline blades
                           more than 1 cm wide. 13. <u>Pluchea</u>
b. Pappus none <u>or</u> pappus present but not capillary.
   c. Carpellate heads axillary, of 1 or 2 flowers in a closed hardened spiny or tuberculate involu-
       cre; staminate flowers on same plant but in separate heads, staminate heads in terminal
       clusters.
       d. Carpellate flowers solitary in an ovoid or top-shaped involucre bearing usually 1 series of
           tubercles or spines. 20. <u>Ambrosia</u>
       d. Carpellate flowers solitary in each of the 2 cavities of an oblong involucre which becomes a
           bur with many spines. 21. <u>Xanthium</u>
   c. Carpellate flowers, if present, not enclosed in a closed hardened involucre.
       d. Heads few-flowered, in glomerules which are subtended by and often exceeded by foliaceous
           bracts; pappus of triangular scales prolonged into terminal bristles. 2. <u>Elephantopus</u>
       d. Heads not in glomerules subtended by foliaceous bracts.
           e. Pappus of 2-6 rigid awns or teeth, mostly downwardly barbed; outer phyllaries some-
               times foliaceous. 34. <u>Bidens</u>
           e. Pappus none or a short cup or crown.
               f. Phyllaries tipped with appendages that are pectinate, lacerate, fringed, toothed, scar-
                   ious or hyaline, or rarely tipped with stout spines; outer corollas of head sometimes
                   enlarged. 53. <u>Centaurea</u>
               f. Phyllaries not as above.
                   g. Leaves, at least the lower, opposite.
                       h. Blades unlobed, lanceolate to ovate; heads 5 mm wide or less. 19. <u>Iva</u>
                       h. Blades large, lobed, with stipulelike appendages at base; heads more than 5 mm
                           wide; viscid-hairy. 22. <u>Polymnia</u>
                   g. Leaves alternate.
                       h. Heads small, 5 mm wide or less, mostly in spikes, racemes, or panicles;
                           blades undivided and narrow or white-tomentose, or pinnatifid. 44. <u>Artemisia</u>
                       h. Heads larger.
                           i. Receptacle conic; phyllaries much shorter than disk. 41. <u>Matricaria</u>

    i. Receptacle flat or convex; phyllaries little, if any, shorter than disk.
      j. Blades 1-3 times pinnately divided or dissected. 43. <u>Tanacetum</u>
      j. Blades oblong, crenate-serrate, sometimes lobed at base. 42. <u>Chrysan</u>-
        <u>themum</u>
a. Heads radiate. (See third a.)
 b. Pappus capillary (of soft hairs or hairlike bristles), rarely absent in ray-flowers.
  c. Corollas of ray flowers yellow or orange.
   d. Phyllaries imbricated in several series.
    e. Heads about 5 cm wide; blades velvety-pubescent beneath. 18. <u>Inula</u>
    e. Heads much less than 5 cm wide.
     f. Phyllaries dotted with conspicuous glands. 38. <u>Dyssodia</u>
     f. Phyllaries not gland-dotted.
      g. Pappus double, the outer series shorter. 7. <u>Chrysopsis</u>
      g. Pappus single; rays short and relatively few. 8. <u>Solidago</u>
   d. Main phyllaries in 1 series of about the same length, often some much shorter ones at base.
    e. Flowers appearing in early spring before leaves, heads solitary on scaly scapes; leaves
     basal, blades cordate-ovate, white-tomentose on lower surface. 45. <u>Tussilago</u>
    e. Flowers appearing after leaves; cauline leaves present, usually pinnatifid. 48. <u>Senecio</u>
  c. Corollas of ray flowers not yellow or orange.
   d. Ray flowers few, 4-5 (-8).
    e. Heads in a corymb; achenes covered with silky hairs, the upper hairs like a short outer
     pappus. 10. <u>Aster</u>
    e. Heads in a slender panicle. 8. <u>Solidago</u>
   d. Ray flowers more numerous than 8.
    e. Phyllaries obviously imbricated in several series of different lengths. 10. <u>Aster</u>
    e. Phyllaries of approximately the same length <u>or</u>, if of several series of different lengths
     then ray flowers scarcely longer than disk flowers. 12. <u>Erigeron</u>
 b. Pappus, at least of disk flowers, a circle of awns, chaffy or rigid bristles, or scales sometimes
  dissected into bristles, sometimes deciduous. (See third b.)
  c. Corollas of ray flowers white, pink, or blue.
   d. Leaves opposite, ovate to lance-ovate; corollas of ray flowers little exceeding disk-
    corollas, the flat portion about as wide as long. 35. <u>Galinsoga</u>
   d. Leaves alternate, lanceolate or narrower; flat portion of corolla of ray flowers linear.
    9. <u>Boltonia</u>
  c. Corollas of ray flowers yellow or orange, sometimes purple at base; plants gland-dotted.
   d. Leaves opposite, pinnately divided, ultimate segments linear; pappus of chaffy scales parted
    into bristles; phyllaries conspicuously gland-dotted. 38. <u>Dyssodia</u>
   d. Leaves alternate on the stem or all basal.
    e. Leaves all basal; pappus of thin scales. 36. <u>Actinea</u>
    e. Leaves alternate on the stem.
     f. Ray corollas narrowed to base, 3-lobed; phyllaries not glutinous but may be gland-
      dotted. 37. <u>Helenium</u>
     f. Ray corollas linear, not 3-lobed; phyllaries often glutinous; pappus of deciduous lin-
      ear scales. 6. <u>Grindelia</u>
 b. Pappus none, a short cup or crown, or a few teeth, awns, or scales (sometimes deciduous) that
  correspond to angles of achene, sometimes with tiny bristles or scales between.
  c. Receptacle chaffy, sometimes only near the center.
   d. Lower leaves opposite; blades large, lobed, small stipulelike appendages at base of petiole;
    rays yellow or whitish; plants viscid-hairy; achenes thick, forming on only ray flowers.
    22. <u>Polymnia</u>
   d. Without the above set of characters.
    e. Rays pale purple, rarely whitish, 2.5-6 cm long; long slender tips of chaff exceeding
     disk corollas. 29. <u>Echinacea</u>
    e. Rays white or rarely pink, at most 1.5 cm long. (See third e.)
     f. Leaves opposite, lanceolate to linear; seeds forming in both disk and ray flowers.
      27. <u>Eclipta</u>
     f. Leaves alternate.
      g. Leaves pinnately dissected into narrow, often filiform, segments.
       h. Rays white or pink, 2-3 mm long; heads less than 1 cm wide; disk corollas
        white. 39. <u>Achillea</u>

    h. Rays white, 5 mm long or more; heads more than 1 cm wide; disk corollas
      yellow. 40. <u>Anthemis</u>
   g. Leaves not dissected, at most slightly lobed.
    h. Rays 5, minute; stem not winged; outer phyllaries ovate; seeds forming in only
      ray flowers. 25. <u>Parthenium</u>
    h. Rays 1-5, less than 1 cm long; stem winged; outer phyllaries oblong; seeds
      forming in disk and sometimes also in ray flowers. 32. <u>Verbesina</u>
  e. Rays yellow, rarely brown or purple at base.
   f. Stems winged; seeds forming in disk-flowers and sometimes also in ray-flowers;
    seeds winged or wingless. 32. <u>Verbesina</u>
   f. Stems not winged.
    g. Receptacle conic or columnar.
     h. Leaves pinnately divided, segments narrow; plants more or less white-
      tomentose; rays shorter than width of disk; seeds forming in both disk and ray
      flowers. 40. <u>Anthemis</u>
     h. Plants not as above.
      i. Leaves opposite, serrate or dentate; receptacle conic; seeds forming in both
       disk and ray-flowers. 26. <u>Heliopsis</u>
      i. Leaves alternate.
       j. Blades pinnately parted, segments lanceolate to linear; disk gray.
        30. <u>Ratibida</u>
       j. Blades unlobed or, if lobed, segments broader than above; disk purple-
        brown or yellow-green. 28. <u>Rudbeckia</u>
    g. Receptacle flat or convex.
     h. Phyllaries of 2 kinds, the outer unlike the inner.
      i. Pappus of rigid, often barbed, awns; inner phyllaries striate, outer ones
       often foliaceous. 34. <u>Bidens</u>
      i. Pappus absent, a mere border, or 2 short awns or teeth, not of rigid awns.
       j. Rays about 5; leaves simple; seeds forming in ray-flowers only.
        24. <u>Chrysogonum</u>
       j. Rays more than 5; leaves often compound; phyllaries joined at base.
        33. <u>Coreopsis</u>
     h. Phyllaries in several series, all similar; pappus none or of 2 scales.
      i. Phyllaries ovate, blunt or, if acute, then blades perfoliate or pinnatifid;
       achenes flat, winged, forming in ray-flowers only. 23. <u>Silphium</u>
      i. Phyllaries acute, usually much longer than wide; achenes thick, wingless,
       forming in disk-flowers only; pappus deciduous. 31. <u>Helianthus</u>
 c. Receptacle naked.
  d. Plants small; leaves all basal; heads scapose; pappus none. 11. <u>Bellis</u>
  d. Plants not as above.
   e. Blades entire; pappus of several awns and short scales or bristles. 9. <u>Boltonia</u>
   e. Blades, or some of them, toothed, lobed, compound, or dissected.
    f. Blades dissected into narrow segments. 41. <u>Matricaria</u>
    f. Blades toothed to bipinnatifid, the teeth sometimes laciniate, but ultimate segments
     not narrow. 42. <u>Chrysanthemum</u>
a. Heads ligulate; plants with milky juice; flowers bisporangiate.
 b. Pappus none; corollas yellow.
  c. Cauline leaves present; peduncles short and slender. 54. <u>Lapsana</u>
  c. Leaves all basal; peduncles long, branched, enlarged upward. 55. <u>Arnoseris</u>
 b. Pappus of short scales; corollas blue or rarely white. 56. <u>Cichorium</u>
 b. Pappus of scales alternating with bristles; corollas yellow or orange. 57. <u>Krigia</u> (See fourth b.)
 b. Pappus capillary.
  c. Pappus-bristles, of some or all of at least inner flowers of head, plumose or barbellate.
   d. Blades linear-lanceolate, entire; branches of pappus-bristles interwebbed. 61. <u>Tragopogon</u>
   d. Blades toothed or lobed; branches of pappus-bristles not interwebbed.
    e. Cauline leaves present. 60. <u>Picris</u>
    e. Leaves all basal or very near base of stem.
     f. Receptacle chaffy. 58. <u>Hypochaeris</u>
     f. Receptacle naked or hairy. 59. <u>Leontodon</u>

        c. Pappus bristles simple or with some soft white scales.
            d. Achenes muricate above, with slender beaks; leaves all basal; heads solitary, on hollow
                scapes. 62. <u>Taraxacum</u>
            d. Achenes not muricate.
                e. Achenes flattened.
                    f. Involucre ovoid or campanulate; flowers 60 or more in a head; achenes beakless.
                        63. <u>Sonchus</u>
                    f. Involucre cylindric or conic; flowers rarely as many as 50 in a head; achenes with a
                        beak or a thickened summit. 64. <u>Lactuca</u>
                e. Achenes not flattened.
                    f. Corollas white, cream, or pink; heads slender, cylindric, drooping. 66. <u>Prenanthes</u>
                    f. Corollas yellow or orange.
                        g. Pappus white; main phyllaries in 1 series. 65. <u>Crepis</u>
                        g. Mature pappus gray, tawny, or brown; main phyllaries in 1 or more series.
                            67. <u>Hieracium</u>

VERNONIEAE, Ironweed Tribe

1. <u>Vernonia</u> Schreb., Ironweed

Leaves alternate, lanceolate, serrate; phyllaries in several series; receptacle flat or convex, naked; heads discoid, in corymbiform clusters; flowers perfect; pappus of 2 series, the inner long, capillary, the outer of short scales or bristles; corolla dark purple, rarely white.

a. Phyllaries without long threadlike tips, obtuse to acuminate.
    b. Heads with 30 or more flowers; stem and lower surface of blades pubescent; pappus purple to
        brown. V. <u>missurica</u> Raf.
    b. Heads with fewer than 30 flowers; stem and lower surface of blades glabrous to puberulent; pap-
        pus purple.
        c. Heads crowded in the inflorescence, branches erect; stem and leaves glabrous; blades dotted
            beneath. V. <u>fasciculata</u> Michx.
        c. Heads not crowded in the inflorescence, branches spreading; stem and leaves puberulent;
            blades little dotted beneath. V. <u>altissima</u> Nutt.
a. Phyllaries with long threadlike tips. V. <u>noveboracensis</u> (L.) Michx.

2. <u>Elephantopus</u> L., Elephant's-foot

<u>E. carolinianus</u> Willd. Leaves alternate or basal, obovate, crenate, petiole sometimes winged; phyllaries 8; heads discoid, of 2-5 perfect flowers, in small glomerules which are subtended by a few large foliaceous bracts; corolla blue-purple, unequally cleft; pappus of flat triangular scales tapering to rigid awns; achenes truncate, 10-ribbed.

EUPATORIEAE, Boneset Tribe

3. <u>Eupatorium</u> L., Boneset, Thoroughwort

Leaves in ours opposite or whorled; heads discoid, in a corymbiform cluster; phyllaries in 2-several series; receptacle flat, convex, or conic, naked; pappus capillary, usually scabrous; corollas white or whitish, blue, purple, or pink.

a. Phyllaries in 2 or more series of different lengths.
    b. Leaves whorled, ovate to lanceolate; corollas pink-purple or rarely white.
        c. Inflorescence or its divisions flat-topped; heads of 8 or more flowers; corolla pink-purple.
            <u>E. maculatum</u> L., Spotted Joe-Pye-weed
        c. Inflorescence with rounded top; heads of 8 or fewer flowers.
            d. Stem usually solid, somewhat glaucous, nodes usually purple; corolla pale purple to white.
                <u>E. purpureum</u> L., Joe-Pye-weed
            d. Stem usually hollow, glaucous, purplish; corolla pink-purple. <u>E. fistulosum</u> Barratt
    b. Leaves opposite or alternate, rarely whorled; corollas white.
        c. At least the lower leaves with elongate slender petioles. <u>E. serotinum</u> Michx.
        c. Leaves connate-perfoliate, sessile, or short-petioled.
            d. Leaves connate-perfoliate, rugose-veiny beneath, lanceolate, long acuminate at tip; rarely,
                the truncate or rounded leaf-bases not joined. <u>E. perfoliatum</u> L., Common B.
            d. Leaves not connate-perfoliate.

e. Blades tapering to base.
      f. Phyllaries narrow, elongate, with conspicuously scarious acuminate or mucronate tips. E. album L.
      f. Phyllaries not as above.
         g. Leaves often whorled, linear to lanceolate, entire to laciniate-toothed; conspicuous axillary fascicles present. E. hyssopifolium L.
         g. Leaves opposite, lance-elliptic, entire to serrate; without axillary fascicles; strongly 3-nerved. E. altissimum L.
   e. Blades with rounded or truncate bases.
      f. Essentially glabrous below inflorescence; phyllaries blunt; blades lanceolate, long-acuminate. E. sessilifolium L.
      f. More or less pubescent; inner phyllaries acute; blades ovate, blunt to acute, 2 or more lateral veins more prominent than the others. E. rotundifolium L.
         g. Blades truncate at base; prominent lateral veins joining midvein at base. var. rotundifolium
         g. Blades more rounded at base; prominent lateral veins joining midvein a little above base. var. ovatum (Bigel.) Torr. (E. pubescens Muhl.)
a. Main phyllaries in 1 series of about the same length (there may be some shorter ones at base); leaves opposite.
   b. Receptacle conic; corollas blue or violet; phyllaries linear, pointed, well separated above base. E. coelestinum L., Mistflower
   b. Receptacle flat; corollas white, pink, or pink-purple; bracts not so slender.
      c. Corollas pink; blades deltoid-ovate; phyllaries 2-nerved, acute. E. incarnatum Walt.
      c. Corollas white.
         d. Leaves small, blades 3-8 cm long, firm, base rounded to truncate, tip acute or obtuse, crenate or crenate-serrate, petioles mostly less than 1.5 cm long; rare. E. aromaticum L.
         d. Leaves larger, membranous, base rounded to acute, tip usually acuminate, serrate, petioles usually more than 1.5 cm long; common and abundant. E. rugosum Houtt., White Snakeroot

4. Kuhnia L., False Boneset

   K. eupatorioides L. Leaves alternate to opposite, linear to ovate, gland-dotted beneath; heads discoid; phyllaries in several series; receptacle naked, flat or nearly so; flowers bisporangiate; corollas cream-color; pappus bristles plumose, brown.

a. Glabrous or slightly hairy; blades entire or sparingly toothed. var. eupatorioides
a. Usually hairy; blades, except the upper, toothed. var. corymbulosa T. & G.

5. Liatris Schreb., Blazing-star

   Leaves alternate, often dotted, entire, narrow; heads discoid, in spikes or racemes; phyllaries in several series; receptacle naked, flat; flowers perfect; pappus of barbellate or plumose bristles; corollas pink-purple, sometimes white.

a. Pappus barbellate.
   b. Heads cylindric to campanulate, mostly with fewer than 18 flowers; spike crowded; corolla-tube glabrous or nearly so within.
      c. Phyllaries obtuse at tip, appressed; stem in inflorescence glabrous or nearly so. L. spicata (L.) Willd.
      c. Phyllaries acute, tips spreading; stem in inflorescence pubescent with long hairs. L. pycnostachya Michx.
   b. Heads subglobose, usually more than 18-flowered, not crowded; corolla-tube hairy within; phyllaries wide or with roundish tips.
      c. Phyllaries with wide scarious laciniate margins, often purple, glabrous. L. aspera Michx.
      c. Phyllaries without, or with only a narrow, scarious margin, finely pubescent. L. scabra (Greene) K. Schum.
a. Pappus plumose, branches at least 15 times as long as width of hairs.
   b. Phyllaries acute, the median ones with spreading or recurved tips; heads 5-12-flowered. L. squarrosa (L.) Michx.
   b. Phyllaries rounded, mucronate, appressed; heads 10-35-flowered. L. cylindracea Michx.

ASTEREAE, Aster Tribe

6. <u>Grindelia</u> Willd., Gumplant

Leaves alternate, resin-dotted; heads, in ours, radiate; phyllaries imbricate or almost equal; receptacle flat or convex, naked; pappus of 2-several deciduous awns or linear scales; corollas yellow; inner and sometimes also outer disk-flowers not seed-forming; ray-flowers carpellate, seeds forming.

a. Phyllaries scarcely resinous, tips not reflexed; leaves entire or with bristle-tipped teeth.
  <u>G</u>. <u>lanceolata</u> Nutt.
a. Phyllaries resinous, tips reflexed; leaves with blunt teeth.   *<u>G</u>. <u>squarrosa</u> (Pursh) Dunal
  b. Blades oblong to ovate. var. *<u>squarrosa</u>
  b. Blades linear-oblong to lanceolate. var. *<u>serrulata</u> (Rydb.) Steyerm.

7. <u>Chrysopsis</u> Ell., Golden-aster

Leaves alternate, sessile; phyllaries in several series; receptacle flat or convex, naked; pappus double, inner series capillary, outer series of short scales or bristles; heads radiate, corollas yellow; disk flowers bisporangiate, ray flowers carpellate, seeds forming in both.

a. Leaf-blades linear, entire, parallel veined. <u>C</u>. <u>graminifolia</u> (Michx.) Ell.
a. Leaf-blades broader, slightly serrate, pinnately veined. <u>C</u>. <u>mariana</u> (L.) Ell.

8. <u>Solidago</u> L., Goldenrod

Leaves alternate, blades toothed or entire; heads radiate, small, often many, variously clustered; phyllaries in several series; pappus capillary; receptacle naked; corollas yellow (in 1 species, white); disk flowers bisporangiate, ray flowers carpellate, seeds forming in both.

a. Heads in a flat-topped, corymblike inflorescence.
  b. Blades not dotted, basal ones long-petioled; achenes glabrous or hairy only at summit.
    c. Blades ovate-oblong or oval, usually rough; involucre 5-9 mm high; achene 10-15-nerved; drier habitats. <u>S</u>. <u>rigida</u> L., Stiff G.
    c. Blades lanceolate or linear; involucre 4-6 mm high; achene 3-7-nerved or -angled; wet habitats.
      d. Blades oblong-lanceolate, toothed or entire, much shorter upward, flat, not 3-nerved.
        <u>S</u>. <u>ohioensis</u> Riddell, Ohio G.
      d. Blades linear, not much smaller upward, folded, somewhat 3-nerved. <u>S</u>. <u>riddellii</u> Frank
  b. Blades dotted, entire, linear; achenes pubescent; basal and lower cauline leaves deciduous; receptacle fimbriate.
    c. Blades with 3 evident nerves; heads mostly peduncled; along Lake Erie, rare. <u>S</u>. <u>remota</u> (Greene) Friesner
    c. Blades with 3 evident nerves and usually 2-4 weaker ones; heads mostly sessile; general in Ohio. <u>S</u>. <u>graminifolia</u> (L.) Salisb.
a. Heads not in a flat-topped, corymblike inflorescence.
  b. Inflorescence long and narrow, of several to many short axillary clusters, or a more or less cylindrical panicle, branches not secund, not curved downward.
    c. Stem pubescent.
      d. Rays white or cream-color. <u>S</u>. <u>bicolor</u> L.
      d. Rays yellow. <u>S</u>. <u>hispida</u> Muhl.
    c. Stem glabrous, except sometimes in the inflorescence or a short distance below.
      d. Phyllaries with prominent spreading or recurved tips; involucre 7-9 mm high; rays 12-16; basal rosettes conspicuous. <u>S</u>. <u>squarrosa</u> Muhl.
      d. Phyllaries appressed.
        e. Inflorescence of short axillary clusters, all but uppermost shorter than subtending leaves; achene pubescent; blades acuminate.
          f. Blades ovate, sharply and coarsely serrate. <u>S</u>. <u>flexicaulis</u> L., Zigzag G.
          f. Blades lanceolate, serrate; stem glaucous. <u>S</u>. <u>caesia</u> L., Blue-stem G.
        e. Inflorescence a narrow terminal panicle; lateral clusters, except sometimes lowest, longer than subtending leaves; leaf-margins ciliate, basal blades, when present, tapering to winged petioles.
          f. Lower blades elliptic to oblanceolate, petioles clasping 1/2-3/4 circumference of stem; disk flowers 4-8; rays 1-8; lower part of stem often red; bog plants.
            <u>S</u>. <u>uliginosa</u> Nutt.

f. Petioles of lower blades not clasping as described on the opposite page; plants of other habitats; cauline blades lanceolate or oblanceolate.

    g. Cauline blades entire, many, finely reticulate, lateral veins obscure, lowest not broadest above middle; inflorescence not leafy. S. speciosa Nutt.

    g. Lower cauline blades sometimes toothed, slightly broadest above middle; inflorescence somewhat leafy. S. erecta Pursh

b. Inflorescence an ovoid or pyramidal terminal panicle, branches secund, some usually curved downward.

  c. Stem and leaves with short gray pubescence; panicle often 1-sided; blades oblanceolate, crenate-serrate to entire, sometimes faintly triple-nerved, apex acute to obtuse, basal ones persistent; short axillary tufts usually present. S. nemoralis Ait., Gray G.

  c. Without the above set of characters.

    d. One set of lateral nerves of blade obviously stronger than the others; blades lanceolate, acuminate, many, crowded; basal and lower cauline leaves deciduous before anthesis.

      e. Stem, at least below the inflorescence, glabrous, often glaucous; blades hairy on veins beneath (or glabrous in var. leiophylla Fern.). S. gigantea Ait.

      e. Stem puberulent or scabrous (or base glabrous), not glaucous; blades hairy to nearly or quite glabrous beneath. S. canadensis L. (Incl. S. altissima L.), Canada G.

    d. Lateral nerves of blade equally conspicuous or one pair only a little stronger.

      e. Blades pellucid-punctate, anise-scented when crushed, sessile, entire, lanceolate; stem below inflorescence glabrous or hairy in lines. S. odora Ait., Fragrant G.

      e. Blades not as above.

        f. Blades pubescent beneath, or blades rugose-veiny, or both; basal leaves usually deciduous before anthesis.

          g. Blades not rugose-veiny, thin, tapering at base, sharply serrate, hairy beneath, at least on veins; stem below inflorescence glabrous. S. ulmifolia Muhl.

          g. Blades strongly to weakly rugose-veiny, tapering or rounded at base, usually hairy beneath; stem hairy, scabrous, or glabrous; if stem and leaves glabrous, then blades strongly rugose-veiny. S. rugosa Ait.

        f. Blades glabrous beneath, not rugose-veiny; stem below inflorescence glabrous; basal and lower cauline leaves present at anthesis, basal narrowed to margined petioles.

          g. Upper surface of blades very rough, hairs minute, broad-based; stem angled; basal leaves very large. S. patula Muhl., Rough-leaf G.

          g. Upper surface of blades not as above; stem terete or nearly so, or striate.

            h. Lower petioles clasping stem 1/2 - 3/4 of its circumference; rays 1-8; disk flowers 4-8; bog plants. S. uliginosa Nutt.

            h. Lower petioles not clasping; blades acuminate; plants of other habitats.

              i. Basal blades oblanceolate to narrowly elliptic, somewhat serrate, gradually narrowed to petiole; upper cauline blades entire, often with tufts in axils; achenes short-hairy; general. S. juncea Ait., Plume G.

              i. Basal blades ovate to elliptic, rather abruptly narrowed to petiole; only uppermost cauline blades entire; achenes glabrous; rare. S. arguta Ait.

9. Boltonia L'Her.

B. asterioides (L.) L'Her. Leaves entire, narrow; heads radiate; phyllaries somewhat imbricate; receptacle conic or convex, naked; disk flowers bisporangiate, corollas yellow, ray flowers carpellate, corollas white, pink, or blue, seeds forming in both; pappus of short scales or bristles and 2-4 long awns or bristles; achenes flat, wing-margined.

10. Aster L. (Incl. Sericocarpus Nees), Aster

Leaves alternate; heads radiate; phyllaries usually in several series of different lengths; receptacle flat or convex, naked; disk flowers perfect, corollas usually yellow, sometimes becoming purple or brown, ray flowers carpellate, corollas white, pink, purple, or blue, seeds forming in both; pappus of capillary bristles, equal, or with a shorter outer series.

a. At least some of the leaves (basal and lower cauline, usually) having blades with cordate bases and petioles.

  b. Heads in a flat-topped corymb; basal blades cordate-ovate.

    c. Phyllaries and peduncles glandular; rays blue or violet, rarely white; tufts of large toothed cordate basal leaves abundant. A. macrophyllus L.

    c. Phyllaries and peduncles not glandular; rays usually white, or violet in age.
       d. Lower stem-leaves and the large ones of the abundant basal tufts serrate-dentate, sinus rectangular. A. schreberi Nees
       d. Lower stem-leaves coarsely toothed, acuminate, often smaller than those higher on stem, sinus cordate or rounded; stem zigzag in inflorescence; basal tufts usually absent. A. divaricatus L.
  b. Heads in an elongate panicle; rays usually blue or violet.
    c. Blade or widened base of petiole cordate-clasping; stem and leaves usually pubescent; bracts many on branches of rather stiff panicle; phyllaries minutely pubescent, tips rhombic. A. undulatus L.
    c. Neither blade not petiole-base cordate-clasping.
       d. Subulate or filiform tips of phyllaries spreading, green areas narrow and inconspicuous; petioles of lower stem-leaves winged; panicle often elongate, branches ascending.
         e. Stem glabrous or pubescent in lines. A. sagittifolius Wedemeyer
         e. Stem and under surface of blades closely gray-pubescent. A. drummondii Lindl. (A. sagittifolius var. drummondii (Lindl.) Shinners)
       d. Tips of phyllaries appressed, rhombic green areas conspicuous.
         e. Cauline blades entire or nearly so, firm.
           f. Petiole-length and blade-size abruptly smaller upward; only lowermost blades cordate at base; inflorescence many-bracted. A. azureus Lindl.
           f. Petiole-length and blade-size gradually smaller upward; most of the blades cordate at base. A. shortii Lindl.
         e. Cauline blades sharply serrate, thin or somewhat fleshy.
           f. Blades pubescent beneath, thin, sharply toothed; petioles usually not winged. A. cordifolius L.
           f. Blades glabrous, somewhat fleshy, very smooth, the cauline often narrowed to winged petioles, sometimes nearly entire. A. lowrieanus Porter. In var. lanceolatus Porter, cauline blades lanceolate.
a. None of the leaves having blades with cordate bases and petioles.
  b. Cauline blades sessile, clasping stem with auricled or cordate bases.
    c. Phyllaries glandular; blades entire or nearly so.
       d. Rays 40 or more, purple; branches of inflorescence glandular-pubescent; heads large; leaves crowded. A. novae-angliae L. New England A.
       d. Rays fewer than 40, blue, pink, or lavender; branches of inflorescence glandular or not.
         e. Blades strongly cordate-clasping at base, appearing almost perfoliate. A. patens Ait.
           f. Blades firm, usually not narrowed just above base. var. patens
           f. Blades thin, the larger slightly narrowed just above base. var. phlogifolius (Muhl.) Nees
         e. Blades somewhat clasping at base but not as above. A. oblongifolius Nutt.
    c. Phyllaries not glandular; blades entire to serrate.
       d. Blades sharply serrate, acuminate, usually with a narrow portion just above base; stem zigzag; rays pale blue or lavender. A. prenanthoides Muhl. Crooked-stem A.
       d. Blades entire or somewhat serrate, not narrowed just above base.
         e. Blades linear or linear-lanceolate, long-tapering at tip; phyllaries with slender green midvein; ligules pale blue or lavender; rare. A. junciformis Rydb.
         e. Blades wider; phyllaries not as above; achenes glabrous or nearly so.
           f. Stem and leaves glabrous, glaucous; phyllaries with rhomboid green tips; rays pale blue or lavender. A. laevis L., Smooth A.
           f. Stem and leaves not glaucous; stem usually stout and purple, hairy, the hairs ordinary or broad-based, to glabrous; phyllaries loose, slender and elongate, the outer ones green; rays purple to lavender. A. puniceus L.
  b. Blades sessile or petioled, not clasping or, if slightly clasping, then without cordate or auricled bases.
    c. Phyllaries rigid, margins of the subulate green tips inrolled; branches stiff, ascending to spreading; rays usually white. A. pilosus Willd.
       d. Stem and sometimes the leaves pubescent.
         e. Blades linear to narrowly lanceolate, upper subulate-tipped. var. pilosus
         e. Blades lanceolate, upper not subulate-tipped. var. platyphyllus (T. & G.) Blake
       d. Stem and leaves glabrous or nearly so.
         e. Heads many; branches of inflorescence spreading. var. demotus Blake
         e. Heads few; branches of inflorescence ascending. var. pringlei (Gray) Blake

c. Phyllaries flat, not as above.
  d. Rays absent; blades linear or nearly so, glabrous with ciliate margins; inflorescence open or spikelike; rare. A. brachyactis Blake
  d. Rays present, sometimes short.
    e. Rays hardly exceeding pappus, rolled at tip, more numerous than disk-flowers; blades linear to lanceolate, entire, rather fleshy; rare. A. subulatus Michx.
    e. Rays not as above, usually fewer than disk-flowers.
      f. Pappus in 2 series, the outer shorter.
        g. Blades linear, 1-nerved; stem and disk-flowers red; rays violet, rarely white; achene with long white hairs. A. linariifolius L.
        g. Blades lanceolate or wider, pinnately veined; rays usually white or cream; inner pappus-bristles clavate.
          h. Heads many, in compact corymb; involucre 3-5 mm high. A. umbellatus Mill.
          h. Heads scattered; involucre 5-7 mm high. A. infirmus Michx.
      f. Pappus in 1 series.
        g. Long silky hairs of achene with appearance of an outer pappus; rays 5-8.
          h. Blades linear to lance-oblong, longitudinally 3-nerved, entire; pappus white or pale. A. solidagineus Michx. (S. linifolius (L.) BSP.)
          h. Blades lanceolate to obovate, pinnately veined, some or all dentate; pappus becoming red-brown. A. paternus Cron. (S. asteroides (L.) BSP.)
        g. Hairs of achene, if present, shorter than above; rays usually more than 8.
          h. Inflorescence corymbiform; achenes sometimes glandular; rare.
            i. Blades toothed, crowded above; achene glandular. A. acuminatus Michx.
            i. Blades entire or nearly so.
              j. Disk white; outer phyllaries acute. A. ptarmicoides (Nees) T. & G
              j. Disk not white; phyllaries blunt, spreading. A. surculosus Michx.
          h. Inflorescence not corymbiform; achenes not glandular.
            i. Phyllaries blunt, squarrose, tips mucronate; heads many, small; bracts in inflorescence many, small; rare. A. ericoides L.
            i. Phyllaries usually acute, not squarrose.
              j. Heads small, 1 cm wide or less, along one side of elongate branches of inflorescence; rays white, lavender, or pink, or becoming pink.
                k. Limb of disk-corolla goblet-shaped, lobes 1/2 - 3/4 its length, soon recurved, purplish; inflorescence-branches often irregular in length, some very long; blades lanceolate to lance-elliptic, usually hairy on midvein beneath. A. lateriflorus (L.) Britt.
                k. Limb of disk-corolla funnel-shaped, lobed less than halfway; inflorescence-branches long, recurving, with many small leaves or bracts; blades narrow, glabrous beneath. A. vimineus Lam.
              j. Heads mostly larger than above; inflorescence-branches not or little one-sided; limb of disk-corolla lobed less than halfway.
                k. Stem tall, erect, furrowed and angled; involucre 7-10 mm high; basal leaves large, rough, coarsely toothed, petioles winged; rays blue or purple, 1-2 cm long. *A. tataricus L. f.
                k. Plants not as above.
                  l. Blades mostly entire, with revolute edge, firm, strongly reticulate beneath, areoles isodiametric; rays usually blue or lavender. A. praealtus Poir.
                  l. Blades not as above; if strongly reticulate beneath, then the areoles not isodiametric.
                    m. Heads long-peduncled, peduncles and inflorescence-branches with many small bracts; phyllaries mostly with rhombic green tips; rays usually blue or lavender; rare. A. dumosus L.
                    m. Peduncles usually short; inflorescence not conspicuously bracted; green tips of phyllaries not or seldom rhombic.
                      n. Slender bog plants; blades entire, linear or nearly so; involucre 5-8 mm high; rays lavender, blue, or white; rare. A. junciformis Rydb.
                      n. Not bog plants; blades mostly serrate, glabrous on lower surface; involucre 3-6 mm high; common and abundant. A. simplex Willd.

o. Rays white; involucre 4 mm high or more.
   p. Blades of main stem 1-4 cm wide. var. <u>simplex</u>.
   p. Blades of main stem elongate, 3-12 mm wide. var.
      <u>ramosissimus</u> (T. & G.) Cronq.
o. Rays lavender to white; involucre 3-4 mm high. var.
   <u>interior</u> (Wieg.) Cronq.

11. <u>Bellis</u> L. , European Daisy

*<u>B</u>. <u>perennis</u> L. Small plant; leaves basal or nearly so; heads radiate, one per scape; phyllaries in 1 or 2 series, equal; receptacle naked, conic; disk flowers bisporangiate, corollas yellow; ray flowers carpellate, corollas white or pink; pappus none.

12. <u>Erigeron</u> L. , Fleabane

Heads radiate or appearing discoid; phyllaries narrow, almost equal, or in several unequal series; receptacle flat or convex, naked; pappus capillary, sometimes lacking in ray-flowers, sometimes an additional row of shorter bristles or scales; disk-corollas yellow, ray-corollas white, pink, or blue.

a. Heads about 5 mm wide; ray-corollas equaling or slightly longer than disk; disk-flowers not more than about 20.
  b. Rays white; stem tall, erect, usually unbranched below inflorescence. <u>E</u>. <u>canadensis</u> L. , Horse-weed
  b. Rays purplish; stem low, diffusely branched. <u>E</u>. <u>divaricatus</u> Michx.
a. Heads 1 cm wide or more; ray-corollas much longer than disk-corollas; disk-flowers many.
  b. Cauline blades sessile or clasping; stem soft; pappus of ray-flowers elongate, capillary.
    c. Ray-flowers fewer than 100, ligules 1 mm wide or more, usually blue or white; heads few or solitary, 2.5 cm wide or more. <u>E</u>. <u>pulchellus</u> Michx. , Robin's-plantain
    c. Ray-flowers 100 or more, ligules 0.5 mm wide or less, pink to white; heads 1 to many, usually less than 2.5 cm wide. <u>E</u>. <u>philadelphicus</u> L. , Philadelphia F.
  b. Cauline blades narrowed to base; stem firm; pappus of ray-flowers none or of short scales or bristles; rays usually white.
    c. Cauline blades lanceolate, many, all but the upper sharply toothed; hairs of stem, except near top, spreading; basal blades elliptic to subcircular, up to 6 cm wide. <u>E</u>. <u>annuus</u> (L.) Pers. Daisy F. , White-top
    c. Cauline blades linear to lanceolate, entire or slightly toothed, few; hairs of stem usually appressed; basal blades oblanceolate, not more than 2.5 cm wide, entire or toothed.
      <u>E</u>. <u>strigosus</u> Muhl. , Daisy F. , White-top

INULEAE, Elecampane Tribe

13. <u>Pluchea</u> Cass. , Marsh-fleabane

<u>P</u>. <u>camphorata</u> (L.) DC. Tall; leaves alternate, blades elliptic to ovate, toothed; heads discoid; receptacle flat, naked; outer flowers of head carpellate, seeds forming, corollas filiform; central flowers of head bisporangiate, few, seeds usually not forming; corolla pink; pappus bristles rough.

14. <u>Filago</u> L.

*<u>F</u>. <u>germanica</u> (L.) Huds. , Cotton-rose. White-woolly; leaves alternate, entire, linear or nearly so; heads discoid, in small dense sessile cluster, sometimes subtended by branches ending in similar clusters; central flowers of head perfect, bractless, pappus capillary; outer flowers carpellate, bracted, seeds forming, outermost without pappus.

15. <u>Antennaria</u> Gaertn. , Pussy-toes, Everlasting

Woolly; diecious; with basal rosettes and stolons; heads discoid, solitary or in close clusters; phyllaries scarious, in several series, those of staminate heads with broad white petaloid tips; pappus capillary, in staminate flowers scanty, bristles often club-shaped, in carpellate flowers copious; receptacle naked, flat or convex.

a. Rosette leaves with only a midnerve, lateral nerves absent or evident only at base. <u>A</u>. <u>neglecta</u> Greene

b. Stolons elongate, procumbent, leaves few and small, eventually with terminal rosettes.
   var. neglecta
b. Stolons ascending at tip, leafy throughout. var. attenuata (Fern.) Cronq. (A. neodioica Greene)
a. Rosette leaves with 3-7 nerves.
   b. Heads solitary; stolons slender, leaves, except terminal ones, scalelike. A. solitaria Rydb.
   b. Heads clustered; stolons leafy and somewhat ascending. A. plantaginifolia (L.) Hook., Plantain-
      leaved E.
      c. Basal leaves and those of stolons bright green above and glabrous or nearly so from the first.
         var. arnoglossa (Greene) Cronq. (A. parlinii Fern.)
      c. Basal leaves and those of stolons dull green and pubescent above, glabrate when old.
         d. Involucre of carpellate heads 7 mm high or less. var. plantaginifolia
         d. Involucre of carpellate heads 8 mm high or more. var. ambigens (Greene) Cronq.
            (A. fallax Greene)

16. Anaphalis DC., Pearly Everlasting

A. margaritacea (L.) Benth. & Hook. Diecious or imperfectly so; blades lanceolate or linear, entire; white-woolly; heads discoid; phyllaries in several series, scarious, dry, mostly white; receptacle convex or flat, naked; pappus capillary; carpellate flowers with filiform corollas; staminate flowers with coarser tubular corollas, sometimes a few at center of carpellate heads.

17. Gnaphalium L., Cudweed

Woolly; leaves alternate, entire; heads discoid; phyllaries dry and scarious, white or colored, in several series; receptacle flat or convex; pappus capillary, bristles sometimes thickened at tip; outer flowers of head carpellate, corollas filiform; inner flowers bisporangiate, few, corollas coarser.

a. Base of pappus-bristles united, deciduous in a ring; involucre brown or purple; inflorescence spikelike; blades spatulate. G. purpureum L., Purple C.
a. Base of pappus-bristles not united, deciduous separately.
   b. Low, branching; involucre brown, 3-4 mm high; heads in many glomerules subtended and over-
      topped by leaves; phyllaries little imbricated. G. uliginosum L.
   b. Erect; involucre 5 mm high or more; heads in corymbiform or paniculate clusters; phyllaries
      white or yellow, imbricated.
      c. Blades clasping, decurrent on stem. G. macounii Greene
      c. Blades narrowed to base, not clasping, not decurrent. G. obtusifolium L.

18. Inula L., Elecampane

*I. helenium L. Large, coarse; stem pubescent; blades large, alternate, elliptic or ovate, vel-vety beneath, cauline clasping, basal petioled; heads radiate, large; phyllaries in several series, outer herbaceous; receptacle flat or convex, naked; ray flowers many, carpellate, slender; disk flowers per-fect; all corollas yellow; pappus capillary, scabrous.

HELIANTHEAE, Sunflower Tribe

19. Iva L., Marsh-elder

Heads discoid, small, green; phyllaries short and broad, in 1-3 series; receptacle small, chaffy; central flowers of head staminate, marginal ones few and carpellate; anthers almost separate; pappus none; achenes obovoid.

a. Heads in spikes, each head subtended by a lanceolate bract; phyllaries 5 or fewer. *I. ciliata Willd.
a. Heads in panicles, without bracts; phyllaries 10. *I. xanthifolia Nutt.

20. Ambrosia L., Ragweed

Monecious or rarely diecious; receptacle flat, with slender chaff; flowers monosporangiate, the two kinds in different heads; staminate heads small, many-flowered, discoid, in spikes or racemes, phyllaries united in a 5-12-lobed involucre, corolla 5-toothed; carpellate flowers below the staminate, in axils of leaves or bracts, without corolla, solitary in a closed globose-ovoid or top-shaped spiny or tubercled involucre; anthers almost separate; pappus none.

a. Blades entire or hastate; staminate heads sessile; staminate involucre zygomorphic. A. bidentata Michx.
a. Blades deeply lobed or cleft, or rarely some unlobed but not entire.
    b. Leaves opposite, blades palmately 3-5-lobed or some merely toothed; tall, very coarse plants. A. trifida L., Greater R.
    b. Leaves opposite or alternate, blades pinnatifid or bipinnatifid.
        c. Pubescent; blades thin, usually twice pinnatifid; fruiting involucre with beak about 1 mm long or more and 5-7 sharp spines about 0.5 mm long. A. artemisiifolia L., Roman R.
        c. Somewhat hoary; blades thick, usually once pinnatifid; fruiting involucre with beak and 1-4 tubercles about 0.5 mm long. *A. psilostachya DC. var. coronopifolia (T. & G.) Farw.

21. Xanthium L., Cocklebur

Leaves alternate, blades lobed, toothed, or repand; monecious; flowers monosporangiate, the two kinds in different heads; staminate heads discoid, small, many-flowered, in short spikes or racemes at end of branches, involucre of separate phyllaries, receptacle chaffy, corolla tubular, anthers separate; involucre of carpellate flowers closed, covered with hooked prickles, 2-loculed, 1 flower in each, corolla none; fruiting involucre (bur) with 1 achene in each locule; pappus none.

a. Blades lanceolate, tip and base acute, a 3-pronged yellow spine in axil of each. *X. spinosum L., Spiny C.
a. Blades broadly ovate, base cordate or truncate, without axillary spines. X. strumarium L., Common C.

22. Polymnia L., Leafcup

Tall, viscid-pubescent; blades large, thin, lobed; heads radiate or discoid; phyllaries 5; receptacle flat, chaffy, outer bracts large, partly enclosing the achenes; ray flowers carpellate, seeds forming, corolla white or yellow; disk flowers perfect, seeds not forming; pappus none.

a. Rays 5, white, very short to about 1 cm long, or none; petioles not winged nearly to base; achene 3-angled, 3-ribbed. P. canadensis L.
a. Rays 10-15, yellow, 1-2 cm long, rarely none; petioles winged to base or nearly to base; achene many-striate. P. uvedalia L.

23. Silphium L., Rosin-weed

Tall, coarse; juice resinous; heads radiate, large; phyllaries in a few series; receptacle flat, chaffy; ray flowers in 2-3 series, carpellate, seeds forming, corollas yellow; seeds not forming in disk flowers; achenes broad, flattened dorsally, winged at the edges, notched at top; pappus none or a tooth at top of each wing.

a. Leaves connate-perfoliate, opposite; stem square. S. perfoliatum L., Indian-cup
a. Leaves not connate-perfoliate.
    b. Lower leaves very large, margins toothed or sometimes pinnatifid; stem-leaves much smaller, bractlike; achene about 1 cm long. S. terebinthinaceum Jacq., Prairie-dock
    b. Stem leafy.
        c. Blades except uppermost pinnately or bipinnately parted; leaves alternate. S. laciniatum L., Compass-plant
        c. Blades toothed or entire, lanceolate; leaves whorled, or sometimes opposite or alternate. S. trifoliatum L.

24. Chrysogonum L.

C. virginianum L. Stem long-pubescent; leaves opposite, long-petioled, blades crenate; heads radiate, few or solitary on slender peduncles; receptacle flat, chaffy; phyllaries in 2 series of about 5 each, the outer foliaceous; ray flowers few, carpellate, seeds forming, corolla yellow; disk flowers perfect, seeds not forming; achene flattened dorsally; pappus a 2-3-toothed crown.

25. Parthenium L.

Leaves alternate; heads small, radiate; phyllaries in 2 to a few series; receptacle conic or convex, chaffy; ray-flowers about 5, carpellate, seeds forming, corollas short, broad, white, inconspicuous; achenes not forming in disk-flowers; pappus none or of 2-3 scales or awns; corolla persistent on achene.

a. Leaf-blades pinnatifid or bipinnatifid. *P. hysterophorus L.
a. Leaf-blades crenate-dentate. P. integrifolium L.

26. Heliopsis Pers., Ox-eye

H. helianthoides (L.) Sweet   Stem glabrous or scabrous; leaves opposite, blades ovate to lance-ovate; heads large, peduncled, radiate; phyllaries in a few series, nearly equal; receptacle conic, chaffy; ray-flowers carpellate, seeds usually forming, corollas yellow; disk-flowers bisporangiate, seeds forming; pappus none, or a few teeth, or a crownlike border.

27. Eclipta L.

E. alba (L.) Hassk. Stem spreading, sometimes rooting at nodes; leaves opposite, blades lanceolate or narrower, slightly serrate; phyllaries in 1 or 2 series; receptacle flat; chaff slender; heads small, radiate; ray-corollas white, minute; disk-flowers bisporangiate, seeds forming; achenes 3-4-angled or compressed; pappus none or a minute crown or 2 short teeth.

28. Rudbeckia L., Coneflower

Usually rough; leaves alternate, blades undivided, lobed, or pinnatifid; heads large, long-peduncled, radiate; receptacle conic or convex, chaff partly enclosing achenes; phyllaries in 2 or 3 rows; ray-flowers neutral, corollas yellow, orange, or partly purple; disk-flowers bisporangiate, seeds forming; pappus none or a short crown.

a. Disk-corollas green or yellow; style tips blunt; lower blades or some of them cleft or pinnatifid. R. laciniata L.
a. Disk-corollas purple or brown; blades 3-lobed or unlobed.
  b. Pappus none; pales with bristle tips; style-tips subulate, elongate. R. hirta L., Black-eyed Susan
    c. Basal and cauline blades ovate, coarsely toothed. var. hirta
    c. Basal blades oblanceolate, entire or finely toothed; cauline blades linear to oblanceolate. var. pulcherrima Farw. (R. serotina Nutt.)
  b. Pappus a crown; pales without bristle tips, under surface glabrous or pubescent, edges ciliate or not; style tips blunt and short.
    c. Pales glabrous, ending in a long cusp, lower cauline blades often 3-lobed. R. triloba L.
    c. Pales glabrous or pubescent, not ending in a long cusp; sometimes ciliate; blades not lobed. R. fulgida Ait.
      d. Basal blades narrowly elliptic, lanceolate, sessile, or petioles winged; rays 1-3 cm long. var. fulgida (R. tenax Boynt. & Beadle)
      d. Blades mostly ovate.
        e. Blades nearly entire or finely dentate, petioles hardly winged; rays 1-3 cm long. var. umbrosa (Boynt. & Beadle) Cronq. (R. umbrosa Boynt. & Beadle)
        e. Blades mostly coarsely dentate; rays 2.5-4 cm long. var. sullivantii (Boynt. & Beadle) Cronq. (R. speciosa Wenderoth)

29. Echinacea Moench, Coneflower

E. purpurea (L.) Moench, Purple C. Leaves mostly alternate, blades ovate or narrower; heads radiate, large, solitary or few, peduncled; phyllaries in 2-4 series; receptacle conic, chaffy, pales with elongate slender tip, longer than disk-flowers; achenes not forming in ray-flowers, corollas reddish-purple, up to 6 cm long; disk-flowers bisporangiate, purple, seeds forming; pappus a short dentate crown.

30. Ratibida Raf., Prairie Coneflower

Leaves alternate, blades pinnately parted; heads radiate, long-peduncled; phyllaries few; receptacle columnar, chaffy; ray-flowers neutral, corollas yellow or partly or wholly brown-purple; disk-flowers bisporangiate, seeds forming, corollas gray-green; pappus a crown, 1 or 2 teeth, or none.

a. Disk ellipsoid, length less than twice width; pappus none; achenes smooth; style-branches lanceolate, acute. R. pinnata (Vent.) Barnh.
a. Disk cylindric, length usually more than twice width; pappus of 1 or 2 teeth; achenes ciliate on inner margin; style-branches short and blunt. R. columnifera (Nutt.) Woot. & Standl.

31. Helianthus L., Sunflower

Coarse; heads large, radiate, peduncled; phyllaries in several series; receptacle flat to conic, chaffy; ray-flowers neutral, corollas yellow; disk-flowers bisporangiate, seeds forming; pappus of 2 or more thin chaffy scales.

a. Lobes of disk-corollas yellow.
 b. Stem leafy only in lower half, or phyllaries appressed and without long recurved tips; blades trinerved.
  c. Blades of middle and upper stem small and bractlike, of lower stem long-tapering to base.
    H. occidentalis Riddell
  c. Upper leaf-blades not conspicuously smaller than lower; phyllaries closely appressed and imbricated, broad, shorter than disk. H. laetiflorus Pers.
 b. Stem leafy; phyllaries with spreading tips.
  c. Disk small, usually 4-7 mm in diameter; blades thin, ovate-lanceolate, trinerved, lower surface resin-dotted, mostly opposite on main stem; stem glabrous; rays about 1 cm long, 5-8.
    H. microcephalus T. & G.
  c. Disk and rays larger.
   d. Blades lanceolate, tapering to base, mostly alternate, not or weakly trinerved.
    e. Stem glabrous below inflorescence, glaucous.
     f. Blades lanceolate, lower surface pubescent; lobes of disk-corollas glabrous or nearly so. H. grosseserratus Martens
     f. Blades linear-lanceolate; lobes of disk-corollas white-pubescent. H. X kellermani Britt. (grosseserratus X salicifolius)
    e. Stem pubescent, scabrous, or hirtellous; blades scabrous on upper surface.
     f. Disk 2-3 cm wide; blades often folded lengthwise, curved; heads in a raceme; petioles short. H. maximiliani Schrad.
     f. Disk smaller; blades not folded, not curved; heads in a corymb or panicle.
       H. giganteus L.
   d. Blades broader or, if lanceolate, then sessile or short-petioled, not tapering to base, mostly opposite, usually trinerved.
    e. Blades mostly opposite, widest at base or just above, rounded or cordate at base, sessile or nearly so (petiole to 1 cm long in H. hirsutus).
     f. Blades clasping; lower surface of blades and of phyllaries gray-pubescent; stem densely villous. H. mollis Lam.
     f. Blades not or scarcely clasping, sessile or short-petioled; stem glabrous to pubescent but not densely villous.
      g. Stem glabrous below inflorescence; leaves divaricate, sessile or nearly so, scabrous above, more or less pubescent below. H. divaricatus L.
      g. Stem usually pubescent; leaves ascending, sometimes with petioles to 1 cm long.
        H. hirsutus Raf.
    e. Petioles evident, often margined; blades tapering to base.
     f. Leaves mostly opposite, ovate-lanceolate, pale and usually soft-pubescent beneath, harsh above, thick, hard, entire or shallowly dentate. H. strumosus L.
     f. Leaves alternate above, median and lower or only lower opposite, usually ovate, usually sharply serrate or coarsely toothed.
      g. Stem glabrous, at least below; blades thin, smooth or slightly rough; veins slender; disk small, the corollas pubescent; usually in dry open woods. H. decapetalus L.
      g. Stem rough.
       h. Annual; receptacle flat or nearly so; leaves chiefly alternate. H. annuus L.
       h. Perennial, with coarse tuber-bearing rhizome; receptacle convex, blades thick and scabrous, midveins coarse and prominent, larger ones white; usually in moist sunny habitats. H. tuberosus L., Jerusalem Artichoke
a. Lobes of disk-corollas red, purple, or brown, or rarely yellow in cultivated H. annuus.
 b. Blades linear, usually less than 1 cm wide, margins usually revolute; disk about 1 cm wide or less. H. angustifolius L.

b. Blades wider, usually trinerved.
  c. Leaves mostly opposite, blades tapering to base, cauline ones gray-green, long-pointed at tip; receptacle convex. H. laetiflorus Pers.
  c. Blades cordate to truncate at base, mostly alternate, except lowermost; receptacle nearly flat.
    d. Leaf-blades often toothed; phyllaries ovate; pales not white-bearded; achenes glabrous or hairy at summit. H. annuus L.
    d. Leaf-blades mostly entire; phyllaries lanceolate; pales at center of disk white-bearded; achenes hairy all over. H. petiolaris Nutt.

32. Verbesina L. (Incl. Actinomeris Nutt.), Wing-stem

Stem more or less winged; heads radiate or rarely rays wanting; phyllaries few or in a few series; receptacle convex to conic, chaffy; ray flowers sometimes few or wanting; disk flowers bisporangiate, seeds forming; pappus of a few awns.

a. Leaves opposite; rays yellow. V. occidentalis (L.) Walt.
a. Leaves alternate.
  b. Rays white, 1-5; heads many, disk 7 mm wide or less. V. virginica L.
  b. Rays yellow, few to 15; disk wider than above.
    c. Heads 1-few; rays 8-15; phyllaries erect or spreading; blades sessile, soft-hairy beneath. V. helianthoides Michx.
    c. Heads usually many; rays 2-8; phyllaries soon deflexed; blades tapering to short petiolelike base; achenes usually winged. V. alternifolia (L.) Britt. (A. alternifolia (L.) DC.)

33. Coreopsis L., Tickseed

Leaves opposite or the upper alternate; heads radiate, long-peduncled; phyllaries in about 2 series, united at base; receptacle flat or slightly convex, chaff thin; disk-flowers bisporangiate, seeds forming; pappus of 2 awns or teeth, a short crown, or none.

a. Leaf-blades palmately divided or palmately compound, when sessile, appearing simple and whorled; rays yellow; achenes winged.
  b. Leaflets or leaf-segments linear or filiform. C. verticillata L.
  b. Leaflets or leaf-segments lanceolate.
    c. Leaves sessile, opposite, leaflets or divisions 3. C. major Walt.
    c. Leaves, at least the main ones, petioled, leaflets or divisions 3 or sometimes 5. C. tripteris L.
a. Leaf-blades unlobed or pinnately lobed or pinnately divided.
  b. Rays yellow with red-brown bases; disk dark; blades pinnately divided, lobes narrow; achenes wingless. *C. tinctoria Nutt.
  b. Rays yellow; disk yellow; achenes broadly winged.
    c. Cauline leaves on only lower part of stem, blades entire or with 1 or 2 pairs of small lobes. *C. lanceolata L.
    c. Stem leafy nearly to summit, blades mostly divided. *C. grandiflora Hogg

34. Bidens L. (Incl. Megalodonta Greene), Bur-marigold

Leaves opposite; heads radiate or discoid; outer phyllaries often foliaceous, inner of different appearance, usually striate; receptacle flat or nearly so, chaffy; disk flowers bisporangiate, seeds forming; corollas yellow; pappus of 2-6 usually rigid awns or teeth, usually barbed, often retrorsely.

a. Aquatic; submersed leaf-blades finely dissected, emersed ones serrate. B. beckii Torr. (M. beckii (Torr.) Greene)
a. Terrestrial or, if sometimes in or near water, then leaf-blades not finely dissected.
  b. Leaf-blades serrate, unlobed, or rarely 3-5-lobed.
    c. Tip of achene convex; rays large and showy or wanting; leaves sessile or connate-perfoliate; heads often nodding in fruit.
      d. Rays absent or present and to 1.5 cm long; achenes at maturity with pale thickened margin; heads usually nodding in fruit; chaff yellow-tipped; outer phyllaries usually exceeding disk. B. cernua L.
      d. Rays to 3 cm long; achenes at maturity without pale thickened margin; chaff red-tipped; outer phyllaries seldom exceeding disk. B. laevis (L.) BSP.

229

c. Tip of achene flat or concave; rays absent or small and scarcely exceeding disk; leaves more or less petioled; heads erect.

    d. Mature achenes toward center of head somewhat 4-angled, with prominent midrib; outer phyllaries mostly 2-6, usually not twice as long as head. B. connata Muhl.

    d. Mature achenes toward center of head flat, midrib obscure; outer phyllaries mostly 6-10, 2 or more times as long as head. B. comosa (Gray) Wieg.

b. Leaves pinnately compound or pinnately divided.

  c. Heads discoid or rays short and rudimentary.

    d. Leaf-blades 2-3 times pinnately divided; achenes linear, 4-angled, 2-3 times as long as phyllaries; heads few-flowered. B. bipinnata L., Spanish Needles

    d. Leaf-blades, or some of them, pinnately compound, segments broader than linear; achenes flat.

      e. Outer phyllaries glabrous, usually 4 (3-5); upper leaves often simple, lower and middle ones 3-foliolate; heads about 6 mm wide or less. B. discoidea (T. & G.) Britt.

      e. Outer phyllaries 5 or more, hairy on margin; heads larger.

        f. Outer phyllaries 5-10, margins sparsely hairy; disk-corollas orange; achenes dark brown or blackish, the inner ones 7-10 mm long. B. frondosa L.

        f. Outer phyllaries 10-15 or more, margins quite hairy; disk-corollas yellow; achenes pale brown, green, or dark, inner ones 9-17 mm long. B. vulgata Greene

  c. Heads radiate, rays exceeding disk.

    d. Achenes cuneate, without thin brittle margin; outer phyllaries 6-8, smooth or ciliate, usually not exceeding disk. B. coronata (L.) Britt.

    d. Achenes obovate, with thin brittle, sometimes erose, margin.

      e. Outer phyllaries 8-12, margins smooth or moderately ciliate, mostly shorter than inner. B. aristosa (Michx.) Britt.

      e. Outer phyllaries 12-25, margins coarsely ciliate, mostly longer than inner. B. polylepis Blake

35. Galinsoga R. & P.

Annuals; leaves opposite, margins of blades dentate or entire; heads small, radiate; phyllaries in 2 series, ovate; receptacle conic, chaffy; ray-flowers 4 or 5, little longer than disk, corollas white or pink, carpellate, seeds forming; disk-flowers bisporangiate, corollas yellow; pappus of fimbriate or laciniate scales, sometimes wanting in ray-flowers.

a. Pappus of ray-flowers wanting, of disk-flowers awnless. *G. parviflora Cav.

a. Pappus of both ray-flowers and disk-flowers awn-tipped. *G. ciliata (Raf.) Blake

HELENIEAE, Sneezeweed Tribe

36. Actinea Juss.

A. acaulis (Pursh) Spreng. Low plants; leaves all basal, oblanceolate to linear, 1-nerved; heads radiate, 3-4 cm wide, solitary on leafless peduncles; phyllaries in 2-3 series; receptacle convex, naked; corollas yellow; ray-flowers carpellate, disk-flowers perfect, seeds forming in both; pappus of several thin scales.

37. Helenium L., Sneezeweed

Leaves alternate, gland-dotted; heads usually radiate; phyllaries in 1-3 series; receptacle convex or conic, naked; ray-corollas yellow or brownish, 3-lobed; disk-flowers bisporangiate, seeds forming, corollas yellow or brown-purple, lobes glandular-hairy; pappus of several thin scales.

a. Leaves linear-filiform, not decurrent on stem. *H. amarum (Raf.) Rock (H. tenuifolium Nutt.)

a. Leaves broader, decurrent on stem.

  b. Disk corollas yellow. H. autumnale L.

  b. Disk corollas purple or brown. H. flexuosum Raf. (H. nudiflorum Nutt.)

38. Dyssodia Cav.

D. papposa (Vent.) Hitchc. Strong-scented; gland-dotted; leaves opposite, pinnatifid or bipinnatifid, segments linear; heads small; phyllaries gland-dotted, in 2 series, inner united; receptacle pitted;

corollas yellow; ray-flowers few, carpellate, seeds forming, corollas scarcely exceeding phyllaries; pappus of scales parted into bristles.

ANTHEMIDEAE, Chamomile Tribe

39. Achillea L., Yarrow

*A. millefolium L. Leaves alternate, blades pinnately dissected; heads small, radiate; phyllaries in several series; receptacle chaffy; ray-flowers few, carpellate, seeds forming, corollas white or pink; disk 2-4 mm wide, the flowers bisporangiate, seeds forming; pappus none.

40. Anthemis L., Chamomile

Leaves alternate, blades pinnatifid or pinnately dissected; heads radiate or discoid; phyllaries nearly equal or in several series; receptacle conic or convex, chaffy at least in the middle; disk flowers bisporangiate, seeds forming, corollas yellow; pappus none or a small crown.

a. Ligules yellow; tubes of disk-corollas flattened; seeds forming in ray-flowers; pappus a short crown. *A. tinctoria L.
a. Ligules white with yellow base; disk-corollas with spur on one side of achene. *A. mixta L.
a. Ligules white.
  b. Strong-scented; leaf-blades usually tripinnatifid; pappus none; achenes rough-tuberculate; receptacle chaffy only at center; no seeds forming in ray-flowers. *A. cotula L., Dog Fennel
  b. Odorless; blades once or twice pinnatifid; pappus a minute border; achenes smooth, about 10-nerved; seeds forming in ray-flowers. *A. arvensis L.

41. Matricaria L., Chamomile

Leaves alternate, blades pinnately dissected, segments narrow; heads radiate or discoid; phyllaries in a few series; receptacle conic or hemispheric, naked; ray-flowers carpellate, when present, seeds usually forming, corollas white; disk-flowers bisporangiate, seeds forming, corollas yellow; pappus none or a short crown.

a. Heads discoid; disk-corollas 4-lobed; phyllaries shorter than disk; receptacle conic; plants fragrant. *M. matricarioides (Less.) Porter, Pineapple-weed
a. Heads radiate; disk-corollas 5-lobed.
  b. Heads 3-4 cm wide; receptacle hemispheric; achenes strongly 3-ribbed; plants scentless or almost. *M. maritima L. var. agrestis (Knof) Wilmott
  b. Heads about 2 cm wide; receptacle conic; achenes faintly 3-5-ribbed; plants fragrant.
    *M. chamomilla L.

42. Chrysanthemum L.

Leaf-blades serrate, dentate, incised, or dissected, alternate; heads radiate or discoid; phyllaries in few to several series; receptacle flat or convex, naked; ray-flowers carpellate, seeds forming, corollas white in species listed below; disk-corollas yellow; pappus none or a cup of short scales.

a. Heads 4-6 cm wide, solitary. *C. leucanthemum L., Oxeye Daisy
a. Heads smaller, in clusters.
  b. Leaf-blades bipinnatifid; ray-flowers present. *C. parthenium (L.) Bernh., Common Feverfew
  b. Leaf-blades crenate or serrate or with a pair of lateral lobes at base; ray-flowers absent or very short; fragrant when crushed. *C. balsamita L.

43. Tanacetum L., Tansy

*T. vulgare L. Strong-scented; leaves alternate, blades twice pinnatifid, rachis winged and toothed or pinnatifid; heads many, discoid; phyllaries in several series; receptacle flat or convex, naked; corollas yellow; flowers bisporangiate, seeds forming; pappus a short crown.

44. Artemisia L., Wormwood, Mugwort

Usually aromatic, often tomentose; leaves alternate; heads small, discoid, panicled; phyllaries in a few series; receptacle small, naked or pubescent; outer flowers of head carpellate, inner bisporangiate, seeds forming in all or in only the outer; corollas green or yellow; pappus none.

231

a. Leaf-blades white-tomentose or silvery-pubescent on one or both surfaces.
  b. Blades silvery-pubescent; receptacle villous. *A. absinthium L.
  b. Blades white-tomentose; receptacle glabrous.
    c. Blades 2-3-pinnatifid, ultimate segments blunt and not more than 1 mm wide; somewhat shrubby. *A. pontica L.
    c. Blades entire to once pinnatifid (or sometimes twice), ultimate segments wider.
      d. Upper surface of blades green, glabrous or glabrate; blades pinnatifid, segments acuminate, entire or incised. *A. vulgaris L.
      d. Upper surface of blades white-tomentose; blades entire, toothed, or pinnately lobed. A. ludoviciana Nutt.
a. Leaf-blades glabrous or pubescent, not white-tomentose, not silvery-pubescent.
  b. Seeds not forming in central flowers; blades 2-3 pinnatifid, segments linear or filiform. A. caudata Michx.
  b. Seeds forming in all flowers; leaf-blades 1-2 pinnatifid.
    c. Heads sessile or nearly so; panicle narrow, of spikelike or glomerate erect branches; involucre 2-3 mm high. *A. biennis Willd.
    c. Heads peduncled, drooping; panicle open; involucre 1-2 mm high; plant sweet-scented. *A. annua L.

SENECIONEAE, Groundsel Tribe

45. Tussilago L., Coltsfoot

*T. farfara L. Low, with rhizomes; leaves basal, long-petioled; blades round-ovate, cordate at base, shallowly lobed and dentate, white-woolly beneath; heads radiate, 2-3 cm wide, solitary on a scaly white-woolly scape, appearing before leaves; principal phyllaries equal; receptacle naked, flat; corollas yellow; ray-flowers many, carpellate, seeds forming; seeds not forming in disk-flowers; pappus soft, capillary, copious.

46. Erechtites Raf., Fireweed

E. hieracifolia (L.) Raf. Stem rather soft, striate; blades alternate, sometimes clasping, lanceolate or oblong, serrate to lobed; heads discoid; involucre cylindric above widened base, main phyllaries equal; receptacle flat, naked; seeds forming at least in marginal flowers; pappus capillary, soft, white, copious.

47. Cacalia L., Indian-plantain

Tall, sometimes with milky juice; leaves alternate; heads discoid; main phyllaries equal; receptacle naked, flat, with sometimes a central projection; flowers bisporangiate, corollas white or pink, deeply 5-lobed; pappus soft, white, capillary.

a. Blades pinnately veined, broadly or narrowly triangular, the lower ones hastate; flowers and phyllaries many more than 5. C. suaveolens L.
a. Blades palmately veined or longitudinally ribbed; flowers and main phyllaries 5.
  b. Blades entire or nearly so, longitudinally 5-7-ribbed. C. tuberosa Nutt.
  b. Blades shallowly lobed and/or toothed, palmately veined.
    c. Stem glaucous; lower surface of blades glaucous. C. atriplicifolia L.
    c. Stem not glaucous, conspicuously grooved; blades green on both sides. C. muhlenbergii (Sch. Bip.) Fern.

48. Senecio L., Groundsel, Ragwort, Squaw-weed

Leaves alternate; heads radiate or discoid; main phyllaries about equal; receptacle naked, flat, often honeycombed; ray-flowers when present carpellate, seeds forming; disk-flowers perfect, seeds forming; corollas yellow; pappus soft, white, capillary.

a. Cauline leaves not diminishing rapidly upward, extending usually to the inflorescence; blades all pinnatifid or only coarsely toothed.
  b. Blades lyrate-pinnatifid, terminal lobe largest and rounded; rays showy. S. glabellus Poir.
  b. Blades not lyrate-pinnatifid; rays absent or short.
    c. Phyllaries glabrous, the smaller ones (and sometimes some of the larger ones) black-tipped; corollas usually all tubular. *S. vulgaris L.

    c. Phyllaries minutely pubescent, not black-tipped; rays present but short and inconspicuous.
      *S. sylvaticus L.
a. Cauline blades diminishing rapidly upward, those near the inflorescence small; basal or rosette
    blades undivided, cauline, except uppermost, often pinnatifid.
    b. Leaves, stems, and phyllaries (or their bases) more or less persistently white-woolly.
      c. Basal blades crenate or crenate-serrate, sometimes pinnatifid at base, white-woolly or gla-
        brate beneath, some of them to 4 cm long or more, ovate- to lance-oblong; phyllaries usually
        white-woolly at base. S. plattensis Nutt.
      c. Basal blades dentate, white-woolly beneath, smaller than above, elliptic or obovate; phyllaries
        white-woolly. S. antennarifolius Britt.
    b. Leaves, stems, and phyllaries glabrous or nearly so at maturity, except sometimes for scattered
    tufts of wool in axils.
      c. Basal blades ovate to circular, cordate to broadly rounded at base, petiole long and not winged;
        achene glabrous. S. aureus L., Golden R.
      c. Basal blades tapering to base, often to winged petioles.
        d. Basal blades obovate to circular, usually tapering at base to winged petioles, crenate
          across rounded tip, teeth sometimes deeper toward base; leaf tufts at ends of stolons
          abundant. S. obovatus Muhl.
        d. Basal blades usually not as above; stolons short or absent.
          e. Heads usually 20 or more, small; stem woolly at base; phyllaries tapering from above
            middle to tip; achenes hispid, at least on angles. S. smallii Britt.
          e. Heads usually fewer than 20; phyllaries tapering to apex from base; achenes glabrous,
            rarely hispid. S. pauperculus Michx.

CYNAREAE, Thistle Tribe

49. Arctium L., Burdock

    Large, coarse; leaves alternate, petioled, blades ovate, large, tomentose beneath; heads dis-
coid, large, globose; phyllaries in many series, rigid, tipped with hooked spines; receptacle bristly;
flowers bisporangiate, corollas purple to white; pappus of many short scabrous bristles.

a. Heads mostly sessile or short-peduncled, each branch of inflorescence racemelike; phyllaries gla-
    brous or somewhat tomentose; petioles of lower leaves mostly hollow. *A. minus (Hill) Bernh.
    Common B.
a. Peduncles mostly longer than heads.
    b. Involucre tomentose; petioles of lower leaves usually hollow; rare in Ohio. *A. tomentosum
    Mill.
    b. Involucre glabrous or nearly so; petioles of lower leaves usually solid. *A. lappa L., Great B.

50. Carduus L., Plumeless Thistle

    Coarse spiny herbs; leaves alternate, decurrent on stem, pinnatifid; heads large, discoid; phyl-
laries in several series, spine-tipped; receptacle flat or convex, bristly; corollas purple, pink, or
white, tube slender, lobes long and narrow; pappus of many capillary bristles.

a. Heads large and showy, 3-6 cm wide, solitary at ends of branches; peduncles naked above; middle
    and outer phyllaries 2-8 mm wide. *C. nutans L.
a. Heads usually not more than 2 cm wide, often clustered; stem leafy to or almost to the head; outer
    phyllaries rarely 2 mm wide. *C. acanthoides L.

51. Cirsium Mill., Thistle

    Coarse, spiny; leaves alternate and basal, blades dentate to pinnatifid; heads discoid, usually
large; phyllaries in many series, usually but not always spine-tipped; receptacle flat to convex, bristly;
flowers bisporangiate, or rarely monosporangiate; corollas pink, purple, yellow, or white, tube slender, lobes long and narrow; pappus of plumose capillary bristles.

a. Leaf-blades decurrent on stem in long spiny wings; phyllaries spine-tipped. *C. vulgare (Savi)
    Tenore, Common or Bull T.
a. Leaf-blades not or very slightly decurrent on stem.
    b. Flowers monosporangiate or some perfect ones present; rhizome extensive; involucre 1-2 cm
    high; phyllaries not or scarcely spine-tipped. *C. arvense (L.) Scop., Canada T.

b. Flowers bisporangiate; heads larger.
  c. Leaf-blades glabrous or pubescent, but not or only thinly tomentose, beneath, more or less green on both sides.
    d. Phyllaries not spine-tipped or only mucronate, with glutinous band; heads rather small, involucre 2-3.5 cm high, more or less tomentose. C. muticum Michx.
    d. Outer phyllaries spine-tipped; heads larger, involucre 4 cm high or more.
      e. Root solid; flowers fragrant; biennial. C. pumilum (Nutt.) Spreng.
      e. Root hollow, thickened; perennial. C. hillii (Canby) Fern.
  c. Leaf-blades densely white-tomentose beneath, glabrous or hispid above.
    d. Peduncles bearing only small bracts, without a circle of leaflike spiny bracts below involucre; heads small; involucre 1.5-2 cm high. C. carolinianum (Walt.) Fern. & Schub.
    d. Peduncles leafy, with a circle of prickly leaflike bracts below the involucre.
      e. Cauline blades unlobed or shallowly lobed. C. altissimum (L.) Spreng.
      e. Cauline blades deeply pinnatifid. C. discolor (Muhl.) Spreng.

52. Onopordum L., Scotch Thistle

*O. acanthium L. Large, coarse, white-tomentose, spiny; stem broadly winged by decurrent leaf-bases; leaves alternate, blades dentate to pinnatifid; heads discoid, large, solitary at ends of branches; phyllaries in many series, spine-tipped; receptacle honeycombed, sometimes bristly; flowers bisporangiate; corolla purple-pink to white, tube slender, lobes long and narrow; pappus of reddish barbellate capillary bristles.

53. Centaurea L., Star-thistle

Leaves alternate, blades entire to pinnatifid; phyllaries in many series, appressed, usually with fimbrillate, dentate, pectinate, or spiny, sometimes dark-colored, appendages; receptacle densely bristly, nearly flat; heads discoid, seeds forming in central flowers, marginal flowers sometimes with corollas enlarged and raylike, seeds not forming; pappus of bristles or scales or none.

a. Phyllaries tipped with long spines; stems winged; corollas yellow. *C. solstitialis L.
a. Phyllaries not spine-tipped.
  b. Leaf-blades deeply pinnatifid, segments long and narrow; phyllaries pectinate.
    c. Outer phyllaries pectinate only at tip, strongly ribbed; involucre less than 1.5 cm high.
      *C. maculosa Lam.
    c. Outer phyllaries pectinate 2/3 of way to base; involucre more than 1.5 cm high.
      *C. scabiosa L.
  b. Leaf-blades unlobed or some of them lobed but not deeply pinnatifid.
    c. Phyllaries entire, erose, lacerate, or toothed, not pectinate; if teeth are long and regular, then they are ascending; appendages dark or light.
      d. Marginal corollas not enlarged; inner phyllaries narrow with slender plumose tips, outer phyllaries wide with almost entire hyaline tips; plant much branched; pappus well developed. *C. repens L.
      d. Marginal corollas enlarged and raylike; inner phyllaries without plumose tips, outer phyllaries toothed or lacerate or, if entire, then brown, concave, with scarious margin.
        e. Phyllaries with ascending narrowly-triangular teeth, the outer toothed to base, the inner only at tip; pappus present; plants evidently white-tomentose. *C. cyanus L., Cornflower
        e. Phyllaries with broad brown concave scarious tips, margins entire or lacerate, middle ones sometimes bifid; pappus none. *C. jacea L.
    c. Phyllaries pectinate, the teeth at least as long as width of central portion; appendages dark.
      d. Pectinate appendage of a phyllary of outer few rows constituting only about 1/3 of whole phyllary; marginal corollas enlarged. *C. vochinensis Bernh.
      d. Pectinate appendage of a phyllary of outer few rows constituting about 2/3 of whole phyllary.
        *C. nigra L.
        e. Appendages dark; marginal corollas not enlarged. var. *nigra
        e. Appendages light; marginal corollas enlarged. var. *radiata DC.

234

54. Lapsana L., Nipplewort

*L. communis L. Leaves alternate, blades ovate, entire to pinnatifid; heads several to many; outer phyllaries short, inner in 1 series, about equal, calyxlike; receptacle naked; corollas all ligulate, yellow; pappus none.

55. Arnoseris Gaertn.

*A. minima (L.) Schweigg. & Koerte Small; leaves basal, blades oblanceolate, mostly toothed; heads solitary at ends of upwardly-thickened branches of scape; principal phyllaries in 1 series, about equal; corollas all ligulate, yellow; pappus none.

56. Cichorium L., Chicory

*C. intybus L. Basal and cauline leaves lanceolate, blades entire to pinnatifid, sometimes runcinate, cauline alternate, somewhat clasping; heads 2-4 cm wide, single or 2-3 together in upper axils; inner phyllaries erect, the outer shorter; receptacle naked or fimbrillate; corollas all ligulate, blue or rarely pink or white; pappus a crown of small scales.

57. Krigia Schreb., Dwarf Dandelion

Leaves all at base of stem or some cauline ones present also; blades entire to pinnatifid, obovate to linear; phyllaries few to many, in about 2 rows; receptacle naked; corollas all ligulate, yellow or orange; pappus double, the outer row of chaffy scales, the inner row of capillary bristles.

a. Flowering stem branched, bearing more than 1 head and 1-3 clasping leaves; pappus-bristles 15 or
   more. K. biflora (Walt.) Blake
a. Flowering stem unbranched, bearing 1 head; leaves all at or near base of stem.
  b. Pappus-bristles about 5-7. K. virginica (L.) Willd.
  b. Pappus-bristles 15 or more. K. dandelion (L.) Nutt.

58. Hypochaeris L., Cat's-ear

*H. radicata L. Leaves all basal or some lower cauline ones present, pubescent, oblanceolate, toothed or pinnatifid; heads 3-4 cm wide, at ends of stout scapes; phyllaries in several series; receptacle chaffy; corollas all ligulate, yellow; pappus of bristles, plumose or the outer ones barbellate; achene beaked.

59. Leontodon L., Hawkbit

Leaves basal, lanceolate or oblanceolate, entire to pinnatifid; heads terminal on branched or unbranched scapes; phyllaries of different lengths; receptacle fimbrillate, villous, or honeycombed; corollas all ligulate, yellow.

a. Scape sometimes branched, bracted; pappus of equal plumose bristles. *L. autumnalis L.
a. Scape unbranched, usually not bracted; pappus of 2 series unequal in length, or pappus of inner
   flowers different from that of outer flowers.
  b. Pappus of marginal flowers of short scales, that of inner flowers of plumose bristles or of both
    plumose bristles and shorter scales. *L. leysseri (Wallr.) G. Beck
  b. Pappus of 2 series, the inner of broad-based plumose bristles, the outer of short bristles.
   *L. hispidus L. (L. hastilis L.)

60. Picris L.

Stems leafy; leaves entire to pinnatifid; corollas yellow, all ligulate; phyllaries in 2 or more series; pappus of plumose bristles.

a. Achenes beakless or nearly so; outer phyllaries narrow. *P. hieracioides L.
a. Achenes with slender beak; outer phyllaries ovate or lance-ovate. *P. echioides L.

61. Tragopogon L., Goat's-beard

Glabrous, succulent; leaves alternate, entire, linear or lanceolate, parallel-veined, clasping; heads large, solitary at ends of branches; phyllaries in 1 series, nearly equal; corollas all ligulate; pappus of plumose interwebbed bristles; achenes beaked.

a. Corollas purple, shorter than the phyllaries; peduncles hollow and enlarged below the heads.
    *T. porrifolius L., Salsify.
a. Corollas yellow.
  b. Corollas shorter than phyllaries; peduncles enlarged below the heads in flower. *T. dubius Scop.
    (T. major Jacq.)
  b. Corollas equaling or longer than phyllaries; peduncles not enlarged below the heads or enlarged a little in fruit. *T. pratensis L. Yellow G.

62. Taraxacum Zinn, Dandelion

*T. officinale Weber   Leaves basal, lanceolate, pinnatifid; heads solitary on hollow naked scapes, several scapes on one plant; inner phyllaries erect, those of outer row shorter; receptacle naked, honeycombed; corollas all ligulate, yellow; pappus capillary; achenes obovoid, muricate, slenderly beaked when mature.

63. Sonchus L., Sow-thistle

Leaves alternate, toothed to pinnatifid; heads many-flowered, in a corymbose inflorescence; phyllaries in few to several series, involucre in fruit swollen at base; corollas all ligulate, yellow or orange; pappus white, capillary; achenes flattened, not beaked.

a. Heads 3-5 cm wide; corollas bright yellow or orange; perennial; achenes transversely wrinkled and longitudinally striate.
  b. Peduncles and phyllaries with gland-tipped hairs. *S. arvensis L.
  b. Peduncles and phyllaries without gland-tipped hairs. *S. uliginosus Bieb.
a. Heads not more than 2.5 cm wide; corollas pale yellow; annual; achenes longitudinally striate.
  b. Auricles rounded; leaves spinulose-dentate; achenes not transversely wrinkled. *S. asper (L.) Hill
  b. Auricles pointed; leaves with soft spiny teeth; achenes transversely wrinkled. *S. oleraceus L.

64. Lactuca L., Lettuce

Leaves alternate, entire to pinnatifid; heads with few to many flowers, in a panicle; involucre narrow, conic or cylindric, phyllaries in 2 or more series of unequal lengths; receptacle naked; corollas all ligulate; pappus soft, capillary; achenes flat, sometimes beaked.

a. Achene-beak slender, slightly shorter than to about twice as long as the body; corolla yellow; pappus white.
  b. Achene-body very flat, oval or oblong, with one prominent median rib on each face (sometimes 2 or more obscure ones).
    c. Stem or midrib of blades beneath usually hairy; involucre 12-22 mm long; achene (including beak) 7-9 mm long; rare. L. hirsuta Muhl.
    c. Only rarely hairy, often glaucous; involucre 10-15 mm long; achene (including beak) less than 7 mm long; common. L. canadensis L.
      d. Leaves, except sometimes the lowermost, entire or toothed, unlobed.
        e. Leaves lanceolate or narrowly ovate, usually entire. var. canadensis
        e. Leaves oblanceolate or obovate, usually toothed. var. obovata Wieg.
      d. Leaves, except sometimes the upper ones, pinnatifid.
        e. Unlobed leaves or lobes of lobed leaves linear or linear-lanceolate, margins untoothed or nearly so. var. longifolia (Michx.) Farw.
        e. Unlobed leaves or the lobes lanceolate to ovate. var. latifolia Ktze.
  b. Achene-body oblanceolate, with 5-7 ribs on each face; leaves sagittate-clasping, entire, toothed, or lobed.
    c. Leaves pinnatifid or unlobed; margins spinulose-toothed; under surface of midvein usually spinulose. *L. scariola L., Prickly L.
    c. Leaves linear and entire or with some linear entire lobes; both the margins and the conspicuous white midrib smooth or nearly so. *L. saligna L., Willow L.

a. Achene-beak absent or short and thick, achene-body with 3 or more ribs on each face; corolla blue (rarely white or yellow); pappus white, gray, or brown.
  b. Pappus white; corolla bluish (or white). L. floridana (L.) Gaertn.
    c. Leaves lyrate or runcinate-pinnatifid. var. floridana.
    c. Leaves elliptic, unlobed, toothed. var. villosa (Jacq.) Cronq.
  b. Pappus gray or brown; corolla blue, white, cream, or yellow. L. biennis (Moench) Fern. Typical L. biennis has bluish to white flowers and lobed leaves; forma aurea (Jennings) Fern. has yellow or whitish flowers and lobed leaves; forma integrifolia (T. & G.) Fern. has unlobed leaves.

65. Crepis L., Hawk's-beard

Leaves entire to pinnatifid, the cauline alternate, usually much smaller upward; heads in panicles or corymbs; principal phyllaries in 1 or 2 series, often becoming thickened at base; corollas ligulate, yellow; pappus soft, white, capillary.

a. Phyllaries glabrous on both sides, the midnerve thick and raised; outer phyllaries ovate, short; receptacle glabrous; achenes pale, the outer ones scabrous. *C. pulchra L.
a. Phyllaries pubescent on the outer side, the outer ones slender.
  b. Receptacle ciliate; longer (inner) phyllaries finely pubescent on the inner side; achenes dark.
    *C. tectorum L.
  b. Receptacle glabrous; phyllaries glabrous on the inner side. *C. capillaris (L.) Wallr.

66. Prenanthes L., Rattlesnake-root

Leaves basal and cauline, variable in shape, alternate; heads slender, cylindric, usually few-flowered; principal phyllaries in 1 row; corollas ligulate; pappus of capillary bristles, often yellow-orange, or reddish.

a. Clusters of heads short, little diverging from the main stem, the whole inflorescence long and slender; upper cauline leaves sessile or somewhat clasping; phyllaries pubescent; petioles of basal leaves winged.
  b. Corollas pink; stem and lower leaf surface glabrous and glaucous. P. racemosa Michx.
  b. Corollas cream-color; stem and leaves rough-hairy. P. aspera Michx.
a. Clusters of heads more or less divergent from the main stem; at least the lower cauline leaves petioled.
  b. Flowers 12-35 per head; basal leaves with winged petioles; phyllaries 12-15; corollas cream-color. P. crepidinea Michx.
  b. Flowers 5-16 per head; phyllaries 5-10.
    c. Heads of 5-7 flowers; involucre of 5 light green bracts. P. altissima L.
    c. Heads of 8-16 flowers; involucre of 6-10 bracts.
      d. Involucre usually with a few hairs; pappus slightly exceeding phyllaries; short outer phyllaries lanceolate. P. serpentaria Pursh
      d. Involucre glabrous, generally papillate, equaling the pappus; short outer phyllaries ovate. P. alba L.

67. Hieracium L., Hawkweed

Mostly hairy, often glandular, hairs sometimes stellate; leaves entire or toothed, simple, alternate, sometimes all basal; phyllaries in 2-3 series; corollas ligulate, yellow to red-orange; pappus gray or brownish, of capillary bristles.

a. Lower stem leafy; basal rosette usually not present at anthesis (except in H. gronovii).
  b. Principal phyllaries imbricated in 2 or 3 series; cauline leaves many, dentate to almost entire.
    H. canadense Michx.
  b. Principal phyllaries in 1 series with some additional small basal ones.
    c. Inflorescence an elongate panicle, usually 3 or more times as long as wide, not leafy bracted; blades obtuse, pubescent, mostly below middle of stem. H. gronovii L.
    c. Inflorescence not elongate, leafy bracted.
      d. Inflorescence open, spreading, peduncles slender; stem glabrous above; blades glabrate and glaucous beneath, acute. H. paniculatum L.
      d. Inflorescence rather stiff; peduncles stout; stem pubescent, often glandular; blades pubescent beneath, obtuse. H. scabrum Michx.

237

a. Basal rosette present at anthesis; cauline leaves none or few.
   b. Corolla red-orange; stolons elongate.   *H. aurantiacum L., Orange H.
   b. Corollas yellow; plants stoloniferous or not.
      c. Blades obovate to elliptic, entire to undulate-dentate; often a few cauline leaves present; not stoloniferous; inflorescence open, paniculiform to corymbiform.
         d. Basal blades often with purple veins; peduncles not stellate, sometimes glandular; stem glabrous except near the heads.   H. venosum L.
         d. Blades without purple veins; peduncles stellate-tomentose and with black gland-tipped hairs; rare.   H. trailii Greene
      c. Blades oblanceolate or narrower; stem-leaves small or none; stem hairy; heads solitary or inflorescence compact, corymbiform.
         d. Blades glabrous or only sparsely hairy above, glaucous.
            e. Stoloniferous; rhizome usually elongate.   *H. floribundum Wimm. & Grab.
            e. Without stolons; rhizome short and thick.   *H. florentinum All.
         d. Blades hairy on both sides, not glaucous.
            e. Stolons short and stout; heads several to many.   *H. pratense Tausch
            e. Freely stoloniferous; heads 1-4.   *H. pilosella L.

1. Leaves needlelike, scalelike, awl-shaped, or flat and narrowly linear.   2.
1. Leaves with expanded blades.   12.

    2. Leaves needlelike, at least some of them in fascicles or clusters on dwarf branches.   3.
    2. Leaves not in fascicles or clusters.   4.

3. Leaves persisting more than 1 season, all of them in fascicles of not more than 5 on dwarf branches.   Pinus
3. Leaves deciduous at end of 1 season, those on dwarf branches in clusters of more than 5, those on elongate branches not in clusters.   Larix
    4. Leaves all awl-shaped or of 2 kinds, some scalelike, some awl-shaped, sometimes in whorls of 3, scalelike ones 4-ranked.   Juniperus
    4. Leaves all of one kind.   5.

5. Leaves scalelike, persistent more than 1 season.   6.
5. Leaves flat and narrowly linear or short and needlelike.   7.

    6. Leaves opposite, 4-ranked, flat; branches flat and fanlike.   Thuja
    6. Leaves alternate; twigs and leaves woolly-pubescent.   Hudsonia

7. Leaves deciduous at end of 1 season, some scattered on twigs, others spreading in 1 plane on deciduous featherlike dwarf branches.   *Taxodium distichum (L.) Rich.
7. Leaves persistent more than 1 season, not on deciduous dwarf branches.   8.

    8. Leaves 4-sided or rarely flattened, not petioled above the sterigmata, which are conspicuous after leaves have fallen.   *Picea
    8. Leaves flat, either not on sterigmata or petioled above sterigmata.   9.

9. Leaf-scar conspicuous, circular, level with stem-surface; leaf-bases not decurrent on stem.   *Abies
9. Leaf-scar on sterigma, or leaves decurrent on stem, or both.   10.

    10. Leaves on any segment of stem of 2 lengths, those on upper side shorter than those spreading laterally, white-striped beneath, sometimes appressed on upper side of stem.   Tsuga
    10. Leaves on all sides of stem of approximately the same length.   11.

11. Leaf-bases decurrent; leaves lighter green or yellow green beneath, sometimes appearing 2-ranked.   Taxus
11. Leaf-bases not decurrent or inconspicuously so; leaves with white or gray stripes beneath.   *Pseudotsuga

—12— LEAVES WITH EXPANDED BLADES

    12. Leaves dichotomously veined, some crowded on thick dwarf branches, some on elongate branches.   Ginkgo
    12. Leaves not dichotomously veined, not on typical dwarf branches.   13.

13. Leaves opposite or whorled.   14.
13. Leaves alternate.   47.

    14. Leaves simple.   15.
    14. Leaves compound.   42.

15. Blades round-ovate, large, to 1.5 dm long or more, sometimes 3 at a node.   16.
15. Blades not as above.   17.

    16. Leaves usually 3 at a node, base rounded to cordate, usually a gland in some vein-axils below; long slender fruits often present.   Catalpa
    16. Leaves 2 or 3 at a node, base cordate; fruits, if present, ovate capsules.   Paulownia

17. Small evergreens with decumbent or prostrate stems.   18.
17. Plants not as above.   19.

    18. Low evergreen shrub with decumbent stems; blades linear to linear-oblong, to about 2.5 cm long, serrate or entire.   Pachystima
    18. Scarcely woody; stems slender, trailing; blades round-ovate, 1-2 cm long; red berries often present.   Mitchella

239

19. Blades entire.   20.
19. Blades crenate, serrate, dentate, or lobed.   31.

    20. Growing attached to branches of trees, partially parasitic; small evergreen shrub, blades thick, obovate or oblanceolate.   Phoradendron
    20. Growing in soil, not parasitic.   21.

21. Leaves and twigs covered with silvery or brown peltate scales.   Shepherdia
21. Leaves and twigs without such scales.   22.

    22. Leaves and stems aromatic; shrubs.   Calycanthus
    22. Not aromatic.   23.

23. Blades with black or translucent dots, usually not more than 1 cm wide; shrubs.   24.
23. Blades without such dots, usually wider.   25.

    24. Each leaf with 2 glands at base.   Ascyrum
    24. Leaves without glands at base, or a single gland between leaf-bases.   Hypericum

25. Petioles connected by stipules or stipule-scars; shrubs of wet places; leaves 2 or 3 at a node, usually 1 dm long or more.   Cephalanthus
25. Petioles not connected by stipules; leaves 2 at a node.   26.

    26. Upper lateral veins curving forward and lying parallel to midvein or actually converging somewhat toward tip of blade.   Cornus
    26. Upper lateral veins not as above.   27.

27. Petioles meeting around stem or connected by a ridge.   28.
27. Petioles not meeting around stem, not connected by a ridge.   29.

    28. Vein in base of petiole 1; upper leaves not perfoliate; shrubs.   Symphoricarpos
    28. Veins in base of petiole 3; upper leaves sometimes perfoliate; shrubs or vines, rarely evergreen.   Lonicera

29. Blades ovate, with rounded, truncate, or cordate base.   Syringa
29. Blades narrower than ovate or tapering to base.   30.

    30. Blades usually not more than 6 cm long, tardily deciduous; small black fruits often present. Ligustrum
    30. Blades usually 1 dm long or more, narrowed to base.   Chionanthus

31. Blades palmately veined or with at least 3 main veins from top of petiole.   32.
31. Blades pinnately veined or with only a midvein visible.   35.

    32. Stipules present.   33.
    32. Stipules absent.   34.

33. Lobes entire or with 1-3 large teeth on each side; petiole without glands.   Acer
33. Lobes or unlobed blades with several to many teeth; petiole often glandular; blades sometimes dark-dotted beneath.   Viburnum

    34. Blades not brown-dotted beneath; petiole without glands; trees and shrubs; fruits, if present, samaras.   Acer
    34. Blades sometimes brown-dotted beneath; petiole often glandular; flat-topped clusters of small white flowers or of drupes often present.   Viburnum

35. Twigs with prominent ridge decurrent from middle of the line connecting bases of petioles; blades lanceolate to ovate, acuminate; low upright shrubs.   Diervilla
35. Twigs without such a ridge.   36.

    36. Blades mostly entire, slightly crenate or few-toothed.   Symphoricarpos
    36. Blades evidently toothed or lobed.   37.

37. Blades doubly serrate, at least some of them shallowly lobed.   Acer
37. Blades not doubly serrate, not lobed.   38.

    38. Bark of ripe twigs green; twigs mostly 4-sided or 4-lined, sometimes cork-winged; blades finely serrate; small trees or decumbent or upright shrubs.   Euonymus
    38. Bark of ripe twigs not green, or other characters not as above.   39.

39. Lateral veins, or those more prominent than the others, 1-3 pairs, branching from lower part of midvein. 40.
39. Lateral veins more numerous. 41.

    40. Teeth of blade-margin many, small, gland-tipped; upper lateral veins curving upward, becoming parallel toward tip of blade; lateral buds evident, appressed; twigs often thorn-tipped. Rhamnus
    40. Teeth of blade-margin few; lateral buds hidden or almost hidden under petiole-base; twigs not thorn-tipped. Philadelphus

41. Blades finely serrate to dentate, if dentate, then lateral veins obviously ending in teeth; blades narrowly to rounded ovate; stipules present or absent. Viburnum
41. Blades dentate or coarsely serrate, ovate; outer bark readily separating from inner on year-old stems; petioles long; stipules absent. Hydrangea

    42. Leaves with 2 leaflets and a terminal tendril; vines. Bignonia
    42. Leaves palmately 5-7-foliolate. Aesculus
    42. Leaves with 3 leaflets. 43.
    42. Leaves with more than 3 leaflets, pinnately compound. 44.

43. Shrubs or small trees; leaflets finely serrate; with stipules or stipule-scars. Staphylea
43. Shrubs or trees; leaflets lobed or coarsely toothed; without stipules. Acer

    44. Vines; stems with aerial rootlets; leaflets dentate. Campsis
    44. Trees and shrubs. 45.

45. Shrubs with large pith; conspicuous stipels sometimes present. Sambucus
45. Trees; stipels absent. 46.

    46. Leaflets lobed or coarsely toothed, on earliest leaves of season usually 3; year-old stems green. Acer
    46. Leaflets entire or serrate, more than 3; year-old stems not green. Fraxinus

—47— LEAVES ALTERNATE

47. Leaves linear or lanceolate, parallel veined, or so thick that veins are not visible. 48.
47. Leaves not as above. 49.

    48. Leaves rigid, thick, mostly basal, evergreen. Yucca
    48. Leaves not rigid and thick, scattered along stem; a grass. Arundinaria

49. Leaves compound. 50.
49. Leaves simple. 73.

    50. Twigs with spines, prickles, stiff bristles, or thorns. 51.
    50. Twigs without spines, prickles, stiff bristles, or thorns. 55.

51. Without spines, prickles, or thorns, but with stiff bristles on twig; leaflets oblong, stalked, entire; pulvini present. Robinia
51. Without spines, prickles, or bristles, but with usually branched thorns on stem; early leaves once compound, later ones wholly or partly decompound; leaf-margin obscurely crenate-serrate; pulvini present. Gleditsia
51. With a pair of stiff prickles or spines in position of or approximately in position of stipules; herbaceous stipules absent. 52.
51. With spines or prickles on stem or leaves or both, but not paired in position of stipules or, if so, then herbaceous stipules present. 53.

    52. Leaflets oblong, stalked; pulvini present; stipular spines present, but spines and prickles lacking elsewhere. Robinia
    52. Leaflets ovate-oblong, sessile or nearly so; pulvini absent; stiff prickles in position of stipules and on rachis and petiole; aromatic. Xanthoxylum

53. Leaves decompound, very large; leaves and thick branches prickly. Aralia
53. Leaves once compound. 54.

    54. Stipules adnate to petiole for most of their length. Rosa
    54. Stipules adnate to petiole for small part of their length, or free. Rubus

55. Leaves 3-foliolate. 56.
55. Leaves palmately 5-7-foliolate; vines. Parthenocissus
55. Leaves with more than 3 leaflets, pinnately compound or decompound. 58.

    56. Vines, leaflets all stalked, margins coarsely toothed or entire; poisonous to touch. Rhus
    56. Shrubs or small trees; leaves with strong odor when crushed. 57.

57. Leaves fragrant, margins of leaflets coarsely dentate or crenate; axillary bud partly concealed by petiole base. Rhus
57. Leaves gland-dotted, margins of leaflets obscurely crenate or serrate to entire; axillary buds concealed under petiole base. Ptelea

    58. Leaflets less than 1 cm wide, rounded at tip and mucronate, margins obscurely crenate-serrate; leaves even-pinnate, the early ones compound, the later ones wholly or partly twice compound. Gleditsia
    58. Leaves not as above. 59.

59. Leaflets entire. 60.
59. Leaflets toothed only at base, a gland under each tooth. Ailanthus
59. Leaflets toothed along all or most of margin. 67.

    60. Vines; leaflets 7-many. Wisteria
    60. Not vines. 61.

61. All the leaflets compound except a few basal ones; ultimate leaflets lance-ovate; leaves even-pinnate, very large. Gymnocladus
61. None of the leaflets compound. 62.

    62. Rachis of leaf winged. Rhus
    62. Rachis of leaf not winged. 63.

63. Stipules membranous, sheathing; leaflets 3-5 (7), usually not more than 2.5 cm long. Potentilla
63. Stipules, if present, not membranous and sheathing; leaflets either more numerous or larger or both. 64.

    64. Leaflets ovate, obviously alternate on rachis; axillary buds in cavity in petiole base; pulvini present; trees. Cladrastis
    64. Without the above set of characters. 65.

65. Leaflets acuminate at tip, usually 8 cm long or more; without pulvini; poisonous to touch. Rhus
65. Leaflets rounded at tip and often mucronate, less than 8 cm long; pulvini present. 66.

    66. Lateral buds evident; leaflets gland-dotted beneath. Amorpha
    66. Lateral buds invisible, under base of petiole, and under leaf-scar membrane after petiole abscises; leaflets not gland-dotted beneath. Robinia

67. Vines; leaflets stalked. Ampelopsis
67. Not vines. 68.

    68. Stipules or stipule-scars present. 69.
    68. Stipules absent. 71.

69. Stipules elongate, adnate to the petiole for more than half their length. Rosa
69. Stipules not adnate to petiole or adnate for less than half their length. 70.

    70. Leaflets long-acuminate; shrubs. Sorbaria
    70. Leaflets obtuse or acute; trees or shrubs. Pyrus

71. Lateral veins of leaflet approximately the same number as the marginal teeth; twigs and petioles with long dense velvety pubescence or glabrous and glaucous. Rhus
71. Lateral veins of leaflet obviously fewer than the marginal teeth; trees; crushed leaves with pungent fragrance. 72.

    72. Leaflets usually more than 9 (to 17 or more), the terminal smaller than those farther down the rachis; pith diaphragmed and chambered, circular. Juglans
    72. Leaflets rarely as many as 9, the 3 terminal largest; or, if leaflets more numerous and the terminal not the largest, then buds yellow; pith solid and continuous; pith 5-lobed or -angled. Carya

73. Leaves evergreen. 74.
73. Leaves deciduous. 87.

    74. Plants scarcely woody, stems slender and trailing, sometimes with erect or ascending branches from rhizomes or above-ground stems. 75.
    74. Plants woody; shrubs with erect or rarely prostrate stems, or trees. 79.

75. Leaf-blades entire. 76.
75. Leaf-blades crenate or serrate. 78.

    76. Blades mostly not more than 1.5 cm long, margins sometimes revolute. 77.
    76. Blades 3 cm long or more, ovate or oblong, rounded at base; twigs and petioles hirsute. Epigaea

77. Blades glabrous beneath, length usually twice the width. Vaccinium (Cranberries)
77. Blades appressed-strigose beneath, length usually less than twice width. Gaultheria

    78. Blades elliptic to almost circular, margin crenulate; plant with odor of wintergreen. Gaultheria
    78. Blades oblanceolate or lanceolate, serrate; leaves sometimes almost in whorls. Chimaphila

79. Lower surface of leaf-blades covered with small peltate scales; blades 1.5-5 cm long, margins crenulate; low shrubs. Chamaedaphne
79. Leaf-blades without such scales. 80.

    80. Margins of blades entire. 81.
    80. Margins of blades toothed. 86.

81. Stipule-scars meeting around twig; blades stiff, 10 cm long or more, elliptic to obovate, shining above. Magnolia
81. Stipule-scars absent or not meeting around twig. 82.

    82. Blade-margins revolute. 83.
    82. Blade-margins not revolute. 84.

83. Blades densely tomentose beneath. Ledum
83. Blades glabrous and glaucous beneath, or pubescent with fine erect white hairs when young. Andromeda

    84. Stem prostrate; blades 1-3 cm long, oblanceolate, rounded or blunt at tip, base tapering. Arctostaphylos
    84. Stem erect; shrubs or small trees; blades longer. 85.

85. Blades mostly less than 8 cm long, coriaceous; winter buds naked or with 2 small scales. Kalmia
85. Blades mostly more than 8 cm long, thick-coriaceous; winter buds with several scales. Rhododendron.

    86. Blades dark green, lustrous, coriaceous, with a few large spine-tipped teeth; stems without thorns. Ilex
    86. Blades crenate-serrate, teeth not spine-tipped; stem with slender thorns. Cotoneaster pyracantha

87. Stipules or stipule-scars meeting or nearly meeting around twig. 88.
87. Stipules or stipule-scars absent or not nearly meeting around twig. 90.

    88. Stipules or stipule-scars (plus leaf-scar) extending 3/4 the distance around twig or more; blades toothed; bark smooth, light gray. Fagus
    88. Stipules or stipule-scars essentially meeting around twig. 89.

89. Blades entire, sometimes auricled at base, about 8 cm to 6 dm long or more. Magnolia
89. Blades lobed, apex formed by tip of lobe; axillary bud enclosed in cavity in petiole base. Platanus
89. Blades more or less lobed, apex truncate or emarginate; margins of lobes entire. Liriodendron

    90. Blades palmately veined or with at least 3 main veins from top of petiole. 91.
    90. Blades pinnately veined or with only a midvein visible. 111.

91. Vines. 92.
91. Not vines. 96.

92. Blades ovate, with 3 or more strong longitudinal veins; tendrils on the petiole. Smilax
92. Blades and tendrils not as above. 93.

93. Blades almost circular, shallowly lobed, petiole not quite at margin; without tendrils; stem slender. Menispermum
93. Blades not as above; petiole at margin; tendrils present or absent, opposite the leaves. 94.

94. Pith brown, with woody partitions at nodes of older branches; bark shreddy; tendril-tips without disks. Vitis
94. Pith greenish or white, without woody partitions at nodes; bark not shreddy. 95.

95. Pith continuous, greenish; blades 3-lobed, those of basal shoots usually 3-foliolate; tendrils ending in disks. Parthenocissus
95. Pith with crosswise fissures, white; blades unlobed or 3-lobed; tendrils mostly in inflorescence, tips slender. Ampelopsis

96. Blades usually 3 kinds on same tree—unlobed and entire, 2-lobed, and 3-lobed, lobes entire, rarely all unlobed; twigs and leaves aromatic; bark green. Sassafras
96. If lobed and unlobed blades on same tree, then lobes toothed. 97.

97. Blade-margin entire. 98.
97. Blade-margin toothed or lobed. 99.

98. Blades round-ovate, cordate; main veins from top of petiole usually 7 (5-9). Cercis
98. Blades ovate; main veins from top of petiole 3. Celtis

99. Blades broadly ovate, unlobed or with 2 or more lobes, lobes toothed, often more than 1 kind on same tree; sap milky. 100.
99. Blades not as above; sap not milky. 101.

100. Twigs gray-green, pubescent; blades densely soft-pubescent beneath; rare. Broussonetia
100. Twigs brown or gray, usually glabrous; blades glabrous to somewhat pubescent beneath, green. Morus

101. Stems with typical lateral thorns; blades sometimes lobed. Crataegus
101. Stems without typical lateral thorns, but spines or prickles may be present. 102.

102. Blades merely toothed. 103.
102. Blades, at least most of them, lobed. 106.

103. Blades 2-ranked, often inequilateral at base. 104.
103. Blades not 2-ranked, equilateral or nearly so at base. 105.

104. Blades ovate or lance-ovate, longer than wide; pith usually in transverse plates. Celtis
104. Blades round-ovate, about as wide as long; pith not as above. Tilia

105. Blades 3-ribbed, with 2 strong lateral lengthwise veins; small shrubs. Ceanothus
105. Blades not 3-ribbed; pith 5-angled or -lobed; top of petiole often flattened; trees. Populus

106. Blades white-tomentose or felted beneath, margin with small lobes or coarse teeth. Populus
106. Blades sometimes hairy beneath but not felted or tomentose. 107.

107. Trees; blades shaped like a star (with a missing point), usually with 5 lanceolate serrate lobes; pith 5-angled or lobed. Liquidambar
107. Shrubs or small trees; blades not as above. 108.

108. Blades usually 1 dm wide or more, sinuses wide, main lobes triangular; twigs and petioles with brownish glandular hairs or bristles. Rubus
108. Blades smaller. 109.

109. Blades about as wide as long, terminal lobe not noticeably longer; prickles or bristles sometimes present. Ribes
109. Blades, or many of them, longer than wide, terminal lobe longer than lateral ones; without prickles or bristles. 110.

110. Blades cuneate at base; lobes irregularly large-toothed. Hibiscus
110. Blades cuneate to subcordate at base; lobes with finely toothed large crenations; outer bark readily separating. Physocarpus

111. Blades wavy-margined or toothed, or some or all of them lobed. 112.
111. Blades with entire margins. 149.

    112. Leaves 2-ranked. 113.
    112. Leaves in more than 2 ranks. 121.

113. Blades wavy-margined, inequilateral at base, broadly ovate. <u>Hamamelis</u>
113. Blades toothed or some or all of them lobed. 114.

    114. Lateral veins the same number as marginal teeth, each vein ending in a tooth. 115.
    114. Lateral veins (main ones, not branches) fewer than total number of marginal teeth both large and small, if 2 sizes are present. 116.

115. Blades elliptic-lanceolate or -oblanceolate; teeth large, long-pointed; pith 5-angled or -lobed. <u>Castanea</u>
115. Blades broadly elliptic or ovate; teeth small; pith circular; bark smooth, light gray. <u>Fagus</u>

    116. Trees and shrubs; pith 3-sided, small; if trees, twigs with wintergreen flavor or bark white, yellowish, or pinkish tan, and exfoliating, or bark white and not exfoliating; if shrubs, growing in bogs, blades round-ovate. <u>Betula</u>
    116. Plants not as above. 117.

117. Lateral veins obscure near leaf-margin or, if prominent, then leaf rounded at apex and lower part of margin entire. <u>Amelanchier</u>
117. Lateral veins prominent, obviously ending in teeth; tip of blade acute or acuminate, serrations extending to base or nearly so. 118.

    118. Shrubs; blades round-ovate, acuminate, doubly serrate, base usually cordate; petioles and twigs usually glandular-bristly. <u>Corylus</u>
    118. Trees; if petioles and twigs glandular, then blades lanceolate to lance-ovate. 119.

119. Bark smooth, dark gray; trunk with projecting ridges. <u>Carpinus</u>
119. Bark rough; trunk without such ridges. 120.

    120. Blades usually decidedly inequilateral at base, firm, doubly serrate; when equilateral at base, then usually each larger serration with only 1 smaller serration; bark not scaling off. <u>Ulmus</u>
    120. Blades, or most of them, equilateral at base, thin; bark finely furrowed, sometimes scaling off in narrow strips. <u>Ostrya</u>

121. Unlobed, 2-lobed, and 3-lobed blades on same tree, margins of lobes and of unlobed blades entire; twigs and leaves aromatic. <u>Sassafras</u>
121. Blades not of the above sort, <u>or</u> plants vines. 122.

    122. Vines. 123.
    122. Not vines. 124.

123. Some blades unlobed, margins entire, others lobed. <u>Solanum</u>
123. Blades not lobed, margins finely serrate. <u>Celastrus</u>

    124. Shrubs with fragrant, deeply-pinnatifid, linear-lanceolate or linear-oblong blades. <u>Myrica</u>.
    124. Blades not at the same time fragrant, deeply-pinnatifid, and linear-lanceolate or -oblong. 125.

125. Shrubs with simple or branched spines subtending short leafy branches; blades with spine-tipped serrations. <u>Berberis</u>
125. Spines not subtending branches. 126.

    126. Small trees; blades elliptic, with prominent scattered hairs on midvein beneath; leaves sour; large panicles of small white flowers or of capsules often present. <u>Oxydendrum</u>
    126. Without the above set of characters. 127.

127. Buds and leaves clustered at tips of branches; blades lobed or toothed, when toothed, usually widest above middle; pith 5-lobed or -angled. <u>Quercus</u>
127. Buds and leaves not clustered at tips of branches <u>or</u> pith not 5-lobed. 128.

    128. Pith small, 3-sided. 129.
    128. Pith not 3-sided. 130.

129. Buds stalked; blades blunt, rounded, or emarginate at apex; woody, conelike carpellate aments persistent. Alnus
129. Buds not obviously stalked; blades acute to acuminate at apex, or, in 1 species, a small bog shrub, blades almost circular; carpellate aments not woody and conelike. Betula

    130. Side veins straight and parallel, each ending in a marginal tooth, teeth and veins the same number; teeth slender and pointed; blades pointed at apex and base; pith 5-lobed or -angled. Castanea
    130. Without the above set of characters. 131.

131. Blades obovate, tip and base pointed, entire except sometimes for a few teeth near tip; pith solid, with diaphragms. Nyssa
131. Blades and pith not as above. 132.

    132. With typical lateral thorns; stipules large, half circular, crenate. Chaenomeles
    132. With typical lateral thorns; stipules, if present, small. Crataegus
    132. Without typical lateral thorns, but some branches may end in thorns. 133.

133. Leaves with glands on midvein, base of blade, or petiole. 134.
133. Leaves without such glands, but buds, twigs, and blades resin-dotted, fragrant. Myrica
133. Leaves with neither glands nor resin dots in positions stated above. 137.

    134. Glands toothlike, along upper side of midvein. Pyrus
    134. Glands of various kinds on petiole or near base of blade. 135.

135. Blades usually lanceolate or narrower; winter buds with 1 scale; twigs often with brittle zone at base; stipules usually present. Salix
135. Blades usually wider than above; winter buds with more than 1 scale. 136.

    136. Blades ovate to deltoid, base cuneate to cordate; pith 5-angled or -lobed; summit of petiole sometimes flattened. Populus
    136. Blades ovate to lanceolate; pith not distinctly 5-lobed; branches sometimes ending in thorns; twigs usually with bitter odor when crushed. Prunus

137. Winter buds with 1 scale; twigs often with brittle zone at base; stipules often conspicuous; blades usually lanceolate or narrower. Salix
137. Winter buds with more than 1 scale, or naked, or woolly; twigs without brittle zone. 138.

    138. Mature blades stellate-pubescent or densely woolly beneath. 139.
    138. Mature blades glabrous or pubescent, but not as above; young blades rarely woolly beneath. 141.

139. Blades with dense white or brown tomentum beneath; small shrubs. Spiraea
139. Blades stellate-pubescent beneath; shrubs or small trees. 140.

    140. Blades sometimes pubescent only on veins beneath; pith chambered. Halesia
    140. Blades sparingly toothed, their under surface and young stems densely stellate-pubescent; pith not chambered. Styrax

141. Blades elliptic to oblong, to about 4 cm long, mucronate, glabrous, margins almost entire; shrub, mostly northern, of wet woods and bogs; bark gray; flowers and red drupes on long slender peduncles. Nemopanthus
141. Plants not as above. 142.

    142. Blades ovate to deltoid, base cuneate to cordate; pith 5-lobed or -angled; summit of petiole sometimes flattened; trees. Populus
    142. Without the above set of characters. 143.

143. Blades rounded, truncate, or cordate at base. 144.
143. Blades narrowed to base or, if somewhat rounded, with gland-tipped teeth. 145.

    144. With branches ending in thorns; pubescent or glabrous; leaf-margins finely to coarsely toothed or lobed; buds ovoid. Pyrus
    144. Without branches ending in thorns; leaf-margins finely and evenly serrate, basal part of margin sometimes entire; buds elongate. Amelanchier

145. Stipules or stipule scars present. 146.
145. Stipules and stipule scars absent. 147.

146. Blades usually widest above middle; buds often superposed; marginal teeth with slender points; ripe fruit red. Ilex
146. Blades usually widest about middle; buds not superposed; marginal teeth gland-tipped, usually incurved; ripe fruit black, unripe sometimes red. Rhamnus

147. Twigs granulate or blistered or white- or pale-speckled. Vaccinium
147. Twigs not as above. 148.

148. Buds elongate, pointed, closely appressed; blades obovate or oblanceolate, pubescent beneath; panicles of small globose capsules sometimes present; rare shrub. Lyonia
148. Buds not as above; blades oblanceolate, elliptic, or rhombic, often crowded; branches slender, sometimes arching; fruits, if present, are aggregates of small follicles. Spiraea

—149— LEAVES ALTERNATE, SIMPLE, DECIDUOUS, PINNATELY VEINED, ENTIRE

149. Leaves and twigs covered with brown or silvery peltate scales. Elaeagnus
149. Leaves and twigs without such scales. 150.

150. Slightly woody vines; blades ovate; fruit a red berry. Solanum
150. Not vines, or, if rarely so, then stems with thorns. 151.

151. Stems with thorns or spines. 152.
151. Stems without thorns or spines. 154.

152. Spines subtending short lateral branches; small shrubs. Berberis
152. Without spines; thorns present. 153.

153. Trees; sap milky; blades glossy. Maclura
153. Spreading shrubs or vines; sap not milky. Lycium

154. Leaves and buds clustered at ends of twigs; pith 5-angled or -lobed; trees. Quercus
154. Leaves and buds not clustered at ends of twigs or pith not 5-lobed. 155.

155. Leaves or twigs or both fragrant. 156.
155. Leaves and twigs not fragrant. 158.

156. Mature stems green; a pair of larger lateral veins from near base of blade. Sassafras
156. Stems not green; lateral veins about equal in size. 157.

157. Blades acuminate or acute at tip, not crowded toward end of twigs. Lindera
157. Blades rounded or emarginate at tip, crowded toward end of twigs. Cotinus

158. Bark in thick polygons separated by longitudinal and cross furrows; petioles loosely jointed; veins in base of petiole 1 or confluent; trees. Diospyros
158. Plants not as above. 159.

159. Blades stellate-pubescent or resin-dotted beneath. 160.
159. Blades neither stellate-pubescent not resin-dotted beneath. 161.

160. Blades stellate-pubescent beneath, not resin-dotted. Styrax
160. Blades resin-dotted beneath, not stellate-pubescent. Gaylussacia

161. Leaves 2-ranked. 162.
161. Leaves not 2-ranked. 163.

162. Base of petiole not surrounding and covering axillary bud; blades to 1.5 dm long or more; twigs not very flexible. Asimina
162. Base of petiole surrounding and covering axillary bud; blades less than 1 dm long; twigs very flexible, enlarged below base of each season's growth-segment. Dirca

163. Pith with diaphragms, solid between diaphragms; leaves often in a mosaic at ends of short branches. Nyssa
163. Pith without diaphragms. 164

164. Lateral veins rather prominent, either curving forward and paralleling midvein or diverging and straight and parallel to near margin. 165.
164. Lateral veins not as described above. 166.

165. Lateral veins curving forward and paralleling midvein in upper part of leaf, or converging; blades pale beneath, minutely pubescent. Cornus
165. Lateral veins straight and parallel almost to margin, there curving forward and anastomosing; blades green beneath, glabrous. Rhamnus

166. Internodes conspicuously unequal; leaves clustered at tips of twigs, margins ciliate; branches clustered. Rhododendron (Azaleas)
166. Internodes not conspicuously unequal; leaves and branches not conspicuously clustered. 167.

167. Winter bud-scale 1; blades usually 4 or more times as long as wide, or silky or woolly beneath, or with revolute margins. Salix
167. Winter bud-scale more than 1. 168.

168. Slender stipules present; petioles mostly more than 7 mm long; blades mucronate at tip. Nemopanthus
168. Stipules absent; petioles mostly less than 7 mm long or somewhat winged. 169.

169. Blades oblanceolate, long-tapering to base; rare shrub. Daphne
169. Blades not as above. 170.

170. Blades lanceolate; twigs usually 5-lined or -angled; decurrent lines from leaf-bases; climbing or spreading shrubs. Lycium
170. Blades mostly oval, elliptic, or obovate; stems sometimes granulate, blistered, or white- or pale-speckled, or twigs self-pruned. Vaccinium

KEY TO DECIDUOUS WOODY PLANTS IN WINTER CONDITION

In this key, length of leaf-scar is the dimension parallel to the long axis of the stem; width is the dimension at right angle to the long axis of the stem. Thus the leaf-scar of maple is <u>short</u> and <u>wide</u>.

If petiole-base is persistent, vein-scars sometimes are not evident until this base is cut off.

KEY TO SECTIONS

a. Leaf-scars opposite or whorled (rarely subopposite). SECTION I, p. 249
a. Leaf-scars alternate.
  b. Vines, with or without tendrils. SECTION II, p. 251
  b. Not vines.
    c. Branches with prickles, spines, or thorns, or ending in thorns. SECTION III, p. 251
    c. Branches without prickles, spines, or thorns. SECTION IV, p. 252

SECTION I. LEAF-SCARS OPPOSITE OR WHORLED

a. Twigs covered with shining, brown or silvery, peltate scales. <u>Shepherdia</u>
a. Twigs without such scales.
  b. Vein-scars several, in a complete or nearly complete circle or ellipse; trees.
    c. Terminal bud present; leaves opposite. <u>Fraxinus</u>
    c. Terminal bud lacking; leaves opposite or whorled.
      d. Leaf-scars of unequal size, in whorls of 3; pith continuous; long slender fruits often present. <u>Catalpa</u>
      d. Leaf-scars opposite or whorled; pith chambered or its space empty between nodes; ovoid capsules sometimes present; buds superposed. <u>Paulownia</u>
  b. Vein-scar 1, or several crowded in a straight, curved, or C-shaped line, appearing as 1. (See third b.)
    c. Vines.
      d. Tendrils present at ends of persistent petioles. <u>Bignonia</u>
      d. Without tendrils, but often with bands of aerial rootlets below nodes. <u>Campsis</u>
    c. Not vines.
      d. Stipules or stipule-scars present.
        e. Terminal bud present; stipules or stipule-scars minute; twigs often green, often lined from nodes, sometimes cork-winged; small trees or erect or prostrate shrubs. <u>Euonymus</u>
        e. Terminal bud absent.
          f. Twigs with terminal thorns; buds appressed. <u>Rhamnus</u>
          f. Twigs without thorns; buds sunken in bark; leaf-scars sometimes whorled; stipules or their scars nearly or quite connecting leaf-scars. <u>Cephalanthus</u>
      d. Neither stipules nor stipule-scars present.
        e. Vein-scar 1; leaf-scar sometimes on persistent petiole-base.
          f. Twigs with decurrent ridge from middle of leaf-scar, thus appearing rather sharply 2- or 4-edged.
            g. Leaf-scar raised by enlarged node, vein-scar obscure; low, somewhat prostrate, shrubs; twigs very slender. <u>Ascyrum</u>
            g. Leaf-scar not raised by enlarged node; erect shrubs. <u>Hypericum</u>
          f. Twigs not as above.
            g. Buds fusiform, scales many; twigs orange, brown, or greenish. <u>Forsythia</u>
            g. Buds ovoid, scales about 4-6; leaf-scar raised.
              h. Leaf-scars often torn or obscure, connected or partly connected by a line; pith hollow or continuous; stems very slender. <u>Symphoricarpos</u>
              h. Leaf-scars not obscure, not connected; pith continuous. <u>Ligustrum</u>
        e. Vein-scars crowded in a transverse straight, curved, or C-shaped line.
          f. Trees; buds brown to almost black, or gray-hairy, lateral ones rounded at summit and flattened parallel to twig, terminal ones short-conic; twigs glabrous or pubescent, sometimes 4-angled. <u>Fraxinus</u>
          f. Shrubs; buds not as above.
            g. Twigs slender, often hairy; leaf-scar tiny, somewhat raised; small black berry-like fruits in panicles often present. <u>Ligustrum</u>

249

  g. Twigs moderate, glabrous or nearly so; leaf-scar of moderate size.

   h. Vein-scars in a C-shaped line, usually near middle of leaf-scar; buds super-posed; lenticels conspicuous. Chionanthus

   h. Vein-scars in a straight or slightly curved line, usually above middle of leaf-scar; buds not superposed. Syringa

b. Vein-scars 3 or more, separate.

 c. Bark with spicy fragrance; bud or bud-aggregate hairy, scales not distinguishable; leaf-scar deeply or shallowly U-shaped. Calycanthus

 c. Without the above set of characters.

  d. Many twigs ending in thorns; end-bud lacking; buds appressed. Rhamnus

  d. Twigs not ending in thorns; end-bud present or absent.

   e. Buds without scales, stellate-scurfy. Viburnum lantana and alnifolium

   e. Buds with at least 1 scale, sometimes hidden by petiole-base.

    f. Axillary bud hidden or partly hidden by petiole-base, when hidden, later growing through it; vein-scars C-shaped; bud-scales 2, mostly hairy; end-bud lacking. Philadelphus

    f. Axillary bud not as above.

     g. Bud-scale 1 or appearing as 1.

      h. Buds plump, often rounded more on side away from twig; leaf-scars all oppo-site. Viburnum opulus

      h. Buds slender, many of them subopposite or alternate. Salix purpurea

     g. Exposed bud-scales 2 or more.

      h. Exposed bud-scales 2, valvate, or rarely separated near base.

       i. Leaf-scar usually raised at first on persistent petiole-base which is later deciduous; twigs often brightly colored (red, green, yellow); globose ter-minal flower-buds present, or terminal buds slender; twigs and buds often with short hairs appressed parallel to long axis of twig. Cornus

       i. Leaf-scar not raised on petiole-base; terminal bud sometimes long and slen-der, sometimes widened at base but not globose; hairs of twigs absent or usually not as above.

        j. Buds not long and slender, not long-tapering at tip; vein-scars sometimes more than 3; buds and/or twigs downy or glabrous; small trees. Acer spicatum and pensylvanicum

        j. Lateral buds long and narrow, appressed; terminal buds often flask-shaped; or buds not so long and narrow, then rufous- or brown-scurfy, or rarely glabrous; hairs sometimes stellate; vein scars 3; shrubs and small trees. Viburnum cassinoides, rufidulum, lentago, and prunifolium

      h. Exposed bud-scales more than 2.

       i. Twigs with a ridge decurrent from middle of line connecting leaf-scars. Diervilla

       i. Twigs without such a ridge.

        j. Leaf-scars rather large, sometimes fan-shaped; vein scars 3, 5, or 7; lenticels conspicuous; tips of twigs usually dead; pith very large; shrubs. Sambucus

        j. Without the above set of characters.

         k. Terminal bud absent; stipule-scars conspicuous; bark of twigs white-streaked. Staphylea

         k. Terminal bud, remnant of last year's inflorescence, or scar of that in-florescence, present; stipule-scars absent.

          l. Terminal bud ovoid, 1 cm long or more, with 12 or more visible scales (or twig terminated by inflorescence-scar); vein-scars sep-arate and more than 3 or in 3 groups; twigs stout. Aesculus

          l. Terminal bud smaller or with fewer scales; other characters not as above.

           m. Leaf-scars small, raised on petiole-base; pith sometimes ex-cavated; vines and shrubs; vein-scars 3. Lonicera

           m. Leaf-scars not small and inconspicuous, rarely raised on petiole-base.

n. Outer papery brown bark easily separating from inner; vein-scars 3 or, if more, buds brown-woolly; dry corymbose or paniculate clusters of flowers or of capsules often present. Hydrangea

n. Outer bark not easily separating; fruits, if present, samaras or drupes; vein-scars 3; leaf-scars short, often like a short wide V or U.

    o. Visible bud-scales usually more than 6; if fewer than 6, one or more of the following characters present: twigs green, brown, or red-brown; twigs glaucous; twigs with milky sap; lateral buds collateral; leaf-scars meeting at upward-pointing angle. Acer

    o. Visible bud-scales usually not more than 4, rarely 6, the first 2 with bases connected; lateral buds longer than wide, somewhat appressed; none of the characters listed after the colon above present. Viburnum acerifolium, recognitum, dentatum, rafinesquianum, and molle

SECTION II. LEAF-SCARS ALTERNATE; PLANTS VINES

a. Vines with tendrils.

  b. Tendrils attached to persistent petiole base; veins of stem scattered; monocots, usually prickly. Smilax

  b. Tendrils attached to stem, opposite leaf-scars; veins in a circle around central pith; dicots; not prickly; stipule-scars short and wide; vein-scars in an ellipse or a C-shaped line.

    c. Pith brown, a diaphragm at each node; bark of older stems shredding; tendrils twining, not ending in disks. Vitis

    c. Pith white or green, without diaphragms; bark not shredding; buds globose to conic.

      d. Tendrils usually ending in adhesive disks; pith continuous. Parthenocissus

      d. Tendrils twining, not ending in disks; pith eventually in thin plates. Ampelopsis

a. Vines without tendrils; aerial rootlets sometimes present.

  b. Vein-scar 1 or several confluent appearing as 1.

    c. Minute stipules or stipule-scars present; bud-scales mucronate. Celastrus

    c. Stipules and stipule-scars absent.

      d. Wartlike or spinelike projections at side of leaf-scar; bud acute. Wisteria

      d. No projections at side of leaf-scar; bud obtuse, hairy. Solanum (See third d)

      d. No projections at side of leaf-scar; bud an indistinguishable mass. Lycium

  b. Vein-scars more than 1, separate.

    c. Leaf-scar short and wide, a thin line; vein-scars 3; usually prickly. Rosa

    c. Leaf-scar about as long as wide; vein-scars 3-many; not prickly; buds hairy.

      d. Buds naked, single, not covered; with aerial roots; poisonous to touch. Rhus

      d. Buds not naked, superposed, small, the lower under circular leaf-scar. Menispermum

SECTION III. LEAF-SCARS ALTERNATE; NOT VINES;
TWIGS BEARING PRICKLES, SPINES, OR THORNS, OR ENDING IN THORNS

a. Twigs with silvery or brown peltate scales. Elaeagnus

a. Twigs without such scales.

  b. Twigs with prickles or with branched or unbranched spines.

    c. Leaf-scars short and wide, encircling the stout twigs halfway or more; vein-scars about 15. Aralia

    c. Leaf-scars not so wide or vein-scars fewer.

      d. Branched or unbranched spines subtending lateral buds. Berberis

      d. Spines, if present, not subtending buds.

        e. A pair of stipular spines at each node, but none between nodes; buds woolly, at least the lower at each node hidden under leaf-scar; trees. Robinia

        e. Usually 1 or 2 prickles at each node, and sometimes some between nodes.

          f. Shrubs or small trees; buds woolly; a pair of stiff prickles in position of stipules but none between nodes. Xanthoxylum

          f. Shrubs; buds not woolly, or petiole-base persistent.

g. Leaf-scar prominently raised on persistent petiole base. <u>Rubus</u>
g. Leaf-scar not so raised.
  h. Exposed bud-scales about 6; twigs with decurrent ridges from nodes; outer bark shreddy. <u>Ribes</u>
  h. Exposed bud-scales 3 or 4; without decurrent ridges; leaf-scar very short, a thin line. <u>Rosa</u>
b. Twigs with lateral branched or unbranched thorns and/or twigs ending in thorns.
  c. Axillary buds superposed, the lower within the leaf-scar, later breaking through the covering; thorns usually branched, above leaf-scars. <u>Gleditsia</u>
  c. Axillary buds single, above leaf-scar, not covered.
    d. Usually 1 thorn at each node, at side of axillary bud; vein-scars more than 3. <u>Maclura</u>
    d. Lateral thorns fewer than nodes or absent; some or all twigs may end in thorns.
      e. Vein-scar 1; buds an indistinguishable cluster; twigs 5-angled. <u>Lycium</u>
      e. Vein-scars usually 3; buds evident and distinguishable.
        f. Buds rounded at tip or globose, projecting at rather wide angle from stem, with about 6 fleshy, often red, scales; lateral thorns prominent. <u>Crataegus</u>
        f. Buds ovoid to lance-ovoid, acute or bluntly acute, appressed or ascending.
          g. Stipule-scars absent; buds glabrous or hairy, the terminal usually present; twigs glabrous or hairy. <u>Pyrus</u>
          g. Stipule-scars or stipule-vestiges present; terminal bud absent.
            h. Buds appressed, lance-ovoid, often opposite; thorns terminal. <u>Rhamnus</u>
            h. Buds not strongly appressed, ovoid, alternate.
              i. Buds small, scales few; stipule-scars round to half-round. <u>Chaenomeles</u>
              i. Buds moderate in size, scales usually several; stipule-vestiges sometimes present; crushed twigs with odor of bitter almonds. <u>Prunus</u>

SECTION IV. LEAF-SCARS ALTERNATE; NOT VINES; WITHOUT PRICKLES, SPINES, OR THORNS

a. Twigs covered with small branch-scars left by abscission of featherlike dwarf branches; tiny and inconspicuous leaf-scars present also; gymnosperms. <u>Taxodium</u> bald cypress
a. Branch-scars absent or, if rarely present, then leaf-scars conspicuous.
  b. Stipule-scars meeting or extending 3/4 the distance around twigs.
    c. Buds more than 1 cm long, fusiform, the lateral ones oblique to twig; bud-scales many; stipule-scars extending about 3/4 distance around twig. <u>Fagus</u>
    c. Buds with 1 or 2 scales; stipule-scars meeting around twig.
      d. Leaf-scar ringlike, nearly surrounding bud; end-bud lacking; twigs zigzag; bud scale 1. <u>Platanus</u>
      d. Leaf-scar not ringlike, not nearly surrounding bud.
        e. Scales of terminal bud 2, valvate; bud somewhat 2-edged. <u>Liriodendron</u>
        e. Scale of terminal bud 1, bearing a leaf-scar; bud not 2-edged. <u>Magnolia</u>
  b. Stipule-scars absent or not as above.
    c. Leaf-scars not 2-ranked.
      d. Buds superposed, in silky craters; pith pink-brown; vein-scars about 3-5; end-bud lacking; twigs stout. <u>Gymnocladus</u>
      d. Without the above set of characters.
        e. Axillary bud (or the lower, if buds are superposed) hidden under leaf-scar.
          f. Buds glabrous; pith pale or pinkish. <u>Gleditsia</u>
          f. Buds woolly; pith brown; twigs sometimes bristly or glandular. <u>Robinia</u>
        e. Axillary bud replaced or hidden by branch-scars or -vestiges. <u>Hibiscus</u> (See third e)
        e. Axillary bud hidden or almost hidden under persistent petiole-base. (See fourth e)
          f. Twigs aromatic; bark not shredding; petiole-base terete, not 3-nerved. <u>Rhus</u>
          f. Twigs not aromatic; bark shredding; petiole-base flat, 3-nerved. <u>Potentilla</u>
        e. Axillary bud evident, not hidden; petiole-base sometimes persistent.
          f. Buds so hairy that scales, if present, are indistinguishable; pith not diaphragmed, not 5-angled.
            g. Leaf-scar strongly raised, it and vein-scars often obscure; small shrubs.
              h. Vein-scar 1; twigs often woolly. <u>Spiraea</u>
              h. Vein-scars 3 in base of petiole-stub; twigs often bristly. <u>Rubus</u>
            g. Leaf-scar not strongly raised on petiole-base; vein-scars evident.
              h. Leaf-scar ring- or U-shaped, encircling bud or bud-group on 3 sides or more.

252

i. Twigs stout, smooth or densely hairy; pith large; bud single; vein-scars more than 3. <u>Rhus</u>
i. Twigs moderate; buds superposed.
  j. Buds 2 at a node, a low conical mass; vein-scars 3; flat circular winged fruits sometimes persistent. <u>Ptelea</u>
  j. Buds usually 3 at a node, uppermost longest, lowermost shortest; vein-scars 3-7; twigs brown, smooth. <u>Cladrastis</u>
h. Leaf-scar not ring-shaped or U-shaped.
  i. Vein-scars 3 or confluent in a line; stipule-scars present; twigs rather slender. <u>Rhamnus</u>
  i. Vein-scars 5 or more, sometimes grouped or in an ellipse; stipule-scars absent; twigs moderate to stout; some species poisonous to touch. <u>Rhus</u>
f. Bud with scales, the scales distinguishable, or, if not, then pith diaphragmed or 5-angled; scales rarely valvate, rarely leaflike.
g. Bud-scale 1; twigs often with brittle zones. <u>Salix</u>
g. Bud-scales 2 or more.
  h. Vein-scars 2; leaf-scars crowded on dwarf branches or scattered on more elongate twigs. <u>Ginkgo</u>
  h. Vein-scars 1 or several confluent, appearing as 1. (See third h.)
    i. Leaf-scars crowded on dwarf branches or scattered on more elongate twigs; leaf-bases decurrent on elongate twigs; gymnosperms. <u>Larix</u>
    i. Leaf-scars not crowded on typical dwarf branches; angiosperms.
      j. Leaf-scar raised on clasping 3-nerved petiole-base; stipules persistent. <u>Potentilla</u>
      j. Leaf-scar not raised on such a petiole-base.
        k. Twigs green and aromatic, often branching the first year; buds sub-globose. <u>Sassafras</u>
        k. Twigs not green and aromatic.
          l. Twigs covered with silvery or brown peltate scales. <u>Elaeagnus</u>
          l. Twigs not covered with such scales.
            m. Stipules or stipule-scars present.
              n. Buds usually superposed; axillary fruits (small red drupes) usually present; tiny pointed stipules often persistent. <u>Ilex</u>
              n. Buds not superposed.
                o. Buds small, hairy; twigs usually puberulent; disklike or cuplike bases of dry capsules often present; small shrubs. <u>Ceanothus</u>
                o. Buds not as above, but with ciliate scales; fruits, if present, blackish drupes; small trees. <u>Rhamnus</u>
            m. Stipules and stipule-scars absent.
              n. Terminal bud or remnant of terminal inflorescence present.
                o. Small trees; buds sometimes superposed; pith chambered; twigs sometimes with stellate pubescence. <u>Halesia</u>
                o. Shrubs; buds not superposed; pith not chambered.
                  p. Buds (and branches) crowded toward tip of annual stem-segments; terminal flower-buds much larger than lateral buds. <u>Rhododendron</u> (Azaleas)
                  p. Buds not crowded toward tip of annual stem-segments.
                    q. Twigs sometimes pubescent (always, in wild species), sometimes with decurrent lines from leaf-scars; leaf-scars tiny, often raised; remnant of inflorescence often present; buds often pubescent. <u>Spiraea</u>
                    q. Twigs glabrous, more or less glaucous, without decurrent lines; bud scales ciliate. <u>Nemopanthus</u>
              n. Terminal bud absent.
                o. Trees.
                  p. Bark with deep fissures surrounding polygons; outer bud-scales 2, much overlapping. <u>Diospyros</u>
                  p. Bark not as above; outer bud-scales more than 2.

q. Buds conical-globose, projecting at wide angle; fruits in large terminal panicle. Oxydendrum
q. Buds ovoid, ascending, sometimes superposed; fruits axillary. Halesia
 o. Shrubs.
  p. Buds with 2 outer scales visible, pointed, elongate, appressed, reddish; scales not mucronate, not resin-dotted; buds not of two sizes. Lyonia
  p. Buds usually with more than 2 outer scales, usually of two sizes, flower-buds larger.
   q. Flower-buds resin-dotted. Gaylussacia
   q. Flower-buds not resin-dotted; bud-scales mucronate, the outer often with long sharp points. Vaccinium
h. Vein-scars 3 or more.
 i. Pith diaphragmed; trees.
  j. Pith hollow between diaphragms, brown; buds superposed. Juglans
  j. Pith solid between diaphragms, white; buds sometimes superposed. Nyssa
 i. Pith not diaphragmed.
  j. Vein scars 3 or in 3 groups.
   k. Twigs usually bearing catkins in winter; pith small, 3-sided.
    l. Buds stalked; new staminate and carpellate catkins and old conelike carpellate ones usually present, old ones with persistent woody scales. Alnus
    l. Buds sessile or somewhat stalked; new cylindrical staminate catkins and sometimes also old carpellate ones usually present, scales of old ones deciduous; twigs sometimes with wintergreen flavor; bark sometimes separating in white, pink, tan, or brown layers. Betula
   k. Catkins absent in winter (or present in Myrica); pith not 3-sided.
    l. Twigs spicy-aromatic and dotted with glandular spots.
     m. Stipule-scars present. Myrica (Comptonia)
     m. Stipule-scars absent; gray-white waxy berries present. Myrica
    l. Twigs spicy-aromatic but without glandular spots. (See third l.)
     m. Buds often superposed, often clustered; leaf-scars not conspicuously crowded toward tip of annual stem-segment. Lindera
     m. Buds not superposed, not collateral; leaf-scars conspicuously crowded toward tip of annual stem-segment. Cotinus
    l. Twigs not spicy-aromatic.
     m. Lowest bud-scale situated above middle of leaf-scar; branch- and stipule-scars present; pith 5-pointed. Populus
     m. Lowest bud-scale not above middle of leaf-scar; branch-scars not regularly present; stipule-scars present or absent.
      n. Leaf-scar raised on persistent petiole-base; pith large, brown, crenate or 5-angled; shrubs. Rubus
      n. Without the above set of characteristics.
       o. Stipule-scars present.
        p. Buds superposed; panicles of short resin-dotted oblong pods usually present; shrubs. Amorpha
        p. Buds not superposed; panicles of such pods not present.
         q. Low shrubs; twigs or buds or both pubescent or puberulent; disklike or cuplike bases of capsules often present. Ceanothus
         q. Taller shrubs or trees; fruits of the above sort not present.
          r. Exposed bud-scales 2 or 3; pith 5-pointed; trees, introduced; native species now only sprouts or young trees. Castanea
          r. Exposed bud-scales usually more than 3; trees or shrubs; pith round or only slightly angled.

s. Older bark shreddy; leaf-scars decurrent; buds appressed; fruit clusters (aggregates of follicles) sometimes present. Physocarpus

s. Bark not shreddy; leaf-scars not decurrent; fruits, if present, not aggregates of follicles.

    t. Crushed bark and wood of twigs with odor of bitter almonds. Prunus

    t. Such odor not present.

        u. Leaf-scars short; buds projecting at rather wide angle from stem, scales somewhat fleshy, often reddish; fruits if present, pomes. Crataegus

        u. Leaf-scars half-elliptic; bud-scales ciliate; sometimes remnants of fruits (small drupes) in axils. Rhamnus

o. Stipule-scars absent.

  p. Leaf-scars large, shield-shaped, triangular, or 3-lobed; vein-scars several; terminal bud large, sometimes appearing naked because valvate scales are leaflike, sometimes with overlapping scales; buds sometimes superposed; pith 5-pointed. Carya

  p. Leaf-scars, vein-scars, and twigs not as above.

    q. Outer bark shreddy; leaf-scars decurrent; shrubs. Ribes

    q. Bark not shreddy; leaf-scars not decurrent.

      r. Twigs brown, green, or reddish; lateral buds very small; leaf-scars mostly toward summit of annual stem-segment (more easily seen on 2-year-old twigs). Cornus

      r. Without the above set of characters.

        s. Pith star-shaped; bud-scales appearing varnished; vein-scars conspicuous; branches cork-winged. Liquidambar

        s. Without the above set of characters.

          t. Buds globose- or oblong-ovoid; scales fleshy, often red. Crataegus

          t. Buds ovoid, tip not rounded. (See third t)

            u. Leaf-scar round or half-round. Sorbaria

            u. Leaf-scar short, much wider than long.

              v. Leaf-scar wide, a thin line. Rosa

              v. Leaf-scar about as wide as bud; bud sometimes hairy. Pyrus

          t. Buds elongate, appressed, often red.

            u. Buds long-pointed, about three times as long as wide, often hairy at tip or at edges of scales. Amelanchier

            u. Buds short-pointed; scales often edged with glands. Pyrus (Chokeberry)

j. Vein-scars more than 3, not in 3 groups.

  k. Buds clustered at ends of branches; bud-scales 5-ranked, often many, closely imbricated; pith 5-angled or -lobed. Quercus

  k. Buds not clustered at ends of branches; bud-scales not as above.

    l. Leaf-scars shield-shaped, triangular, or 3-lobed, large.

      m. Terminal bud present.

        n. Pith 5-lobed or -angled; buds sometimes superposed; end-bud with overlapping scales or appearing naked because valvate scales are leaflike. Carya

        n. Pith circular; plants poisonous to touch. Rhus vernix

      m. Terminal bud absent; pith brown; twigs stout; buds hemispheric; vein scars about 9. Ailanthus

    l. Leaf-scars crescent-shaped or short; vein-scars not more than 7.

m. Buds hairy or gummy; stipule-scars absent; leaf-scars not decurrent. <u>Pyrus</u> (Mountain-ash)

m. Buds not hairy, not gummy; stipule-scars present; leaf-scars decurrent; lowest vein-scar largest. <u>Physocarpus</u>

c. Leaf-scars 2-ranked.

  d. Leaf-scar ring-shaped, surrounding or nearly surrounding bud; terminal bud absent; buds hairy.

    e. Bud solitary, short-conic; twigs very flexible, appearing jointed because enlarged at tip of season's growth; base of terminal bud-scar prominent. <u>Dirca</u>

    e. Buds superposed usually in groups of 3; stems not flexible, not appearing jointed. <u>Cladrastis</u>

  d. Leaf-scar not ring-shaped, not nearly surrounding bud.

    e. Buds naked, or naked after abscission of 2 stipular scales, or so hairy that scales are not discernible.

      f. Terminal bud absent; buds superposed, scurfy. <u>Styrax</u>

      f. Terminal bud present.

        g. Stipule-scars present; pith continuous, without diaphragms; terminal bud with 2 stipular scales, or scars of these scales. <u>Hamamelis</u>

        g. Stipule-scars absent; pith with firmer diaphragms, sometimes chambered; lateral buds sometimes superposed. <u>Asimina</u>

    e. Buds with obvious scales.

      f. Vein-scar 1 or several confluent appearing as 1; pith may have cavities; bud scales 2, broad, much overlapping. <u>Diospyros</u>

      f. Vein-scars 3 or more.

        g. Pith 3-angled, small; aments usually present; twigs sometimes with wintergreen flavor; buds often elongate; bark sometimes separating in white, pink, tan, or brown layers. <u>Betula</u>

        g. Pith 5-angled, star-shaped. (See third g.)

          h. Bud-scales several, 4-ranked; buds often of two sizes on twig, the larger usually 4-angled; trunks and larger branches fluted; bark dark gray, smooth. <u>Carpinus</u>

          h. Visible bud-scales 2 or 3; buds not obviously of two sizes, not 4-angled. <u>Castanea</u>

        g. Pith neither 3-angled nor 5-angled.

          h. Stipule-scars absent.

            i. Terminal bud present, usually long-pointed; leaf-scar short. <u>Amelanchier</u>

            i. Terminal bud absent; buds small, obtuse; leaf-scar oval or triangular. <u>Cercis</u>

          h. Stipule-scars present.

            i. Terminal bud present; pith with a diaphragm at each node but without diaphragms between nodes. <u>Broussonetia</u>

            i. Terminal bud absent.

              j. Buds appressed to stem; bud-scales 2-ranked.

                k. Pith chambered, sometimes only at nodes; vein-scars 3 or confluent in 3 groups. <u>Celtis</u>

                k. Pith continuous; vein-scars more than 3, often in an ellipse. <u>Morus</u>

              j. Buds not appressed to stem.

                k. Visible bud-scales 2 or a third visible at tip, usually red or green; buds bulging more on one side than on the other, oblique to twig. <u>Tilia</u>

                k. Visible bud-scales more than 2.

                  l. Bud-scales 2-ranked.

                    m. Vein-scars 3, some of which may be aggregates; buds often hairy, especially at tip. <u>Ulmus</u>

                    m. Vein-scars more than 3, often in an ellipse; buds not hairy. <u>Morus</u>

                  l. Bud-scales not 2-ranked.

                    m. Bud scales 4-ranked, slightly striate; buds usually of 2 sizes on twig, the larger 4-angled; bark dark gray, smooth; trunks and larger branches fluted; no aments present in winter. <u>Carpinus</u>

                    m. Bud-scales not 4-ranked; buds not of two sizes, not 4-angled; aments present in winter.

n. Bud-scales striate; bark finely furrowed and scaly; small trees. <u>Ostrya</u>

n. Bud-scales not striate; buds blunt at apex; shrubs (our species) or small trees. <u>Corylus</u>

Glossary

Acaulescent. Without an aerial leaf-bearing stem.

Accrescent. Continuing to enlarge after the usual time.

Achene. A dry indehiscent fruit with thin pericarp fitting closely around the single seed.

Actinomorphic. Regular; radially symmetrical; descriptive of a flower or set of flower-parts which can be cut through the center into equal and similar parts along 2 or more planes.

Acuminate. Long-tapering to a pointed apex or base.

Acute. Sharp-pointed.

Adnate. United; said of unlike structures, as stamen and petal.

Aggregate fruit. A ripened gynecium of separate carpels.

Alternate. Describing the arrangement of leaves or other structures which occur singly at successive nodes or levels; not opposite or whorled.

Alveolate. Like a honeycomb.

Ament. A slender, usually flexible, often pendent, spike or raceme of monosporangiate apetalous flowers.

Anastomosing. Connected forming a network.

Andrecium, Androecium. The set of stamens of a flower.

Annual. Living through but one growing season; yearly.

Anther. The upper part of the stamen, containing microsporangia and, later, pollen sacs.

Anthesis. The period during which flowers are open.

Apetalous. Without petals.

Appressed. Closely pressed against, as a bud against a stem.

Aquatic. Living in water.

Arachnoid. Cobwebby.

Arcuate. Curved in form of a bow; arching.

Areolate. With areoles.

Areole. A small marked off space, as one formed by anastomosing veins.

Aril. A fleshy appendage of some seeds from the region of the hilum, partially or wholly covering the seed.

Aristate. Tipped by a bristle or awn.

Articulate. Jointed; consisting of segments united at joints.

Ascending. Growing obliquely upward.

Attenuate. Long-tapering, becoming very narrow.

Auricle. An earlike lobe or appendage.

Auricled, Auriculate. Having an auricle or auricles.

Awl-shaped. Tapering from a narrow base to a point.

Awn. A slender bristlelike structure; in grasses, usually a continuation of a nerve of the lemma, the glumes, or rarely the palea.

Axil. Angle formed by one structure with another, as a leaf with a stem.

Axile. In the axis; descriptive of placentae at center of an ovulary.

Axillary. In the axil.

Barbellate. Minutely barbed; said of pappus bristles that have short hairs attached along the side.

Bearded. Bearing long or stiff hairs.

Berry. A fruit with fleshy pericarp.

Biennial. Living through two growing seasons.

Bifid. Two-cleft.

Bipinnate. Twice pinnately compound.

Bipinnatifid. Twice pinnatifid.

Bisporangiate (flower). Having both stamens and carpels.

Blade. The expanded, more or less flat, portion of a leaf, petal, sepal, etc.; the portion of a leaf above the petiole or the sheath.

Bract. A leaf unlike ordinary foliage leaves, usually smaller or of different shape, usually found as part of the inflorescence, subtending a flower or a flower-cluster.

Bractlet. Secondary bract.

Branchlet. An ultimate branch; in woody plants, the stem segment formed during the latest growing season.

Bulb. Bud, often subterranean, with fleshy scales.

Bundle-scar. Vein-scar.

Caducous. Falling very early.

Callosity. More or less protuberant thickening.

Callus. Hardened downward extension of the lemma in grasses.

Calyx. The set of separate or united sepals of a flower.

Campanulate. Bell-shaped.

Canescent. With fine close gray or white pubescence; hoary.

Capillary. Hairlike.

Capitate. Like a head; in a head.

Capsule. Dry dehiscent fruit of 2 or more carpels.

Carpel. Megasporophyll of seed plants; the ovule-bearing structure.

Carpellate. Having only carpels or carpellate flowers.

Cartilaginous. Tough and hard but elastic.

Caruncle. An appendage at the hilum of a seed.

Catkin. Ament.

Caudate. Tailed; tail-like.

Caulescent. Having an obvious stem above ground.

Cauline. Of the stem.

Chaff. Bracts on the receptacle of plants in the Composite Family.

Chartaceous. Having texture of paper.

Cilia. Marginal hairs.

Ciliate. Having cilia.

Circumscissile. Dehiscent crosswise, the top coming off as a lid.

Clasping (leaf). With its base partly or wholly surrounding the stem.

Clavate. Club-shaped.

Claw. Narrow or stalk-like base of a petal or similar structure.

Cleft. See Lobed.

Cleistogamous. Descriptive of a flower the perianth of which remains unopened.

Coalescent. United; said of parts or structures of the same kind.

Coma. Tuft of hairs.

Comate, Comose. With a coma.

Conduplicate. Folded lengthwise.

Cone. A determinate axis bearing sporophylls in regular arrangement.

Confluent. Merging, indistinguishable as individuals.

Connate. United; said of like structures, as stamen and stamen.

Connate-perfoliate. Said of opposite sessile leaves with bases united around the stem and appearing as a single blade pierced by the stem.

Connective. The portion of an anther between the pollen sacs.

Connivent. Not united but with margins meeting.

Cordate. Heart-shaped.

Coriaceous. Leathery.

Corm. A solid fleshy upright underground stem.

Corolla. The set of separate or united petals of a flower.

Corona. A crownlike outgrowth from a corolla or an andrecium.

Corymb. A convex or flat-topped inflorescence, the outer flowers opening first; like a raceme except that the lower pedicels are longer than the upper.

Corymbiform. In the form of a corymb, but not necessarily a true corymb.

Crenate. With rounded teeth or scallops.

Crenulate. Finely crenate.

Culm. The stem of grasses; sometimes, also, the stem of sedges.

Cuneate. Wedge-shaped.

Cuspidate. With a stiff sharp point.

Cyathium. Inflorescence in Euphorbia; see description of that genus.

Cyme. A convex or flat-topped determinate inflorescence, the central flower opening first.

Deciduous. Abscising; descriptive of a plant whose leaves all abscise at end of one growing season.

Decompound. More than once compound.

Decumbent. Lying on the ground, the apex ascending.

Decurrent. Said of one organ extending along the edge of another, as of a leaf blade extending as a wing along the stem.

Deflexed. Bent or turned abruptly downward.

Dehiscence. Process or method of splitting open at maturity of a structure such as a fruit or a pollen sac.

Dehiscent. Splitting open, usually in some regular way.

Deltoid. With shape of an equilateral triangle.

Dentate. With rather coarse teeth projecting at right angle to margin.

Denticulate. Finely dentate.

Diadelphous (stamens). United in 2 groups.

Diaphragmed (pith). With transverse partitions or plates.

Dichotomous. Two-forked.

Diecious, Dioecious. Descriptive of a species having only monosporangiate flowers, the staminate and carpellate flowers on separate plants.

Diffuse. Spreading loosely.

Digitate (leaf). Compound with leaflets arising at one point (top of petiole).

Dimorphic. Of two forms.

Disc, Disk. See Hypogynous Disc.

Discoid. See description of Composite Family.

Dissected. Divided into narrow segments.

Distal. Away from the center or point of attachment.

Divaricate. Greatly divergent.

Divergent. Spreading; separated.

Divided. See Lobed.

Downy. With short fine soft hairs.

Drupe. An indehiscent fruit with outer portion of pericarp fleshy and inner portion of pericarp stony.

Dwarf branch. A short leaf-bearing branch of pine and other conifers.

Echinate. With prickles, the prickles often short and thick.

Ellipse. A surface longer than wide, rounded similarly at both ends, widest at middle.

Ellipsoid. A solid the longitudinal section of which is an ellipse.

Elliptic, Elliptical. In form of an ellipse.

Emarginate. Shallowly notched at apex.

Emersed. Above water.

Entire (margin). Unbroken by indentations.

Ephemeral. Persisting for a short time, as a day or less.

Epigynous (flower). In which the sepals, petals, and stamens are borne above the gynecium.

Epigynous disc, disk. See Hypogynous disc.

Epiphyte. A plant growing attached to another plant but not parasitic on it.

Erose (margin). Appearing as if gnawed.

Even pinnate. With an even number of leaflets.

Evergreen. With green leaves throughout the year.

Excurrent. Extending beyond the margin.

Exfoliating. Peeling off.

Exserted. Protruding; extending beyond surrounding parts, as stamens extending beyond corolla.

Falcate. Scythe-shaped; with curved axis.

Fascicle. A cluster or bundle.

Fertile. Bearing seeds, spores, or gametes.

Fibrillose. Containing fibers or disintegrating into fibers.

Filament. Stalklike part of a stamen.

Filiform. Narrow, threadlike.

Fimbriate. Fringed.

Fimbrillate. Minutely fringed.

Flexuous. Wavy or more or less zigzag.

Floret. Of grasses, consisting of lemma, palea, and flower-parts.

Flower. A determinate sporophyll-bearing stem-tip.

Foliaceous. Leaflike.

Follicle. A dry fruit developing from a single carpel, dehiscing along one suture.

Fruit. A ripened gynecium and such parts as ripen with it; sometimes a ripened flower-cluster (multiple fruit).

Funnelform. With shape of a funnel.

Fusiform. Spindle-shaped; thickest at middle, tapering toward each end.

Galeate. With a hood-shaped or helmet-shaped part.

Gibbous. Enlarged on one side.

Glabrate. Becoming glabrous with age; nearly glabrous.

Glabrous. Without hairs.

Gland. A secreting structure; a body with appearance of a gland.

Glaucous. Covered with a whitish powdery or waxy substance that can be rubbed off.

Globose. Sphere-shaped.

Glomerule. A small compact cluster.

Glumaceous. Resembling or having glumes.

Glume. One of the 2 bracts at base of spikelet of grasses.

Glutinous. Sticky or gummy.

Grain. One-seeded fruit with ovule-coat adnate to pericarp; caryopsis.

Granular, Granulate, Granulose. Covered with small grains or meal.

Gynecium, Gynoecium. The set of carpels of a flower.

Habitat. Place where a plant grows.

Halberd-shaped. Hastate.

Hastate. With shape of an arrow-head but with basal lobes divergent.

Head. A dense globular cluster of sessile or nearly sessile flowers.

Hilum. Scar on a seed where stalk was attached.

Hirsute. Having coarse, rather stiff, hairs.

Hispid. Having bristly hairs.

Hispidulous. Minutely hispid.

Hoary. Canescent.

Hyaline. Thin and translucent or transparent.

Hypanthium. As used in this manual, a cup-like, saucerlike, or tubular structure on which the sepals, petals, and stamens are borne, the structure growing from the receptacle below the gynecium (hypanthium then hypogynous, flower perigynous); also a similar structure above the gynecium of an epigynous flower (hypanthium epigynous). Since origin of the structure is usually not superficially apparent, this term has been chosen arbitrarily and used throughout; in some instances its technical accuracy may be questioned.

Hypogynous (flower). Having calyx, corolla, and andrecium below the gynecium.

Hypogynous disc or disk. A fleshy cushionlike structure growing from the receptacle below the gynecium and above the attachment of the calyx (flower hypogynous); a similar structure within the hypanthium (disk then perigynous) or upon the gynecium of an epigynous flower (disk then epigynous).

Imbricate, Imbricated. Overlapping as do shingles on a roof.

Incised. Cut sharply and irregularly.

Included. Not projecting beyond surrounding parts.

Indehiscent. Not opening.

Indusium. The covering of the sorus in ferns.

Inequilateral. With unequal sides.

Inferior. Arising below other flower-parts; descriptive of ovulary of an epigynous flower.

Inflexed. Turned inward.

Inflorescence. Flower-cluster.

Inserted. Attached.

Internode. Portion of stem between 2 successive nodes.

Involucral. Of an involucre.

Involucrate. With an involucre.

Involucre. A set or circle of bracts below a flower or a flower-cluster.

Irregular. Not regular.

Isobilateral. Can be cut through the center into halves along 2 planes at right angle to each other, but the halves made by one plane unlike those made by the other.

Keel. Projecting ridge; the two lower petals of a papilionaceous corolla.

Laciniate. Cut into narrow pointed segments.

Lanceolate. Lance-shaped; widest above base, tapering to apex, several times as long as wide.

Leaflet. One of the divisions of the blade of a compound leaf.

Legume. A dry fruit of one carpel dehiscent along suture and midrib.

Lemma. In grasses, the lower (outer) of the 2 bracts enclosing a flower.

Lenticular. Lens-shaped; biconvex.

Ligulate. With or resembling a ligule; (corolla), see description of Composite Family.

Ligule. A small flap or appendage; specifically, the appendage at junction of blade and sheath in grasses.

Limb. Expanded part of a sympetalous corolla distal to the tube.

Linear. Long and narrow, the sides parallel or nearly so.

Lobed. Separated by indentations (sinuses) into segments (lobes) longer than teeth. Strictly, relative depth of indentations (and length of segments) is indicated by the succession of terms, lobed, cleft, parted, divided, as follows: lobed, indented less than halfway to base or midvein; cleft, about halfway; parted, more than halfway; divided, most of the way. Loosely, lobed is used to mean any of these degrees of division.

Locule. Cavity of an ovulary or anther.

Loment. A legume divided by transverse constrictions into 1-seeded segments that separate at maturity.

Lyrate. Pinnatifid, the terminal lobe largest.

Marcescent. Withered but persistent.

Margined (petiole, rachis). With a strip of blade along each side.

Membranous, Membranaceous. Thin and pliable.

-merous. Suffix meaning in number of parts.

Midrib. Midvein.

Monadelphous (stamens). United in one group.

Monecious, Monoecious. Descriptive of a species having staminate and carpellate flowers on same plant.

Moniliform. Constricted at regular intervals; like a string of beads.

Monosporangiate. Descriptive of a flower having either stamens or carpels but not both.

Mucronate. With a sharp abrupt point.

Multiple fruit. Formed by ripening of a flower-cluster.

Muricate. With sharp points or prickles.

Nerve. Prominent vein of a leaf or flower-part.

Neuter, Neutral (flower). With neither stamens nor carpels.

Node. Short zone of stem at level where leaf or leaves are borne.

Nodulose. Having little knobs.

Nut. A hard indehiscent 1-seeded fruit.

Ob-. Prefix meaning in an opposite direction; thus obovate is inversely ovate.

Oblique. With unequal sides; not at right angle to; slanting.

Oblong. Longer than wide, the sides nearly or quite parallel and the ends rounded.

Obtuse. Blunt.

Ocrea. A tubular sheath around the stem formed by a pair of united stipules.

Ocreola. Small ocrea in the inflorescence of plants in the Buckwheat Family.

Odd pinnate (leaf). With an odd number of leaflets.

Opposite. Two at a node or at same level of an axis or rachis.

Orbicular. Circular.

Oval. Broadly elliptic.

Ovate. Having the shape of a median longitudinal section through an egg with widest portion below the middle.

Ovoid. Egg-shaped, the widest portion below the middle.

Ovulary. Lowest part of a closed carpel or set of united carpels, in which ovules are borne.

Ovule. Megasporangium in spermatophytes; forerunner of a seed.

Palate. A projection on lower lip of a 2-lipped sympetalous corolla partially or wholly closing the throat.

Palea. The bract subtending the flower of grasses.

Palmately compound. With leaflets all arising at apex of petiole.

Palmately veined. With principal veins diverging from top of petiole.

Panicle. A compound elongate inflorescence.

Paniculate. Like a panicle; in a panicle.

Paniculiform. With the form or appearance of a panicle but not necessarily a true panicle.

Papilionaceous (flower). See description of Pea Family.

Papillate, Papillose. With minute rounded projections.

Pappus. The bristles, hairs, awns, or teeth at top of an achene, as in plants of Composite Family.

Parietal (placenta). On the inner surface of wall of ovulary.

Parted. See Lobed.

Pectinate. Pinnatifid, with divisions narrow, resembling teeth of a comb.

Pedicel. Stalk of a flower in a flower-cluster.

Peduncle. Stalk of a solitary flower or of a flower-cluster.

Pellucid. Transparent.

Peltate. Attached by some part of the surface instead of by the margin.

Pepo. A fruit with hard or leathery rind, as a melon or gourd.

Perennial. Living through more than two growing seasons.

Perfect (flower). Bisporangiate.

Perfoliate (leaf). With base surrounding the stem and appearing as if pierced by the stem.

Perianth. Calyx and corolla together or one of them, if only one is present.

Pericarp. Wall of the fruit.

Perigynium. A sac enclosing the ovulary of a flower of Carex.

Perigynous (flower). Having sepals, petals, and stamens borne on a hypanthium which is free from the ovulary.

Perigynous disc, disk. See Hypogynous disc.

Petal. One member of the set of flower-parts between stamens and sepals.

Petaloid. Like a petal or petals.

Petiole. Stalk of a leaf.

Petiolule. Stalk of a leaflet.

Phyllary. One of the bracts making up the involucre subtending the head of flowers in composites.

Pilose. With long soft hairs.

Pinna. A primary division of a pinnately compound leaf.

Pinnate (leaf). Compound with leaflets along the rachis.

Pinnately compound. Pinnate.

Pinnately veined. With one main vein, the midvein, from which principal lateral veins branch.

Pinnatifid. Pinnately cleft.

Pinnule. A division of a pinna.

Placenta. Ridge or surface within the ovulary on which ovules are borne.

Placentation. Type of arrangement of ovules within the ovulary.

Plaited. Having folds.

Plumose. Like a plume or feather.

Pollinium. A mass of coherent pollen.

Pome. A fruit consisting of ripened ovulary plus hypanthium adnate to ovulary.

Prickle. A sharp pointed outgrowth from epidermis or cortex of stem.

Procumbent. Prostrate, not rooting at nodes.

Prophyll. One of the bracts subtending the flower in some species of Juncus.

Pubescent. Hairy.

Pulvinus. An enlarged base or apex of petiole or petiolule, as in the Pea Family.

Punctate. Dotted.

Pyxis. A capsule with crosswise dehiscence, the top coming off as a lid.

Raceme. Elongate inflorescence with each flower on a pedicel.

Racemiform. With form or appearance of a raceme, but not necessarily a true raceme.

Racemose. In a raceme; like a raceme.

Rachilla. Branch of a rachis; axis of the spikelet of grasses.

Rachis. Axis of a pinnately compound leaf; sometimes used, also, to mean the axis of an inflorescence.

Radiate (head). Having both disk and ray flowers.

Ray. A pedicel of an umbel; a ray flower of the composites.

Ray flower. See description of Composite Family.

Receptacle. Apex of a flower-stalk on which the flower-parts are borne.

Reflexed. Bent backward.

Regular (flower, corolla, calyx). Radially symmetrical; actinomorphic.

Reniform. Kidney-shaped.

Repand. With wavy margin.

Repent. Prostrate on the ground.

Reticulate. Forming a network.

Revolute. Rolled backward or under.

Rhizome. An underground stem.

Rosette. A cluster of radiating leaves usually near or at ground level, separated by very short internodes.

Rotate (corolla or calyx of united petals or sepals). Wheel-shaped, flat, the tube very short or absent.

Rudimentary. Vestigial.

Rugose. With wrinkled surface.

Runcinate. Pinnatifid with segments turned backward.

Saccate. Sack-shaped.

Sagittate. With shape of an arrow-head.

Salverform (corolla or calyx of united petals or sepals). With limb spreading abruptly at right angle to the slender tube.

Samara. An indehiscent winged fruit.

Scabrous. Rough.

Scape. A leafless or nearly leafless stem bearing a flower or a flower-cluster.

Scapose. Borne on a scape; with a scape.

Scarious. Transparent or translucent, thin, dry, not green.

Scurfy. Covered with scalelike particles.

Secund. Twisted or turned to one side.

Seed. A ripened ovule.

Sepal. One member of the outermost set of flower-parts.

Septate. Divided by partitions.

Serrate. Toothed, the teeth pointing forward, toward the apex.

Serrulate. Finely serrate.

Sessile. Without a stalk.

Setaceous. Bristlelike.

Silicle, Silique. See description of Mustard Family.

Sinus. Indentation between lobes or teeth.

Sorus. A cluster of sporangia in ferns.

Spadix. A fleshy spikelike inflorescence.

Spathe. A bract subtending a spadix.

Spatulate. With shape of a spatula; oblong and rounded at apex, tapering to narrow base.

Spike. An elongate inflorescence consisting of an axis bearing sessile flowers.

Spikelet. A secondary spike; specifically, the inflorescence-unit in grasses and some sedges; see description of Grass Family.

Spine. A sharp-pointed structure which in origin is a leaf or part of a leaf.

Spinulose. Having small spines.

Sporangium. A spore-case.

Sporocarp. A globose case containing a group of sporangia of Water Ferns.

Sporophyll. A spore-bearing leaf.

Spur. An elongate sac projecting from a part of a flower.

Squarrose. With spreading or recurved tip.

Stamen. Microsporophyll of seed plants, bearing microsporangia in which microspores and, later, pollen grains develop.

Staminate. Having only stamens or staminate flowers.

Staminode. A sterile structure resembling a stamen or in the position of a stamen.

Standard. The upper (odd) petal of a papilionaceous flower.

Stellate. Star-shaped.

Sterigma. A short projection or peg on which the leaf of some conifers is borne, left on the stem when the leaf abscises.

Sterile. Not forming spores, pollen, or seeds.

Stigma. Uppermost part of a carpel or set of united carpels.

Stipe. Stalk of a structure, as of an ovulary.

Stipel. Stipule of a leaflet.

Stipitate. Borne on a stipe.

Stipules. A pair of appendages, one on either side of base of petiole.

Stolon. A basal horizontal branch rooting at nodes.

Striate. Marked with fine lines.

Strict. Standing upright, straight, sometimes rigid.

Strigose. With appressed or ascending stiff hairs.

Style. The narrow portion of a carpel or set of united carpels between stigma (or stigmas) and ovulary.

Stylopodium. Expanded style-base.

Submersed. Under water.

Subtend. To be below.

Subulate. Awl-shaped.

Succulent. Soft, fleshy, juicy.

Sulcate. Furrowed or grooved.

Superior. Descriptive of the ovulary of a hypogynous flower or of a perigynous flower in which the hypanthium is free from the ovulary.

Superposed. Placed one above the other.

Suture. Line of joining of margins of a carpel.

Sympetalous. Of united petals.

Synsepalous. Of united sepals.

Tendril. A slender coiling segment of leaf or stem.

Terete. Circular in cross section.

Ternate. Divided into 3 segments or leaflets; in 3's.

Terrestrial. Growing on land, not in water.

Thallus. A plant-body not differentiated into stems, leaves, and roots.

Thorn. A sharp-pointed stem or branch of a stem.

Throat. The place where the tube and limb (of a corolla or calyx of united petals or sepals) join.

Tomentose. Densely woolly.

Trichome. A hair.

Trifoliolate. Having 3 leaflets.

Truncate (base or apex). Ending with a straight or nearly straight edge or surface perpendicular to the long axis, as if cut off.

Tuber. A short thick underground stem or part of a stem.

Tubercle. A small knob or projection.

Tunicate (bulb). With coats (leaves) in concentric layers, as an onion.

Twig. The most recently formed segment of a woody stem; branchlet.

Two-ranked. In two vertical rows on the stem.

Umbel. An inflorescence with all the pedicels arising from apex of peduncle.

Umbellate. Like an umbel; in an umbel.

Umbellet. A secondary umbel of a compound umbel.

Umbelliform. With the form or appearance of an umbel but not necessarily a true umbel.

Uncinate. Hooked at tip.

Undulate. Wavy.

Urceolate. Urn-shaped.

Utricle. A one-seeded fruit with loose thin pericarp.

Valvate. Dehiscent by valves; with margins meeting but not overlapping.

Valve. One of the segments into which a capsule splits at dehiscence.

Vein-scar. Scar within a leaf-scar resulting from breaking of a vein during leaf-abscission.

Venation. Arrangement of veins.

Verrucose. With wartlike protuberances.

Versatile (anther). Attached at or near its center to the filament.

Verticil. A whorl.

Verticillate. Whorled.

Vescicle. A small sac or cavity.

Vestige. A small undeveloped or poorly developed structure.

Vestigial. Existing as a vestige.

Villous. With long soft hairs.

Whorl. A group of 3 or more leaves or other structures at a node.

Whorled. In a whorl.

Woolly. Covered with long, somewhat matted, hairs that are not straight.

Zygomorphic (flower, corolla, calyx). Can be cut through the center along only one plane into 2 equal and similar parts that are mirror images of each other; bilaterally symmetrical.

Index

Robinia, 138
Robin's-plantain, 224
Rockrose Family, 156
Rorippa, 118
Rosa, 132
Rosaceae, 125
Rose, 132
Rose Family, 125
Rose-mallow, 154
Rose-moss, 97
Rose-of-Sharon, 154
Rose Pogonia, 77
Rosin-weed, 226
Rotala, 161
Royal Fern Family, 17
Rubiaceae, 206
Rubus, 130
Rudbeckia, 227
Rue-anemone, 105
    False, 106
Rue Family, 143
Ruellia, 205
Rumex, 92
Rush, 65
Rush Family, 65
Russian-olive, 160
Russian-thistle, 95
Rutaceae, 143
Rye, 37

Sabatia, 177
Sage, 192
Sagina, 99
Sagittaria, 26
Salicaceae, 79
Salicornia, 95
Salix, 79
Salsify, 236
Salsola, 95
Salvia, 192
Salvinia, 21
Salviniaceae, 21
Salvinia Family, 21
Sambucus, 211
Samolus, 175
Sandalwood Family, 91
Sand-spurrey, 98
Sandvine, 180
Sandwort, 99
Sanguinaria, 110
Sanguisorba, 131
Sanicula, 166
Santalaceae, 91
Saponaria, 101
Sarracenia, 121
Sarraceniaceae, 121
Sarsaparilla, 164
Sassafras, 109
Satureja, 193
Saururaceae, 79
Saururus, 79
Savory, 193

Saxifraga, 123
Saxifragaceae, 122
Saxifrage, 123
    Golden, 123
Saxifrage Family, 122
Scheuchzeria, 26
Schizachne, 37
Schizaeaceae, 17
Scirpus, 53
Scleranthus, 98
Scleria, 54
Scotch Thistle, 234
Scouring-rush, 15
Scrophularia, 199
Scrophulariaceae, 197
Scurf-pea, 137
Scutellaria, 189
Sea-rocket, 116
Secale, 37
Sedge Family, 50
Sedum, 122
Seedbox, 162
Selaginella, 16
Selaginellaceae, 16
Selaginella Family, 16
Self-heal, 190
Senecio, 232
Senecioneae, 232
Senna, 135
Sensitive Plant, 136
Sericocarpus, 221
Setaria, 49
Seymeria, 202
Shadbush, 128
Sheep-sorrel, 92
Shepherdia, 160
Shepherd's Purse, 116
Sherardia, 207
Shooting Star, 174
Sibara, 120
Sickle-pod, 135
Sicyos, 213
Sida, 154
Side-oats Grama, 44
Silene, 100
Silphium, 226
Silverbell, 176
Simaroubaceae, 144
Sisymbrium, 117
Sisyrinchium, 74
Sium, 168
Skullcap, 189
Skunk Cabbage, 63
Smartweeds, 93
Smilacina, 71
Smilax, 72
Smoke-tree, 148
Snakeroot
    Black, 166
    Sampson's, 177
    Seneca, 144

    Virginia, 91
    White, 219
Snapdragon, 199
Sneezeweed, 230
Sneezeweed Tribe, 230
Snowbell, 176
Snowberry, 210
    Creeping, 173
Snow-on-the-mountain, 146
Soapwort, 101
Solanaceae, 195
Solanum, 195
Solidago, 220
Solomon's Seal, 72
Sonchus, 236
Sorbaria, 127
Sorbus, 127
Sorghastrum, 50
Sorghum, 50
Sorrel-tree, 173
Sour Gum Family, 171
Sourwood, 173
Sow-thistle, 236
Spanish Needles, 230
Sparganiaceae, 23
Sparganium, 23
Spartina, 44
Spatter-dock, 102
Spearmint, 194
Specularia, 213
Speedwell, 201
Spergula, 99
Spergularia, 98
Spermacoce, 208
Sphenopholis, 39
Spicebush, 110
Spider Flower, 112
Spiderwort, 65
Spiderwort Family, 64
Spikenard, 164
Spike-rush, 52
Spindle-tree, 149
Spiraea, 126
    False, 127
Spiranthes, 77
Spirodela, 64
Spleenwort, 19, 20
Sporobolus, 42
Spring Beauty, 97
Spurge, 146
Spurge Family, 145
Spurrey, 99
Squarrosae, 61
Squash, 213
Squawroot, 204
Squaw-weed, 232
Squirrel-corn, 111
Squirrel-tail Barley, 38
Stachys, 191
Staff-tree Family, 148
St. Andrew's Cross, 155

Vitis, 152
Vulpinae, 58

Wahoo, 149
Wake Robin, 72
Waldsteinia, 129
Wallflower, Western, 118
Walnut, 84
Walnut Family, 83
Water Cress, 118
Water-hemlock, 168
Waterleaf, 183
Waterleaf Family, 182
Water-lily, 102
Water-lily Family, 102
Water-lotus, 102
Water-milfoil, 164
Water-milfoil Family, 163
Water-parsnip, 168
Water-pennywort, 166
Water-plantain, 26
Water-plantain Family, 26
Water-purslane, 161
Water-shield, 102
Water Stargrass, 65
Water Starwort, 147
Water Starwort Family, 147
Waterweed, 27
Water-willow, 206
Waterwort, 156
Waterwort Family, 156
Waxweed, Blue, 161
Wayfaring Tree, 211

Wheat, 37
White-top, 224
Whitlow-grass, 114
Whitlow-wort, 98
Whorled Pogonia, 77
Wild Balsam Apple, 213
Wild Bergamot, 192
Wild Crab, 128
Wild Ginger, 91
Wild Hyacinth, 70
Wild Licorice, 207
Wild Rice, 45
Wild-rye, 37
Wild Sweet Potato, 180
Willow, 79
Willow Family, 79
Willow-herb, 162
Wineberry, 130
Winged Pigweed, 94
Wing-stem, 229
Winterberry, 148
Winter Cress, 119
Wintergreen, 171
    Creeping, 173
Witch Hazel, 125
Witch Hazel Family, 124
Wolffia, 64
Wolffiella, 64
Wood Betony, 203
Woodbine, 152
Wood-nettle, 90
Wood Reed, 41
Wood-rush, 67

Woodsia, 18
Wood Sorrel, 142
Wood Sorrel Family, 142
Woodwardia, 20
Wormwood, 231

Xanthium, 226
Xanthorhiza, 107
Xanthoxylum, 144
Xyridaceae, 64
Xyris, 64

Yam, 73
Yam Family, 73
Yarrow, 231
Yellow-eyed-grass, 64
Yellow-eyed-grass Family, 64
Yellow Mandarin, 71
Yellow Rocket, 119
Yellow-root, 107
Yellowwood, 136
Yew, 21
Yew Family, 21
Yucca, 71

Zannichellia, 25
Zea, 50
Zizania, 45
Zizanieae, 45
Zizia, 168
Zosteraceae, 23
Zygadenus, 69
Zygophyllaceae, 143

LEAVES

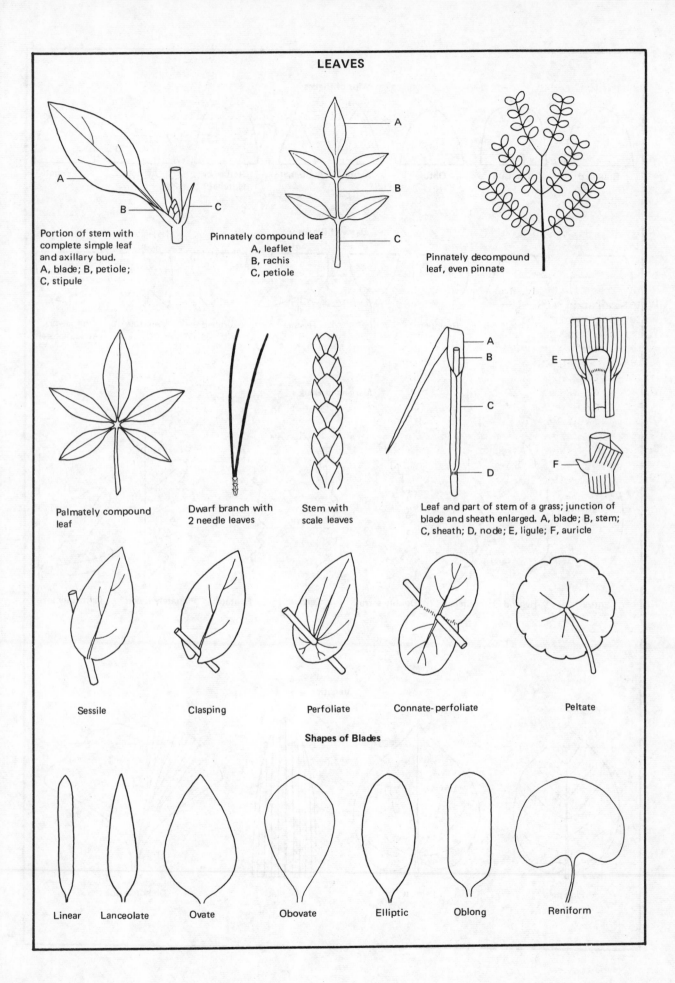

Portion of stem with complete simple leaf and axillary bud.
A, blade; B, petiole; C, stipule

Pinnately compound leaf
A, leaflet
B, rachis
C, petiole

Pinnately decompound leaf, even pinnate

Palmately compound leaf

Dwarf branch with 2 needle leaves

Stem with scale leaves

Leaf and part of stem of a grass; junction of blade and sheath enlarged. A, blade; B, stem; C, sheath; D, node; E, ligule; F, auricle

Sessile

Clasping

Perfoliate

Connate-perfoliate

Peltate

Shapes of Blades

Linear   Lanceolate   Ovate   Obovate   Elliptic   Oblong   Reniform

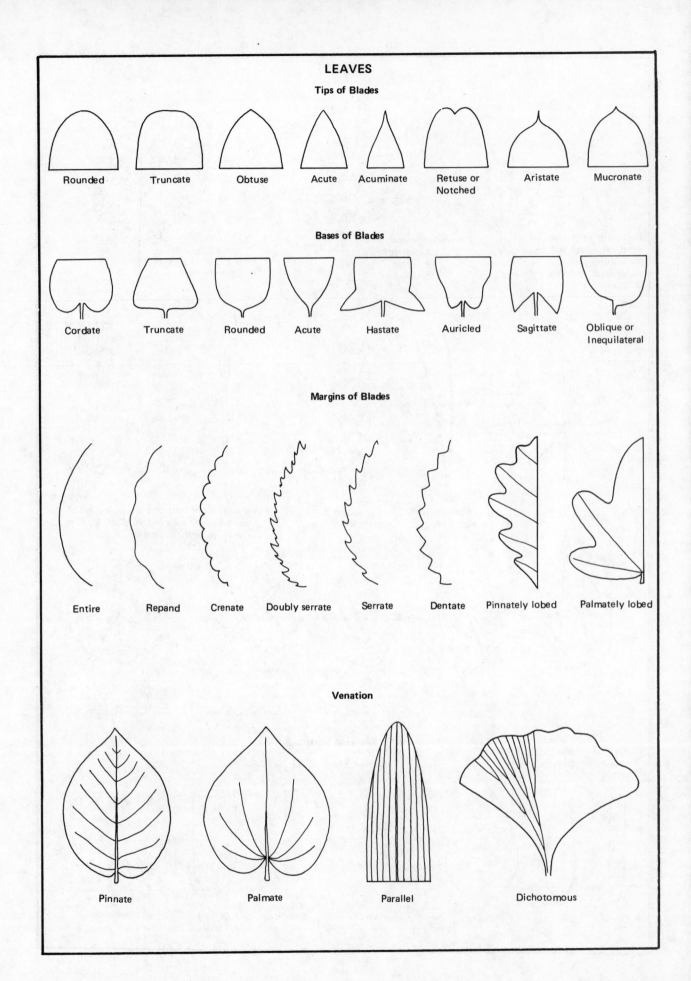

LEAVES

Tips of Blades

Rounded   Truncate   Obtuse   Acute   Acuminate   Retuse or Notched   Aristate   Mucronate

Bases of Blades

Cordate   Truncate   Rounded   Acute   Hastate   Auricled   Sagittate   Oblique or Inequilateral

Margins of Blades

Entire   Repand   Crenate   Doubly serrate   Serrate   Dentate   Pinnately lobed   Palmately lobed

Venation

Pinnate   Palmate   Parallel   Dichotomous

FLOWERS

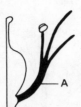

Hypogynous.
A, stamen;
B, petal;
C, sepal

Hypogynous, stamens adnate to petals

Perigynous.
A, hypogynous hypanthium

Epigynous, ovulary partly inferior

Epigynous, ovulary wholly inferior

Epigynous with epigynous hypanthium

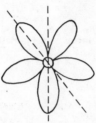

Stamens opposite petals

Stamens alternate with petals

Regular corolla

Zygomorphic corolla

Isobilateral corolla

Gynecia of 2 carpels. A, carpels separate; B-F, carpels united (from united only at base to completely united);
G, stigmas united, styles and ovularies separate

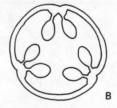

Cross section of ovulary of one carpel

Cross sections of ovularies of 3 united carpels. A, placentation axile; B, placentation parietal; C, placentation free central

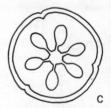

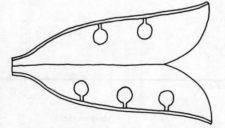
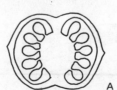

Ovulary of one carpel, split open

Cross sections of ovularies of 2 united carpels. A, placentae axile; B, placentae parietal

FLOWER CLUSTERS

Spike. A, peduncle

Raceme. A, bract; B, pedicel

Panicle

Cyme

Corymb

Head

Cyathium

Umbel

Spadix with spathe

Grass spikelet. A, floret; B, rachilla; C, second glume; D, first glume

Grass floret. A, lemma; B, palea

FERNS

Leaf and rhizome

Sporangia. A, opening transversely; B, opening longitudinally

Leaflet with marginal sori

Portion of blade with sori

Five types of indusia

Marginal sori. A, continuous; B, discontinuous

TWIGS IN WINTER CONDITION

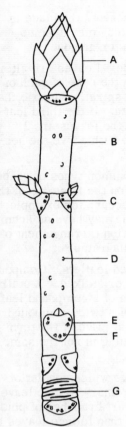

A, terminal bud, with imbricate
scales; B, internode; C, node;
D, lenticel; E, leaf-scar; F, vein-
scar; G, bud-scale scars

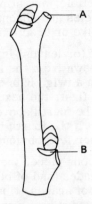

Leaf-scars alternate,
2-ranked. A, scar of
terminal bud; B,
stipule-scar

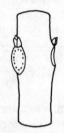

Leaf-scars whorled;
vein-scars in an
ellipse

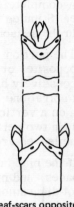

Leaf-scars opposite,
meeting around twig

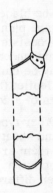

Stipule-rings

Pith 5-pointed

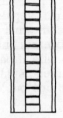

Pith diaphragmed,
hollow

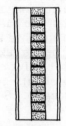

Pith diaphragmed,
solid

Pith continuous

Vein-scar single

Buds superposed,
bud-scales valvate

Buds collateral

Leaf-scar almost
encircling bud

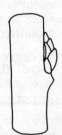

Bud appressed

<u>Note</u>. Terms used but not explained in these pages can be found in the glossary.

Leaves

A complete leaf consists of <u>blade</u>, <u>petiole</u>, and <u>stipules</u>. If the petiole is absent, the blade is <u>sessile</u>. Stipules are present in some species and absent in others; if present, they may abscise early in the season. Stipules may be leaflike or bractlike, or they may be spines or glands.

The portion of a stem at which a leaf (or leaves) is attached is called a <u>node</u>. Leaves are <u>alternate</u>, <u>opposite</u>, or <u>whorled</u>. If a line is drawn from the lowest leaf-base through the base of each of the successively higher alternate leaves on a twig, it is evident that the line is a spiral. Suppose that on this spiral one leaf is designated as the first, the next above as the second, etc. If the third leaf-base is on a vertical line above the first, the fourth above the second, and so on, the leaves are 2-ranked (in two vertical rows on the stem). If the fourth is above the first and the fifth above the second, the leaves are 3-ranked. Several other less obvious types of ranking occur.

If the blade of a leaf is in one continuous piece, the leaf is <u>simple</u>. The common types of shapes, tips, bases, and margins of simple leaf-blades (and of bladelets) are illustrated on the first two of the pages of diagrams. Various combinations of shapes, tips, bases, and margins occur; for example, an ovate blade may have an acute tip, a rounded base, and an entire margin, or it may have an acuminate tip, a cordate base, and a serrate margin. The four general types of <u>venation</u> (arrangement of the principal veins in a leaf-blade) are illustrated on the second of the pages of diagrams.

The blade of a <u>compound</u> leaf is divided into wholly separate segments called <u>leaflets</u>. Compound leaves are <u>pinnately</u> compound or <u>palmately</u> compound. Leaflets may be sessile or stalked. A stalked leaflet consists of <u>bladelet</u> (expanded portion) and <u>petiolule</u> (stalk). If the leaflets of a compound leaf are compounded one or more times, the leaf is <u>decompound</u>. Ultimate leaflets of a twice compound leaf may be called secondary leaflets; of a thrice compound leaf, tertiary leaflets. Use of the term ultimate leaflet (or bladelet) is convenient, especially when the degree of compounding is not the same in all parts of a compound leaf.

Other types of foliage leaves include needle leaves, scale leaves, and leaves which consist of blade and sheath. Many plants have one or more kinds of leaves in addition to ordinary foliage leaves. Among them are bud-scales, spines, scale leaves on rhizomes and at base of dwarf branches of pine, and bracts that subtend flowers and flower-clusters. These are usually smaller than foliage leaves and are often different in shape.

Flowers

In a complete flower of angiosperms there are four sets of parts arranged in concentric cycles or in spirals. These parts, in order, beginning at the base, are: <u>sepals</u>, <u>petals</u>, <u>stamens</u>, and <u>carpels</u>. In the hypogynous flower, these parts are attached to a somewhat enlarged stem-tip, the <u>receptacle</u>. Sepals taken together make up the <u>calyx</u>; petals taken together make up the <u>corolla</u>. <u>Perianth</u> consists of calyx and corolla together, or of either calyx or corolla if only one of them is present. Stamens collectively make up the <u>andrecium</u> (androecium). A stamen usually consists of <u>filament</u> and <u>anther</u>. If the filament is missing, the anther is sessile. Sometimes one or more filaments are present without anthers.

The <u>gynecium</u> (gynoecium) of a flower consists of a single carpel or of two to many separate or united carpels. A carpel usually has three parts: <u>stigma</u>, at the apex; <u>style</u>; and <u>ovulary</u> (ovary), the widened basal portion containing one or more <u>ovules</u>. If the style is absent, the stigma is sessile. The ripened gynecium, together, sometimes, with other structures that ripen with it, is the <u>fruit</u>. If the ovules mature, they become <u>seeds</u>. Often it is only the ovulary of the carpel or carpels that persists until maturity; the style and stigma usually wither before the fruit is ripe. Pistil, a term widely used, has been omitted here because <u>gynecium</u> and <u>carpel</u> more simply convey the intended meaning.

The degree of union of carpels varies all the way from slight to complete. If carpels are only partly united, they are usually united at base and separate above; but rarely the stigmas, or the styles, or both are united and the ovularies are separate. A <u>compound</u> ovulary consists of the united ovularies of two or more carpels. In like manner, a style or a stigma may be compound.

It may help the student to understand the relationship between carpel and compound ovulary if, with diagrams and, if possible, an actual carpel at hand, he carefully thinks through the following explanation. If a young fruit of a plant such as pea (a single carpel) is split along the groove at the placenta and the sides are spread apart, ovules are seen to be attached to the two edges of the carpel. If the two edges are brought together (that is, if the carpel is closed as it was before it was split), the ovulary is 1-loculed and the ovules within it are attached to a parietal (marginal) placenta.

Imagine that the two walls of a carpel meet at the placenta at an angle of 120 degrees, and that three of these carpels are joined in such way that in cross section the whole has the appearance of a pie cut in three pieces, each piece a carpel. The compound ovulary so formed has as many locules as carpels, and the placentation is central (axile). Imagine again that the edges of each of the three carpels are separated as in the split pea-fruit, and that the carpels are joined to one another in a circle with each edge of a carpel meeting an edge of an adjoining carpel. The compound ovulary so formed has one locule; the placentae are parietal (where the edges of the carpels meet) and are equal in number to the carpels. Within a 1-loculed compound ovulary the ovules are sometimes attached to a postlike structure in the center. Such placentation is called free central. Ovules may also be attached basally in the ovulary or suspended from the top.

Variations from the situations described above may occur. A carpel may contain no ovules. In some species some of the ovules regularly fail to develop into seeds. Partitions may be present in the lower part of an ovulary but absent in the upper part; then the ovulary in cross section appears to have one locule above and more than one locule below. A septum may form within a carpel resulting in twice as many locules as carpels.

Because it is sometimes difficult or impossible for the student to decide how many carpels make up a gynecium of a single carpel or of united carpels, the keys have been constructed in this edition in such way that this decision never has to be made. However, the number of carpels can usually be inferred from one or more of the following: number of styles or style-branches; number of stigmas or stigma-lobes; number of locules; and number of placentae.

Except in those plants that have multiple fruits, the term fruit, as used here, means the ripened gynecium of a flower, plus, sometimes, other structures such as receptacle or hypanthium that ripen with it. If the gynecium consists of a single carpel, the fruit is one of several types including achene, follicle, drupe, and others. If the gynecium consists of separate carpels, two or more of which ripen, the fruit is an aggregate of achenes, follicles, drupes, etc. , (whatever the individual carpels become). If the gynecium consists of united carpels, the fruit may be any one of a number of types, among which are capsule, achene, grain, berry, and pome. Some fruits fit none of the named types and are referred to by such terms as berrylike, nutlike, etc. A multiple fruit is a ripened flower-cluster.

A flower is said to be incomplete if it lacks one or more of the four sets of flower-parts. A flower is bisporangiate (perfect) if it has both stamens and carpel or carpels, regardless of whether a perianth is present. A monosporangiate flower has either stamens or carpels but not both; however, vestigial stamens may be present in a carpellate flower and vestigial carpels may be present in a staminate flower. A species is diecious (dioecious) if its flowers are monosporangiate and the two kinds are on different plants. A species is monecious (monoecious) if its flowers are monosporangiate and the two kinds are on the same plant. Note that the adjectives monecious and diecious can not correctly be used to describe flowers. Sometimes both bisporangiate and monosporangiate flowers are found on the same plant.

Flowers can be classified as hypogynous, perigynous, or epigynous. In a hypogynous flower the parts of the perianth are attached to the receptacle below the gynecium; the ovulary is said to be supeior. The stamens may be borne on the receptacle below the gynecium or they may be adnate to the corolla. In a perigynous flower there is a cuplike, saucerlike, or tubular structure attached to the receptacle below the gynecium. In this manual, this structure is called hypanthium (see hypanthium in glossary). Attached to the hypanthium, usually at its rim, are perianth and stamens. The ovulary is superior in a perigynous flower; that is, the hypanthium is attached below the ovulary. In an epigynous flower the other flower-parts are borne at the summit of, or somewhere along the side of, the gynecium. The ovulary is said to be wholly or partly inferior. An epigynous hypanthium may be present, attached to the summit of the ovulary and extending upward, bearing upon its edge the other flowerparts. The stamens may be free from or adnate to the corolla. To decide whether a flower is hypogynous, perigynous, or epigynous, cut the flower lengthwise through the center and look at a cut surface.

Types of symmetry of flowers or of flower-parts (regular, zygomorphic, isobilateral) are illustrated on the third of the pages of diagrams. If all the sets of flower-parts are regular, the flower can be said to be regular. In many flowers, however, the sets of flower-parts do not all have the same symmetry. If, for example, calyx and corolla are regular but stamens and carpels are not, then, in strict accuracy, only the perianth can be called regular.

A peduncle may end in a solitary flower or it may end in a cluster of flowers. The stalk of a flower in a cluster is a pedicel; if the pedicel is absent, the flower is sessile. The term inflorescence, as used here, is synonymous with flower-cluster. The common types of flower-clusters are illustrated on the fourth of the pages of diagrams. A spike consists of a single more or less elongate axis bearing sessile or nearly sessile flowers. A raceme differs from a spike in that each flower is pediceled, the pedicels not greatly different in length. A corymb is like a raceme except that the pedicels are progressively shorter from base to tip of the cluster. Sequence of flowering in these three types is usually, but not always, from base to tip. A panicle is compound and somewhat elongate, usually made up of spikes, racemes, or corymbs, but sometimes of other types of clusters. The cyme which is diagrammed is a dichasium. A simple dichasium consists of three flowers; below the terminal (oldest) flower is a pair of opposite branches, each ending in a flower. In a compound dichasium there is, below each of the younger flowers of the three-flowered unit, a pair of opposite branches, each ending in a flower or a bud. Repeated compounding in this manner results in widening of the inflorescence. Other flower-clusters that are often called cymes have branching patterns different from that of the dichasium.

In an umbel the order of flowering may be from center to periphery or in the opposite direction. Umbels and corymbs may be compound; that is, each ray may bear, instead of a single flower, a secondary inflorescence. A head is similar to an umbel except that the pedicels are short or absent. Spikelet is used mostly to designate the small spikes of grasses and sedges. The spadix, usually with a subtending spathe, is characteristic of the Arum Family. The cyathium is the flower-cluster in Euphorbia. Because some flower-clusters do not fit any of the named types or are difficult to interpret, it is convenient to use the term racemiform (or umbelliform, etc.) to indicate that a cluster has the superficial appearance of, but not necessarily the strict character of, a raceme (or umbel, etc.).

Ferns

Pteridophytes differ from spermatophytes (seed plants) in that they have no seeds. Pteridophytes in the Ohio flora include horsetails, club-mosses, Selaginella, quillworts, and true ferns. In this manual, keys to the pteridophytes are based upon vegetative characters and upon features of sporangia.

Sporangia of true ferns are usually grouped in sori. These sori are on the lower sides of leaf-blades. They may be distributed over the lower blade-surface, or they may be in a continuous or a discontinuous band at blade-margin. A sorus may be naked (without an indusium) or it may have an indusium (a scalelike structure which covers the young sporangia). The indusium may be above (when the blade is lying with its lower side upward), or it may be at least partly beneath, the sporangia. As the sporangia enlarge, the indusium usually becomes withered and may eventually disappear, and sori that were originally separate sometimes become confluent. For a long time special terms have been applied to parts of fern plants. Frond means leaf; stipe, petiole; pinna, primary leaflet of a compound frond; and pinnule, division of a pinna.

Deciduous Woody Plants in Winter

Many of the characters used in keys to woody plants in winter are those of twigs. The term twig, as used here, means the stem-segment that grew during the most recent growing season. The scars of leaves, of veins, and of stipules are among the twig-features useful in identification. In this manual, length of a leaf-scar or of a stipule-scar is measured vertically on the twig; width is measured horizontally around the twig. Thus a wide, short leaf-scar extends far around the twig but only a short distance up and down the twig. This usage differs from that in many other manuals. In some species, at leaf-abscission, a portion of the lower part of the petiole is left on the stem. This petiole-base may remain all winter or may abscise during the winter. Stipule-scars are usually separate at sides of a leaf-scar, but sometimes they appear as a continuous ring around the twig. In some species stipules themselves persist in winter. Most twigs have lenticels (small cork-filled breaks in the epidermis); their number and character may occasionally be distinctive.

Every bud contains a stem-tip. In a reproductive bud, young flower-parts are on the stem-tip or on each of several stem-tips if the young stem in the bud is branched; in a vegetative bud, young leaves are on the stem-tip; and in a mixed bud, both young leaves and young flower-parts are on the stem-tip. Many buds remain dormant; of those that grow, only a flower or a flower-cluster emerges from a reproductive bud; only a leaf-bearing stem, from a vegetative bud; and a stem with both leaves and a flower or flowers, from a mixed bud. Lateral buds usually are located in the axils of leaves; therefore on winter-twigs they are usually just above leaf-scars. If the leaf-scar is U-shaped or ring-shaped, it may encircle or almost encircle the bud. Buds may be solitary at a leaf-scar, or collateral (side by side), or superposed (one above another). Sometimes they are located under the surfaces of leaf-scars or are covered by persistent petiole-bases. Unless otherwise stated, sizes of buds given in the keys are those of the larger ones on the twigs, often the terminal buds. Buds may be without scales (naked) or covered with a single scale or with two or more scales. Bud-scales may be valvate (with edges meeting) or imbricate (with edges overlapping). When a stem emerges from a bud, the bud-scales, if present, eventually abscise leaving scars. Consequently, at the base of each twig of a woody plant with scaly buds, there is a ring of bud-scale scars. These scars remain for a few years and eventually disappear as the bark roughens.

In some woody species each terminal bud regularly abscises before winter leaving a bud-scar (stem-scar). If leaves are opposite, the twig then ends in the two buds at the last node. If leaves are alternate, the last lateral bud usually has the appearance of a terminal bud. It can be distinguished from a true terminal bud by: 1) the presence of a leaf-scar under it; and 2) the presence of the scar of the terminal bud some distance around the twig from the leaf-scar. It is wise to look at several twigs before deciding that the terminal bud has abscised, because a scar of a terminal flower or flower-cluster on some of the twigs may be mistaken for a terminal bud-scar.

Thorns, spines, and prickles are sharp-pointed structures sometimes found on twigs and older stems. A prickle is superficial; it can be separated from the stem by pushing it from the side. In this manual, thorn and spine are applied to different structures. A thorn is a sharp-pointed stem; a spine is a leaf or some part of a leaf such as a stipule or a marginal tooth. If the spine is a leaf, it is located on the stem in the position of a leaf and it subtends a bud or a branch.

Pith of twigs or of somewhat older stems is sometimes distinctive. Pith is usually continuous but it sometimes has cavities. Horizontal diaphragms may be present, sometimes only at nodes; the space between the diaphragms may be filled with pith or it may be hollow. In cross section pith usually appears circular, but in some species it is triangular or star-shaped (5-pointed or 5-angled). It is usually white but it may be some other distinctive color.

Many features other than those of twigs are useful in identification of woody plants in winter. Among the important ones are: growth habit, whether the plant is a tree, a shrub, or a vine; type of branching; characters of bark, such as color, exfoliation, whether smooth or fissured and, if fissured, the pattern of the fissures; persistence of dead leaves; and persistence of fruits.

KEYS TO SPECIES OF SOME GENERA OF DECIDUOUS WOODY PLANTS IN WINTER CONDITION

Populus

a. Trees columnar in form, branches erect; buds resinous, appressed. P. nigra var. italica
a. Trees with spreading branches.
  b. Buds pubescent, not or only slightly resinous.
    c. Twigs glabrous or slightly pubescent; buds 5-10 mm long. P. grandidentata
    c. Twigs pubescent; buds about 6 mm long, ovoid.
      d. Twigs and buds densely white-tomentose. P. alba
      d. Twigs and buds thinly gray-tomentose. P. canescens
  b. Buds not pubescent, or only slightly puberulent at base, sometimes resinous.
    c. Buds elongate, slender, 1.5 cm long or more, length twice width.
      d. Buds very resinous, fragrant, the largest about 2 cm long; twigs brown. P. balsamifera and P. gileadensis
      d. Buds resinous or not resinous, usually shorter than 2 cm; twigs greenish-brown or greenish-yellow, often angled. P. deltoides
    c. Buds mostly less than 1 cm long (if longer, then ovoid), not or only slightly resinous.
      d. Pith orange; twigs light in color; buds mostly less than 1 cm long or rarely to 1.5 cm, ovoid, slightly puberulent at base. P. heterophylla
      d. Pith not orange; twigs brown and glabrous; buds ovoid or lance-ovoid, less than 1 cm long, the lateral appressed. P. tremuloides

Carya

a. Bud-scales valvate.
  b. Buds yellow, densely gland-dotted; outer scales leaflike. C. cordiformis
  b. Outer bud-scales brown, gland-dotted, deciduous; inner scales pale-pubescent. C. illinoensis
a. Bud-scales imbricate.
  b. Outer scales of terminal bud persistent; bark of trunk shaggy, light gray, separating in long
      strips; husk of nut 4-10 mm thick; inner bud-scales enlarging greatly in spring as bud opens.
    c. Terminal bud 1.3-2 cm long; twigs brown or green-, red-, or gray-brown, dark; fruit 3.5-5
      cm long. C. ovata
    c. Terminal bud 2-3 cm long; twigs orange or gray; fruit 4-7 cm long; leaf-rachises and petioles
      usually persistent. C. laciniosa
  b. Outer scales of terminal bud deciduous in autumn or early winter; bark not shaggy.
    c. Terminal bud 1.5-2 cm long; inner scales silky, accrescent in spring; twigs red-brown; fruit
      3.5-5 cm long, husk usually more than 3 mm thick; bark deeply furrowed. C. tomentosa
    c. Terminal bud usually not more than 1 cm long; inner bud-scales slightly accrescent in spring;
      fruit about 3.5 cm long, the husk usually not more than 3 mm thick.
      d. Husk of nut splitting to base along 3-4 sutures; bark sometimes separating. C. ovalis
      d. Husk of nut not splitting to base or, if so, along only 1 suture. C. glabra

Quercus

    In some species of oak, dead leaves remain on the tree far into the winter; if leaves are present,
use the key on p. 87. If immature fruits are found on the twigs in winter, the tree does not belong to
any of the following species in the key, all of which are in the White Oak group: Q. macrocarpa, bicolor, robur, prinus, stellata, muehlenbergii, and alba. Twig characters, upon which this key is
based, are often not sufficient for correct identification of oaks; it is wise to use any additional characters available.

a. Buds 7-10 mm long, woolly or densely and finely pubescent.
  b. Buds strongly angled, yellow-gray, ovoid; twigs glabrous. Q. velutina
  b. Buds slightly angled, red-brown, woolly, conic to fusiform; twigs pubescent. Q. marilandica
a. Buds smaller than above, or glabrous, or pubescent only in the upper part.
  b. Stipules persistent among the buds clustered at stem-tip.
    c. Twigs light brown, glabrous; lateral buds glabrous, the terminal cluster sometimes hairy;
      bark peeling off in curling pieces from branches 2-4 cm in diameter. Q. bicolor
    c. Twigs gray, hairy or glabrous; buds gray, hairy; young stem often cork-ridged. Q. macro-
      carpa
  b. Stipules not persistent among the buds clustered at stem-tip, or not noticeable.
    c. Terminal bud mostly more than 5 mm long; twigs mostly glabrous.
      d. Buds obtuse, broadly ovoid, mostly glabrous, lateral about as large as terminal and
        spreading at wide angle; twigs brown, greenish, or reddish. *Q. robur
      d. Buds bluntly acute to blunt, ovoid, reddish-brown, with white or tawny hairs on upper half
        or more; acorn usually with concentric rings at tip. Q. coccinea (See third d)
      d. Buds acute; other characters not as in the two d's above.
        e. Buds ovoid, hairy at tip or glabrous, dark brown; twigs brown. Q. borealis
        e. Buds ovoid, gray or yellow-brown, glabrous, scales with erose or split edges. Q.
          shumardii (See third e)
        e. Buds slender-conic, light brown, upper part hairy; bark deeply fissured. Q. prinus
    c. Terminal bud mostly 5 mm long or less.
      d. Twigs pubescent; buds more or less pubescent.
        e. Buds red, acute; twigs gray- or rusty-tomentose. Q. falcata
        e. Buds red-brown or brown, rounded or bluntly acute, sparsely pubescent, sometimes only
          on lower half; twigs gray-pubescent. Q. stellata
      d. Twigs glabrous or nearly so.
        e. Buds acute.
          f. Buds 2-3 mm long, glabrous; twigs very slender, glabrous. *Q. phellos
          f. Buds somewhat longer, glabrous or pubescent in upper part.
            g. Buds and twigs yellow- or gray-brown; bud-scales pale-edged. Q. muehlenbergii
            g. Buds and twigs brown; dead leaves usually persistent in winter.
              h. Tree with many small divergent pinlike branches. Q. palustris
              h. Tree without many small divergent pinlike branches. Q. imbricaria

     e. Buds obtuse, ovoid to almost globose.
       f. Bark peeling off in curling pieces from branches 2-4 cm in diameter; stipules often present in terminal bud-cluster; twigs light brown. Q. bicolor
       f. Bark not peeling off as above; stipules absent in terminal bud-cluster.
         g. Twigs gray, purplish, or red-brown, often glaucous; buds red or brown. Q. alba
         g. Twigs and buds yellow- or gray-brown; bud-scales pale-edged. Q. muehlenbergii

Magnolia

a. Terminal bud about 1 cm long, slender, acute, green; twigs slender, green. *M. virginiana
a. Terminal bud and twigs not as above.
  b. Terminal bud glabrous, acuminate, 3-5 cm long; twigs stout. M. tripetala
  b. Terminal bud pubescent.
    c. Terminal bud 3-5 cm long; twigs downy, stout; leaf-scars about as wide as long. M. macrophylla
    c. Terminal bud mostly shorter than 3 cm; twigs glabrous or slightly pubescent.
      d. Hairs of terminal bud pale; leaf-scar crescent-shaped, wider than lateral bud, encircling it about half; native tree. M. acuminata
      d. Hairs of terminal bud gray; leaf-scar somewhat triangular, not crescent-shaped, not wider than lateral bud; planted as an ornamental. *M. soulangeana

Rhus

a. Leaf-scar circular, raised on persistent petiole-base which covers axillary bud; twigs pleasantly aromatic; small amentlike clusters of reproductive buds often present. R. aromatica
a. Leaf-scar, petiole-base, and twigs not as above.
  b. Terminal bud absent; lateral bud a globose hairy mass; leaf-scar ring-shaped, U-shaped, or somewhat shield-shaped, half to almost completely encircling lateral bud; fruits red.
    c. Lateral bud about half encircled by notch of leaf-scar; twigs puberulent. R. copallina
    c. Lateral bud almost completely encircled by leaf-scar.
      d. Twigs densely long-hairy. R. typhina
      d. Twigs glabrous, glaucous. R. glabra
  b. Terminal bud present; leaf-scar shield-shaped or triangular, bud usually in notch at top; fruits white or drab; POISONOUS TO TOUCH.
    c. Vines or spreading and bushy shrubs; aerial rootlets often present. R. radicans
    c. Erect shrubs or small trees. R. vernix

Acer

a. Buds white-pubescent, scales few; twigs glabrous, usually glaucous, green or tinged with red or purple; leaf-scars meeting in a sharp upward-pointing angle. A. negundo
a. Buds and twigs not as above.
  b. Outer bud-scales 2, valvate; buds stalked; shrubs or small trees.
    c. Twigs glabrous; bark white-striped; buds about 8 mm long. A. pensylvanicum
    c. Twigs pubescent; bark not white-striped; buds shorter than above. A. spicatum
  b. Bud-scales imbricate, more than 2 exposed; buds sessile or nearly so.
    c. Buds acute, conical, brown to almost black, about 4-8 pairs of scales visible. A. saccharum and A. nigrum
    c. Buds ovoid to globose, obtuse or bluntly acute or, if acute, then differing from the above in color or in number of scales.
      d. Terminal bud usually 6 mm long or more.
        e. Bud-scales red, or red and green. *A. platanoides
        e. Bud-scales green, or the lower ones somewhat brownish. *A. pseudoplatanus
      d. Terminal bud (or lateral buds, if terminal is missing) less than 6 mm long.
        e. Buds red, often collateral; reproductive buds globose; scales with conspicuous light or brownish cilia; twigs brown to red; terminal bud present; trees.
          f. Twigs with strong, rather unpleasant, odor when broken. A. saccharinum
          f. Twigs without such odor. A. rubrum
        e. Buds not collateral, sometimes reddish but not wholly red unless with a collar; flowers from mixed buds; terminal bud sometimes absent; small trees or shrubs.
          f. Buds with asymmetric pale collar, the terminal usually absent. *A. palmatum
          f. Buds without a collar; terminal bud may be absent from some twigs.

g. Leaf-scars usually meeting around twig; twig dull, grayish-brown, -red, or -green, usually somewhat pubescent; buds usually pubescent. *A. campestre
g. Leaf-scars not meeting, connected by a line; bud-scales glabrous or ciliate.
h. Buds 2-3 mm long; twigs very slender, glabrous. *A. ginnala
h. Buds a little longer; twigs slender, often slightly hairy. *A. tataricum

Fraxinus

a. Twigs 4-angled, mostly glabrous; buds gray- or tawny-pubescent. F. quadrangulata
a. Twigs terete, densely pubescent; buds brownish. F. americana var. biltmoreana and F. pennsylvanica var. pennsylvanica (See third a)
a. Twigs terete, glabrous or nearly so; buds glabrous or nearly so but often rough.
b. Buds brown.
c. Leaf-scar crescent-shaped or at least upper margin concave. F. americana var. americana
c. Leaf-scar circular or semicircular, upper margin straight or notched below bud. F. pennsylvanica var. subintegerrima
b. Buds dark brown to blackish; fruit oblong.
c. Lateral buds appressed, rounded at tip, the terminal sharp-pointed at tip; leaf-scar circular or longer than wide, sometimes notched below bud; twigs yellow-gray. F. nigra
c. Lateral buds divergent, rounded at tip to a slight point, rather plump; leaf-scar semicircular; twigs greenish-gray, widened and flattened at nodes. *F. excelsior, European A.

Viburnum

a. Buds naked; buds and twigs stellate-scurfy.
b. Leaf-scars half-circular, length at least half width. V. alnifolium
b. Leaf-scars shorter. *V. lantana
a. Buds with at least one scale.
b. Outer bud-scales 2, valvate or separate near base, sometimes connate and appearing as one.
c. Bud-scales connate; buds plump, appressed. V. opulus
c. Bud-scales separate; buds oblong or linear, or widened at middle or near base.
d. Buds long and slender, often curved, the vegetative 1 cm long or more, those containing flower-clusters 1.5-3 cm long, widened near base and long-acuminate at tip.
e. Buds light brown, scurfy. V. cassinoides
e. Buds dark brown or red-brown, somewhat scurfy, becoming lead-color. V. lentago
d. Buds shorter than above, the vegetative less than 1 cm long, those containing flower-clusters about 1 cm long, widest at middle, about half as wide as long.
e. Buds densely red-scurfy, the lateral ones blunt; rare. V. rufidulum
e. Buds brown-scurfy, becoming gray; twigs short, spreading at wide angle. V. prunifolium
b. Outer bud-scales more than 2, usually 2 pairs.
c. Bark light gray, on stems 2 years old and older the outer layer splitting and coming off, the brown under layer then exposed. V. molle
c. Bark not as above.
d. Twigs usually pubescent or puberulent; bud-scales often hairy at tip, the outer pair pointed at apex, about 1/4 length of bud. V. acerifolium
d. Twigs glabrous or rarely pubescent; bud-scales ciliate, hairy at tip or glabrous, the outer pair 1/4 - 1/2 length of bud.
e. Lower bud-scales acute at tip. V. dentatum and V. recognitum
e. Lower bud-scales rounded or truncate, often short-mucronate. V. rafinesquianum